GLOBAL RESOURCES AND THE ENVIRONMENT

In the past few decades, sustainability of natural resources and the social and environmental issues that surround them have become increasingly topical. This multidisciplinary book discusses the complex relationships between society, natural resources, and the environment. Major resources including water, agriculture, energy, minerals, and forests are considered, as well as different facets of the environment including climate, landforms, and biodiversity. Each resource is discussed in the context of both environmental and socioeconomic factors affecting their present and future distribution and demand. Presenting a balanced, comprehensive overview of the issues surrounding natural resources and sustainability, this accessible volume will be of interest to policymakers, resource managers, graduate students and researchers in the natural and social sciences.

CHADWICK DEARING OLIVER is the Pinchot Professor of Forestry and Environmental Studies and Director of the Global Institute of Sustainable Forestry at Yale. His research focuses on global issues of landscape management, and he has considerable experience advising public and private forest resource organizations in the United States and abroad. He is the senior author of the highly cited book, *Forest Stand Dynamics*, now in its second edition.

FATMA ARF OLIVER was a statistical consultant for the Boeing Company until her retirement in 2000. She is the author of a number of scientific and technical papers.

T0296906

"Since the United Nations Conference on Environment and Development in Rio during 1992, international agencies, national governments, non-governmental organizations, academic institutions, the media, and the public have recognized the significance of the interactions between social, environmental, and economic needs and problems. Considerable research, development, and application have emerged but interdisciplinary approaches to problems have often been lacking. This book provides a comprehensive and integrative approach to the dynamics of natural resources, the concept of sustainability, the needs and impacts of humans, and the roles of climate and biodiversity. This substantial tome is a major contribution to broader understanding and cooperation, written by outstanding authors, and will be a fundamental resource for policy-makers, managers, and academics for many years."

– Professor Jeff Burley, Oxford Forestry Institute

"In a world with increasing specialization and more well-articulated interest-groups than ever, the Olivers' book provides the opposite of the mainstream and fragmented flows of news and information. An overview and key figures describing the main components of our global production and consumption of resources are presented, analyzed, and conclusions are drawn – not only to describe problems and challenges, but mainly to illuminate potentials and possible solutions to support a highly needed sustainable development of our world society and its expected 9 billion people by 2050. Suggestions for the reader: read it, learn from it, expand your knowledge, and continue to upgrade analytical skills using this important book as your solid platform."

– Professor Palle Madsen, Forest and Landscape College, IGN, University of Copenhagen

"Chad and Fatma Oliver have written a magisterial work – fact-filled, comprehensive, and illuminating. I am in admiration of their accomplishment. There is nothing else like it, and it will fill a real need."

– James Gustave Speth, founder and former President of World Resources Institute, and former Dean of Yale School of Forestry and Environmental Sciences

"*Global Resources and the Environment* is a labor of love and a wonderful achievement. It spans widely across the natural and the social sciences, as well as economics, synthesizing a huge amount of relevant literature for scientists, policymakers, and the general reader. It is also admirably positive, while clear-sighted, in its future predictions: there are still enough resources in the world, provided they are equitably distributed and managed with greater reverence and care. A wholly recommended tour de force."

– Edward Davey, Project Director, World Resources Institute

"This book is an extremely helpful and timely contribution for the current or would-be professional at the international scale. It provides thorough information on the world and by region, and presents it using smart, easy-to-understand figures and graphs that synthesize a broad spectrum of knowledge. Using copious data analysis, this work carefully balances current theories and empirical lessons to propose doable actions and a promising future, but not without paying attention to dead ends and misconceptions."

– Dr. Gerardo Segura, Senior Natural Resources Management Specialist, World Bank

GLOBAL RESOURCES AND THE ENVIRONMENT

CHADWICK DEARING OLIVER

Yale University

and

FATMA ARF OLIVER

Former engineer at the Boeing Company, Seattle

CAMBRIDGE
UNIVERSITY PRESS

University Printing House, Cambridge CB2 8BS, United Kingdom

One Liberty Plaza, 20th Floor, New York, NY 10006, USA

477 Williamstown Road, Port Melbourne, VIC 3207, Australia

314-321, 3rd Floor, Plot 3, Splendor Forum, Jasola District Centre, New Delhi - 110025, India

79 Anson Road, #06-04/06, Singapore 079906

Cambridge University Press is part of the University of Cambridge.

It furthers the University's mission by disseminating knowledge in the pursuit of education, learning and research at the highest international levels of excellence.

www.cambridge.org
Information on this title: www.cambridge.org/9781107172937
DOI: 10.1017/9781316779484

First published 2018

A catalogue record for this publication is available from the British Library

Library of Congress Cataloging in Publication data
Names: Oliver, Chadwick Dearing, 1947– author. | Oliver, Fatma Arf, author.
Title: Global resources and the environment / Chadwick Dearing Oliver, School of Forestry and Environmental Studies, Yale University and Fatma Arf Oliver, Engineer, retired from The Boeing Company, Seattle.
Description: Cambridge, United Kingdom ; New York, NY : Cambridge University Press, 2018. | Includes bibliographical references and index.
Identifiers: LCCN 2017052815| ISBN 9781107172937 (hardback) | ISBN 9781316625415 (paperback)
Subjects: LCSH: Environmentalism. | Natural resources – Management. | Human ecology. | Sustainability.
Classification: LCC GE195 .O55 2018 | DDC 304.2–dc23
LC record available at https://lccn.loc.gov/2017052815

ISBN 978-1-107-17293-7 Hardback
ISBN 978-1-316-62541-5 Paperback

We dedicate this book to our children (*) and their families:
* Elif Kendirli
* Doa Kendeerlee
* Chadwick Cahit Oliver and Julie Hanrahan, and our grandchildren (**)
** Eli Thomas Oliver
** Aisha Ann Oliver
* Renin Hilary Oliver

Contents

Color plates are to be found between pp. 326 and 327

Preface

People often feel anxious when advanced to a new job – when a politician moves to a higher office, a businessman takes on more responsibility, or an athlete turns professional. Over time, they learn to be more comfortable with their new situation and feel less overwhelmed.

The situation itself does not change. Rather, with more detailed knowledge, people learn to organize their tasks in context – what is likely to happen and what is not likely; what is important and needs tending and what is less urgent and can be delayed; and in what sequence things are expected to occur.

Many people are becoming anxious over the environmental situation because they are bombarded at a fast pace with a large number of individual facts and assertions but with little overarching context. The purpose of this book is to put the numerous aspects of the environment, resources, and people into perspective so that readers will feel less over-whelmed. The book is based on the premise that much in-depth knowledge can be presented simply, thoroughly, and succinctly. The intended audience for this book is advancing, mid-career professionals, policymakers, managers, academicians, and anyone else who works with or is interested in environmental/resource issues. Even those with a focus on one issue will benefit from a broader knowledge of the subject as a whole.

If people understand the present conditions, and likely (and unlikely) changes in the environment and resources, they can craft effective economic, social, business, and tech-nical policies.

The book intends to educate the reader thoroughly but, more importantly, give him/her a perspective as knowledge changes. It is not intended for skim-reading. The chapters and sections are relatively independent and may be read selectively.

To the academician, the book complements the trend of increasing specialization. The emphasis on specialization is creating much new knowledge but there is a need to integrate and understand broad perspectives. The changes in a given resource are generally more affected by changes in other resources than by actions taken by specialists in the target resource.

A hope is that this book will be adopted for graduate courses. Students would need to take several courses to learn about the environmental and resource systems covered in this single volume; they would then have less time to take more courses in their chosen specialty.

The book is a monograph, not an edited collection of papers. Consequently, we have been forced to deepen our own understanding of each resource and how it interacts with the others and the environment. Economics is not discussed directly but we set the bounds of reasonable futures within which economic solutions can be constructed and debated.

The interactions among the environment, resources, and people are best understood from the perspective of complex, dynamic systems and sustainability, described in the first section of three brief chapters. Then nine sections follow, of two to four short chapters each, on people, major environmental systems, and major resources. The environmental systems are climates, landforms, and biodiversity. The resources are water, agriculture, energy, rocks/minerals, and forests. Except for water, aquatic systems are only minimally covered and oceanic resources are not covered at all. Soils are covered as part of landforms. Disturbances have been well studied recently, but are treated as one "change" that is part of dynamic systems – not as a separate subject.

The book's figures provide more global information subdivided by world regions that cannot be completely covered in the text. These figures allow the reader to extend his/her understanding of specific regions beyond the text.

Resources and the environment are covered globally because both are global issues; what is done in one part of the world affects all others. Described first are the underlying scientific principles of each resource/environmental topic and what makes each important; then, how it is distributed globally and managed; and finally, what future scenarios and options exist for maintaining, changing, or mitigating it.

The book predicts neither a "gloom and doom" nor a "glowing" environmental/resource future. Our future depends partly on how we manage the present. The book promotes an informed dialogue about future directions for the world's environment and resources. It is not a motivational book, but it does not shy away from pointing out obvious "dead ends" and possible actions when analyses lead to such insights.

The book is a result of our common global perspective. It was inspired and informed by the synergies of Chad's practical experiences and knowledge of resources and Fatma's analytical perspective and interest in systems theories. More than forty years ago we met and married when Chad was in graduate school at Yale University and Fatma was teaching mathematics at the University of New Haven. Our different backgrounds and perspectives – Chad from a small town in South Carolina, USA, and Fatma from Istanbul, Turkey – have stimulated fascinating discussions ever since. We both enjoy traveling and have done it extensively.

Our travels and discussions with local people have made this book better. As colleagues that we visited in other countries learned of our writing the book, they became excited, offering us various insights, informed discussions, and tours with local resource professionals. Beforehand, we would study the area and frame mentally the resources, environment, and people as a system. Subsequent observations and discussions on the visit would show discrepancies between our mental model and our hosts', which led to fruitful discussions and learning. It soon became obvious that energy policies affected agriculture as much as agronomists; that the condition of forests was less impacted by forest managers than by agriculture policies; and that water policies were affected strongly by agriculture, energy, and urbanization. Consequently, it became apparent that one needs to understand all resources to make informed decisions.

This book synthesizes existing knowledge, amalgamates different fields, and comes up with integrating observations. The world's information is now so vast, complicated, and rapidly changing that many things in this book will no doubt be out of date by the time it reaches print – just as many things changed while writing it. Since the book began about ten years ago, the calculated time of a previous Ice Age (not the present one) was changed by over 100 million years; and the time when North and South America joined has been adjusted by a few million years. A colleague at Yale who showed scientifically why Neanderthals could never breed with *Homo sapiens sapiens* has been proven wrong. Locations of different landforms are inconsistent among authorities. This is, and will be, the future of science: no person or generation has perfect knowledge. The future will be about learning details, synthesizing, comparing, measuring, and experimenting; forming thoughtful, flexible mental models; and continuing to learn from and adjust them.

We have tried to use the latest, most accurate data. Some things reported earlier have not been updated with new data. Some recent data have not been consistently posted. Sometimes, an analysis with new data would delay the book but would not show meaningful differences. Consequently, sometimes we show trends that are several years old. Copious online data sets were used. Some of the data is imperfect or incomplete; however, it is preferable to err by commission and use imperfect data rather than to err by omission, for two reasons:

- Out of respect for people who need to make decisions, we prefer to state an educated observation, rather than shy away on the basis of imperfect knowledge or data. Sometimes the data shortcomings are pointed out.
- Fatma's experience with data sets as an engineer has shown that the best way to improve data is to use it.

Hopefully, the reader will appreciate the dynamic nature of knowledge and use this book as a basis to adjust from, rather than as a static authority.

We also hope that environmental and resource knowledge will become more integrative and holistic. This book is designed to encourage that trend.

All units are metric.

Acknowledgments

Many professional colleagues helped shape this book. Some influenced us through general discussions, specific suggestions, and giving us new perspectives. Among them are David M. Smith, Yale University; Daniel Botkin, George Mason University; E. C. Burkhardt, consultant forester; Glenn Galloway, University of Florida; Bruce Lippke, University of Washington; Gerhard Schreuder, University of Washington; John Perez-Garcia, University of Washington; John Stanturf, USA Forest Service; Palle Madsen, Copenhagen University; Roger Sedjo, Resources for the Future; Gerardo Segura, The World Bank; Jim Geary, Megamekanik, Istanbul; Maxwell McCormack, University of Maine; Suha Gursey, Yale University; and Cahit Arf, Fatma's father, Middle East Technical University. The mistakes are our own.

Many others invited us to travel abroad in various capacities. They became interested and very helpful in regard to the global, integrating nature of the book and selflessly spent time with us to make it better. **Africa:** John Kakonga, UNDP; the Forest Service of Liberia; and Wangari Maathai, Kenya. **Alaska:** Bernard Bormann, University of Washington; and Robert Deal, USA Forest Service. **Armenia:** Jeff Masarjian, Armenia Tree Project; and Zack Parisa, SilviaTerra, LLC. **Australia:** Keith Jennings, retired forester, Brisbane; Chadwick C. Oliver, Melany, Queensland. **Bhutan:** Pema Gyametsho, Maungmoe Myint, Kinley Tshering, Younten Phuntsho, Ministry of Agriculture and Forests. **Brazil:** Daniel Piotto, Federal University of Southern Bahia. **Chile:** Christian Salas, University of Frontera, Temuco. **China:** Yaji Song, Yale University; Xuemei Han, Department of Management and Budget, Fairfax County, VA, USA; Jianping Ge, Rumei Xu, Qingxi Guo, Xiaojun Kou, Po Mu, Paul Mou, Tianming Wang, and Liming Feng, all at Beijing Normal University; Yu Tian, Chinese Research Academy of Environmental Sciences; Kezhen Guo, Ministry of Water Resources, Hahot; Shang Hua, Dalian University of Technology; Qingxi Guo, Northeast Forestry University; Ge Sun, Chinese Academy of Forestry; and Chun Fu, Nanchang University. **Ecuador:** Glenn Galloway, University of Florida. **Europe:** Dietlef Kraft, Prince of Thurn and Taxis Forests; Gabriella Nosswitz, formerly Freiburg University; Patrice Harou, University of Nancy. **India:** P. P. Bhojvaid and Mahabir Sharma, Indian Forest Service; Alark Saxena, Yale University; Rajendra Pachauri, TERI, New Delhi. **Japan:** Edwin Miyata, USA Forest Service; Takao Fujimori, Japan Forest Technology Society; Akira Osawa, Kyoto University. **Mexico:**

Gerardo Segura, The World Bank. **Nepal**: Lhakpa Norbu Sherpa, Mt. Everest National Park. **Russia:** Herrick Fox, USA Forest Service. **South Korea:** Jong-Choon Woo, Kaangwon National University; Pil Sun Park, Seoul National University. **Thailand:** Patrick Baker, University of Melbourne. **Turkey**: Melih Boydak, Isik University, Istanbul; Turkish Ministry of Forestry; Ugur Zeydanli, DKM, Ankara; Nuri Ozbagdatli and Bahtiyar Kurt, UNDP, Ankara. **Ukraine**: Sergiy Zibtsev, Dmytro Melnychuk, and Petro Lakyda, National University of Life and Environmental Sciences, Kiev.

Chad taught an international silviculture class at the University of Washington, where the students and professors gave him very helpful input.

At the Yale School of Forestry and Environmental Studies, Chad also taught a class with the same title as this book. Fatma sat in on the classes and researched specific subjects. The class gave many insights, data, ideas, and corrections that are in the book. Shilo Tilleman-Dick generated the idea for Tables 6.1 and 6.2 as a class project. He graciously allowed me to expand and modify it and use the result here. Maungmoe Myint offered many helpful suggestions. Jason Rauch gave insights into minerals. And Vaaruni Eashwar provided insights and information about organic farming.

Chad also taught mid-career international short courses at Yale, reaching people in a variety of sectors from business to policy ENGOs from throughout the world. Their questions and comments helped give a different perspective.

Chad's faculty colleagues, graduate students, and the Yale university staff have been especially helpful. We enjoyed and learned a lot in discussions with PhD students Xuemei Han, Alark Saxena, and Kris Covey, as well as with staff at the Global Institute of Sustainable Forestry and Ucross Laboratory: Mary Tyrrell, Charles Bettigole, Sabrina Szeto, Henry Glick, and others.

Many colleagues gave us written and oral comments on the different chapters. David Evans, Yale University; Alark Saxena, Yale University; Xuhui Lee, Yale University; Elif Kendirli, New York University; Oswald Schmitz, Yale University; Esteban Rossi, Pontificia Universidad Javeriana, Colombia; James Shaw, Oklahoma State University; Vaaruni Eashwar, Environmental Defense Fund, Washington, DC; Ganesh and Jayashree Eashwar, Madras, India; Kenneth Gillingham, Yale University; Tom Graedel, Yale University; and Tim Gregoire, Yale University, were all very helpful and generous with their expertise. Diana Karwan, University of Minnesota, Twin Cities, had long ago helped with the water chapters; and Aleksandr Onuchin, Siberian Branch of the Russian Academy of Science, gave excellent insights into the relation of forests to water flow.

Several Yale graduate students worked on the book's construction: Ajit Rajiva helped with the figures – especially with some creative ideas – and Benjamin Rifkin and Vaaruni Eashwar helped organize and format the references. Sabrina Szeto made the maps in Chapter 4.

We thank others who are acknowledged in the attributions table for providing figures or quotations.

We both recognize that none of this would have been possible without the excellent educations we gained from high school (Camden High School, Chad; Robert College,

Fatma); undergraduate (The University of the South, Chad; Bosphorus University, Fatma); and graduate schools. And we appreciate the support of our extended families.

We are grateful to Emma Kiddle and Zoë Pruce and the other staff at Cambridge University Press for their help, dedication, and patience; Richard Hallas for his copy editing; and Anubam Vijayakrishnan and his team at Integra Software Services for their care in assembling the book.

As soon as the book is in print, we are sure we will wake up, horrified, to remember all the helpful people we have omitted from this list. To those, we apologize and thank them as a group.

Figure Credits

For more details please see the captions and references that accompany the figures.

We appreciate the many publicly available data sets used for this book. We are also grateful to the following publishers, institutions, and individuals for permission to reproduce material:

American Chemical Society

Figure 27.6: Reprinted with permission from J. Johnson, E. Harper, R. Lifset, and T. Graedel. Dining at the periodic table: Metals concentrations as they relate to recycling. *Environmental Science and Technology.* 2007;41(5):1759–65. Copyright (2007) American Chemical Society.

Table 27.2: Adapted with permission from J. N. Rauch, Global spatial indexing of the human impact on Al, Cu, Fe, and Zn mobilization. *Environmental Science and Technology.* 2010;44:5728–34. Copyright (2010) American Chemical Society.

Jeff Blossom

Figure 4.4f: Used with permission of Jeff Blossom, Harvard University. J. Blossom. *Global oil pipelines: Using Google Earth as a classroom tool*: Harvard Map Collection, Harvard College Library; 2009 [cited June 4, 2017]. Available from: http://worldmap.harvard.edu/data/geonode:global_oil_pipelines_7z9

BP Statistical Review of World Energy

Figures 25.1, 25.2, 25.4, 25.9: Source: Includes data from *BP Statistical Review of World Energy* 2016.

Brinly-Hardy Company

Figure 21.6: Courtesy of Brinly-Hardy Company, Jefferson, IN, USA.

Charles Bettigole

Figure 11.6: Courtesy of Mr. Charles Bettigole, New Haven, CT, USA.

Melih Boydak

Figures 11.8a and 14.7b: Courtesy of Professor Melih Boydak, Isik University, Istanbul.

University Press of Colorado

Figure 14.16: Courtesy of University Press of Colorado [C. D. Oliver. *Mitigating Anthropocene Influences in Forests in the United States*. In: V. A. Sample and R. P. Bixler, editors. *Forest Conservation and Management in the Anthropocene*, 2014].

Columbia Institute for Water Policy

Figure 17.3: Based on R. P. Osborn. *Hanford Nuclear Reservation: Black Rock Groundwater Could Affect Movement of Radioactive Contamination under Hanford*: Columbia Institute for Water Policy; 2007. (www.columbia-institute.org /blackrock/Issues/Hanford.html)

Creative Commons

Figure 4.1: Licensed under a Creative Commons 3.0 Attribution License. (https:// creativecommons.org/licenses/by-sa/3.0/)

Figures 25.5 and 25.6: Licensed under the Creative Commons 3.0 Attribution License: Creative Commons license (CC BY-SA 3.0). The Shift Project, Paris. www .theshiftproject.org

Dora Cudjoe

Figure 11.1a, Also color plate 04a: Courtesy of Ms. Dora Cudjoe, Washington, DC.

Environmental Performance Index

Table 6.1 and 6.2; Figure 6.5a–b: Data used with permission of the *Environmental Performance Index*, Yale University. (http://epi.yale.edu/)

Free Software Foundation, Inc.

Figure 9.2: © 2000, 2001, 2002 Free Software Foundation, Inc. https://commons .wikimedia.org/wiki/File:Ice_Age_Temperature.png#/filelinks

The Fund for Peace

Figure 5.6f, Table 6.2: Data used with permission of The Fund for Peace (http:// fundforpeace.org/fsi/).

Google Earth

Figure 10.11: Courtesy of Google Earth: Map data: SIO, NOAA, US Navy, NA, GEBCO. Image © 2017 CNES.Airbus

The Harvard Forest, Harvard University

Figure 5.4a: Photograph by John Green (1999), courtesy of The Harvard Forest, Harvard University.

The Heritage Foundation

Table 6.2: Data used with permission of The Heritage Foundation, Washington, DC.

International Union for Conservation of Nature

Figure 15.8 and Table 15.3: Data used with permission of IUCN: Data in IUCN 2016. *The IUCN Red List of Threatened Species*. Version 2016–3. www.iucnredlist.org. Downloaded on February 28, 2017.

Island Press

Figure 3.1a, b: From *Panarchy* edited by Lance H. Gunderson and C. S. Holling. © 2002 Island Press. Reproduced by permission of Island Press, Washington, DC. (Figures 2.1 and 3.0).

The Maddison Project

Figure 6.3: Data courtesy of the Maddison Project. www.ggdc.net/maddison/maddison-project/home.htm

Roger Mesznik

Figure 9.3, Also color plate 06: Used with permission of Professor Roger Meszik, New York City.

Natural Earth

Figure 4.4b–e: Made with Natural Earth. Free vector and raster map data @ naturalearthdata.com. Cartography by Sabrina Szeto.

Erik Ndayishimiye

Quotation, Chapter 6: Used with permission from Erik Ndayishimiye, citizen of Rwanda and MEM, 2017, School of Forestry and Environmental Studies, Yale University.

New Economics Foundation

Tables 6.1, 6.2, and Figure 6.4: Happy Planet Index courtesy of the New Economics Foundation: info@neweconomics.org

Rivers of the World Project

Figure 4.4a: *Rivers of the World Atlas*. Rotterdam, The Netherlands: 2010. Based on Maps 2.1–2.7.

Science

Figure 3.3: From E. Ostrom. A general framework for analyzing sustainability of social-ecological systems. *Science*. 2009;325(5939):419–22. Reprinted with permission from AAAS.

Mariana Sarmiento

Figure 12.7a, b: Courtesy of Ms. Mariana Sarmiento, Bogota, Colombia.

Springer-Verlag

Figure 14.1a, b: From Springer *Vegetatio* "Vegetation science concepts I. Initial floristic composition, a factor in old-field vegetation development," 4, 1954, pages 414–415, F. E. Egler. With permission of Springer.

Stockholm International Peace Research Institute

Figure 25.3: Data used with permission of Stockholm International Peace Research Institute, Solina, Sweden. www.sipri.org

Taylor and Francis Group

Figures 1.3, 2.6: Reproduced from C. D. Oliver, 2014. "Forest Management and Restoration across Temporal and Spatial Scales." *Journal of Sustainable Forestry* 33. Supplemental Issue 2014 (www.tandfonline.com/loi/wjsf20) by permission of Taylor and Francis Group, LLC, a division of Informa plc. Permission conveyed through Copyright Clearance Center, Inc.

Figure 14.13a–f: Reproduced from C. D. Oliver, A. Osawa, and A. Camp. "Forest dynamics and resulting animal and plant population changes at the stand and landscape levels." *Journal of Sustainable Forestry.* 1997;6(3–4):281–312. (www.tandfonline.com/loi/wjsf20) by permission of Taylor and Francis Group, LLC, a division of Informa plc. Permission conveyed through Copyright Clearance Center, Inc.

Figures 24.7, 24.8, 28.6: Reproduced from Journal of Sustainable Forestry (www.tandfonline.com/loi/wjsf20), Taylor and Francis Group, LLC, a division of Informa plc. Open access (C. D. Oliver, N. T. Nassar, B. R. Lippke, and J. B. McCarter. Carbon, fossil fuel, and biodiversity mitigation with wood and forests. *Journal of Sustainable Forestry.* 2014;33(3):248–75.)

Figures 14.15b, 30.1, 30.2a–b: Reproduced from *Restoration of Boreal and Temperate Forests*, CRC Press (www.crcpress.com); 2004. pp. 31–59) and from *Restoration of Boreal and Temperate Forest, second edition.* (2015) pp. 37–68 by permission of Taylor and Francis Group, LLC (www.tandfonline.com), a division of Informa plc. Taylor and Francis, publishers. Permission conveyed through Copyright Clearance Center, Inc.

Transparency International

Figure 5.6e, Table 6.2: Source: Transparency International. Licensed under CC-BY-ND 4.0.

US Energy Information Agency

Figures 25.5, 25.6: Source: US Energy Information Administration, via: www.tsp-data-portal.org, accessed May 2016.

US National Oceanic and Atmospheric Administration

Figure 8.2: *Full Mauna Loa CO$_2$ record*: US Department of Commerce, National Oceanic & Atmospheric Admin., Earth System Research Laboratory, Global Monitoring Division; 2016 [cited July 6, 2016]. Available from: www.esrl.noaa.gov/gmd/ccgg/trends/full.html

United Nations Food and Agriculture Organization

Figures 7.1a, b; 7.7; Also color plate 03: Source: Food and Agriculture Organization of the United Nations, UN FAO. *Global Forest Resources Assessment 2000*. Rome. Reproduced with permission.

Figure 29.1, Also color plate 12: Open source: UNFAO. *Global Forest Resources Assessment (fra) 2000 main report* (Food and Agriculture Organization of the United Nations; 2001. UNFAO GeoNetwork). www.fao.org/geonetwork/srv/en/main .home

Wikimedia Commons

Figure 7.10: Courtesy of Wikimedia Commons: https://commons.wikimedia.org/wiki/ File:Global_tropical_cyclone_tracks-edit2.jpg

Figure 9.2: Courtesy of Free Software Foundation, Inc. https://commons.wikimedia.org /wiki/File:Ice_Age_Temperature.png#filelinks

Photographs and figures not otherwise acknowledged above were taken or created by the authors.

Part I

Introduction, Dynamic Systems, and Change

1

Introduction

1.1 Resources, Environment, and People

This book is about resources viewed in the context of their environment and people. It is not an economics, policy, or business text; instead, it gives the status and trajectory of the environment and resources so that better informed economic, business, and other policies can be made. Resources will be discussed in five categories: water, agriculture, energy, minerals, and forests. Three categories of the environment will be discussed: climate, landforms, and biodiversity. The general model for this book is depicted in Figure 1.1.

Figure 1.1 has the shortcomings of all models. It is a simplification of the underlying network structure. It appears here as a hierarchy to emphasize that upper levels are initially formed by lower levels. Once formed, both upper and lower levels can influence each other, even to the point of annihilation; consequently, the system transforms to a network [1]. These concepts will be discussed in Chapter 2.

People will be discussed separately, since they are both a resource and part of the environment [2, 3]. Their capabilities as manual or intellectual labor and problem solvers, and their creativity, place them as a major resource. But they are also a significant factor in changing the environment in all three categories discussed.

"Raw materials" are unaltered physical components used by people. People commonly modify raw materials into small and large tools and infrastructures (axes, looms, roads, bridges) known as "building blocks" (Chapter 2). The concept of building blocks can be expanded to include intangible skills and institutions as well as tangible entities developed from raw materials.

Exactly what constitutes a resource is somewhat arbitrary, but resources will be considered here as tangible entities that people use to survive and prosper. Discussions of intangible intellectual property, economics, policy, pure ideas, and technologies are beyond the scope of this book.

The use of raw materials and building blocks change over time and space; therefore, the demand for a resource – and what is considered a resource – changes. For example, copper for telephone wires was a widely used resource but is being replaced by photo-optics and then by wireless cell phones.

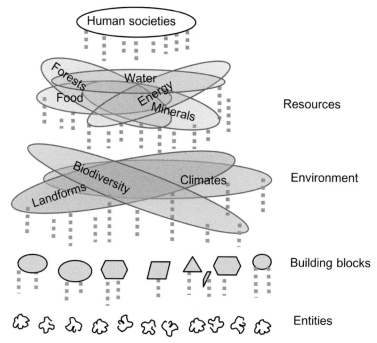

Figure 1.1 General model discussed in this book. (A black and white version of this figure will appear in some formats. For the color version, please refer to the plate section. Color plate 1.)

The human population more than doubled from 3 to 7.3 billion people from 1960 to 2015, and is expected to exceed 9 billion people by 2050 (Figure 5.11). The urban environment is expected to triple by 2050 to 9% of the Earth's land surface [4], possibly at the expense of forests. Farmland may increase slightly, both with population growth and with more people demanding a protein-rich meat diet. The desert areas at low latitudes will probably expand with global warming, but the upper latitudes will probably not warm enough to make the current boreal, tundra, and polar lands more habitable.

There are optimistic and pessimistic trends. The agriculture area has remained nearly stable since 1960 (Figure 21.5), and people are receiving more food (Figure 22.10) despite a doubling of the world's population (Figure 5.11). The world fertility rate halved from 5.0 to 2.5 births per female between 1960 and 2015 (Figure 5.8). The effects of the fertility decline will be felt during the next few decades. If the trend continues, the population may soon stabilize and even decline to a less precarious level.

1.2 Focus of the Book

The book's focus is detailed discussions of each resource with respect to properties, availability, global distribution, and management. Extensive use of currently available data supports these discussions. The attempt is to provide an in-depth understanding of

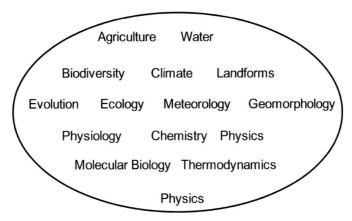

Figure 1.2 A conceptual hierarchy of scientific disciplines from more fundamental (bottom) to more applied [9].

each resource and the environment. As more data become available the graphs and the charts can be updated. Interactions and changes over time and scale, human-induced or otherwise, will be emphasized.

The challenge will be to keep the present and impending large numbers of people from damaging the environment and resources that sustain them – and thus dramatically constraining themselves.

Resource management is an art and a science. The need is to anticipate and manage change so that the human species and those aspects of the environment and resources vital for its survival are maintained – most notably, clean air, clean water, food, and biodiversity. The other very important issue is to promote equity – inter- and intra-generational rights to productive lives for all people [5, 6]. Quick, short-sighted, emotional fixes are to be avoided at all costs.

The book is optimistic. It stands midway between environmental ruin and technological heaven [7]. Although depletion and misuse generate very serious challenges, the world currently has sufficient food, energy, water, and minerals; an abundance of wood; and biodiversity can be stabilized [8]. Despite heartbreaking violence in several places, the pace of advancement in renewable resources shows people can overcome challenges. Optimism is not only warranted, but necessary to give people the confidence to undertake the changes that are needed for everyone to have.

Changes in climate and many other features always have occurred and always will. Scientists now have at least some understanding of how these changes might take place and how dire ones can be adapted to.

Each field of learning exists within a hierarchy of disciplines (Figure 1.2), with physics as the most fundamental discipline. Other fields (disciplines) emerge from fundamental principles as one ascends in the hierarchy. Fundamental fields are used to "explain" or "understand" the behavior of fields immediately above [9]. For example, understanding of climates and biodiversity helps the understanding of agriculture; understanding of evolution and ecology helps the understanding of biodiversity; understanding of physiology helps the understanding of evolution and ecology; and so on. Theoretically, a learned person would be able to comprehend all resources by understanding physics – and even comprehend economics and human behavior. In practice, an expert in each field tries to maintain a strong awareness of fields immediately above and below their specialty, so they can understand how new information can be incorporated.

1.3 Recurrent Themes

Several recurrent themes emerged as the book was being written.

"Social drivers" will mean new behaviors, ideas, institutions, or employment that form the core of a larger population's varying lifestyle [10, 11]. Social drivers will be discussed throughout this book.

Small, resource-based cities constructed in rural areas could offer relief from overly crowded large cities. Their core employment could be scientifically and technically knowledgeable people using new, small-scale, technical tools to manage farms and forests in place of costly large machinery and broad-scale uniform practices. Local secondary manufacture would add another source of labor. As small cities, a threshold population would exist to provide amenities that attract secondary employment in services and entertainment that would make people not want to migrate to large cities.

1.4 Socioenvironmental Systems

Systems are groups of interrelated entities of any kind, with several characteristics:

- The properties or behavior of each subgroup of the group affects the properties or behavior of the whole group;
- The properties and behavior of each subgroup depend on the properties and behavior of at least one other subgroup;
- The properties and behavior of each subgroup are affected by the properties and behavior of the whole group;
- The way that one subgroup affects the whole group depends on the properties and behavior of at least one other subgroup.

Every possible subgroup of a system has all four characteristics. Each has an effect and none can have an independent effect on the whole system; consequently, subgroups cannot be treated independently. A system has some properties or behaviors that none of its subgroups has. This behavior is referred to as "emergence." The system loses these properties and behaviors when it is taken apart. The subgroups of a system may themselves be systems, and every system may be a subgroup of a larger system (Chapter 2) [12, 13].

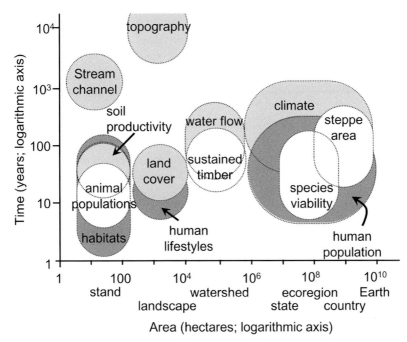

Figure 1.3 Different spatial scales and rates of change for some properties of resources and the environment [19, 20]. Reproduced with permission of Taylor and Francis Group.

Social, economic, ecological, and environmental systems interact so closely that it is appropriate to address "socioenvironmental" systems rather than distinct "human" and "natural" systems [14–18]. This recognition is appropriate because nearly all of the world is influenced by – and influences – people. However, combining human and environmental systems is a dramatic change in our embedded way of thinking. It means that "natural" has little meaning and cannot be used as a baseline "ideal" against which to measure human impact. Avoiding human impact is not a sure way of preventing harm to the environment. Socioenvironmental systems are complex (Chapter 2).

Each resource, the environment, and people are continuously changing because of innate processes, external disturbances, and human needs. They change at different rates and scales, and a wide variety of conditions can materialize at the various times and spaces (Figure 1.3) [20]. Together, these changes can hinder the return of the socioenvironmental systems to a previous condition, even if this return were desirable.

1.5 Spatial and Temporal Scales

Resources exist at a variety of spatial scales (Figure 1.3). We address resources from very small scales, such as the individual fruit, mineral, or person, to intermediate scales, such as the landscape or ecosystem to the entire Earth [20]. People have broadened the scale of where they obtain resources. Presently, many resources are transported worldwide.

The rise of complex societies has been attributed to the need to coordinate resources across larger areas [21]. The early trade of resources over large areas has become more prevalent as transportation and communication have become rapid and inexpensive.

This global sharing of resources has allowed many more people to live more comfortably than they otherwise could. People travel farther and more regularly and are increasingly linked into an international network of connections that make communication very rapid – from accessing immediate business-oriented information, to sharing creative ideas and health remedies, to connecting with distant friends. People rarely die of some common diseases that once were fatal.

Problems have also arisen with expansion to the global level, including moving plants and animals to foreign areas where they become pests; exploiting resources in one region to profit people in another; pollution from one region spreading to others; and bad investment decisions in one region causing economic problems in another.

People and resources are not uniformly distributed in the world, nor are concentrations of people matched with concentrations of needed resources; however, people have a strong ability to sustain themselves based on what is available.

Both the distributions and changes of people and resources have patterns – some more obvious than others. Analyses of global inventory data over the past decades are helping elucidate the patterns.

The 180+ world countries are commonly grouped by regions ("country groups") of similar resources, cultures, and environments to facilitate understanding of the patterns. The United Nations uses 22 country groups; however, this book further divides some United Nations groups and uses 32 country groups (Figure 1.4).

1.6 Organization of the Book

The book is divided into eleven parts of one to four chapters each. An introductory section sets the stage by describing complex systems and their implications for sustainability and change. The second part discusses human societies and people's abilities to adapt, make choices, cooperate, and so form and sustain themselves. Human societies are perhaps the most complex systems.

Then, sequential sections discuss the three environmental factors of climates, landforms, and biodiversity, followed by five parts on resources: water, agriculture, energy, minerals, and forests. The organizations of the sections vary. Generally, each section first describes the properties and distribution of its subject. Then, it discusses how its subject changes and can be managed. During these discussions, the effect of each subject on the other resources, environmental aspects, and people is expounded. The last section considers an overall perspective and the integration of all resources.

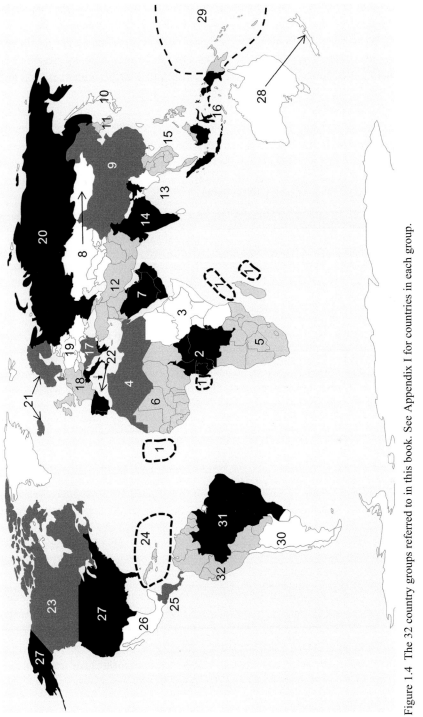

Figure 1.4 The 32 country groups referred to in this book. See Appendix I for countries in each group.

Near the ends of most sections are suggestions for possible actions that could be taken, with their merits and shortcomings. Some actions may be mutually exclusive, and some will be easier to implement than others. These suggestions were developed by the authors, by students at the School of Forestry and Environmental Studies and the Jackson Institute of Global Affairs at Yale University, and by others. It is hoped that the suggestions will stimulate readers to develop creative ways to manage the environment and resources.

The book incorporates maps, graphs, and statistical analyses and makes use of recently emerging large data sets. It also discusses trade-off issues [22].

References

1. Y. Bar-Yam. General Features of Complex Systems. *Encyclopedia of Life Support Systems (EOLSS).* (UNESCO, EOLSS Publishers, 2002).
2. J. L. Simon. *The Ultimate Resource 2.* (Princeton University Press, 1998).
3. A. Goudie. *The Human Impact on the Natural Environment*, sixth ed. (Blackwell, 2006).
4. K. C. Seto, B. Güneralp, L. R. Hutyra. Global Forecasts of Urban Expansion to 2030 and Direct Impacts on Biodiversity and Carbon Pools. *Proceedings of the National Academy of Sciences.* 2012;109(40):16083–8.
5. C. D. Oliver. Sustainable Forestry: What Is It? How Do We Achieve It? *Journal of Forestry.* 2003;101(5):8–14.
6. G. Brundtland, M. Khalid, S. Agnelli, et al. *Our Common Future.* (Oxford University Press, 1987).
7. Y. N. Harari. *Sapiens: A Brief History of Humankind.* (Random House, 2014).
8. E. Von Weizsacker, K. Hargroves, M. Smith, C. Desha, P. Stasinopoulos. *Factor Five: Transforming the Global Economy through 80% Improvements in Resource Productivity.* (Earthscan/Routledge, 2009).
9. C. D. Oliver, B. C. Larson. *Forest Stand Dynamics*, update ed. (John Wiley, 1996).
10. G. Björklund, R. Connor, A. Goujon, et al. Chapter 2. Demographic, Economic and Social Drivers, in *The UN World Water Development Report 3: Water in a Changing World.* (UNESCO, Earthscan, 2009): pp. 29–40.
11. E. Dugarova, P. Utting. *Emerging Issues: The Social Drivers of Sustainable Development.* (UN Research Institute for Social Drivers, 2013).
12. R. L. Ackoff. Science in the Systems Age: Beyond IE, OR, and MS. *Operations Research.* 1973;21(3):661–7.
13. D. Meadows. *Thinking in Systems: A Primer.* (Chelsea Green Publishing, 2008).
14. F. Berkes, C. Folke. *Linking Social and Ecological Systems.* (Cambridge University Press, 1998).
15. J. Liu, T. Dietz, S. R. Carpenter, et al. Coupled Human and Natural Systems. *AMBIO: A Journal of the Human Environment.* 2007;36(8):639–49.
16. P. Matson, W. C. Clark, K. Andersson. *Pursuing Sustainability: A Guide to the Science and Practice.* (Princeton University Press, 2016).
17. M. Cote, A. J. Nightingale. Resilience Thinking Meets Social Theory: Situating Social Change in Socio-Ecological Systems (SES) Research. *Progress in Human Geography.* 2012;36(4): 475–89.
18. A. Hornborg. Introduction: Conceptualizing Socioecological Systems. In: C. Crumley, ed. *The World System and the Earth System: Global Socioenvironmental Change and Sustainability since the Neolithic.* (Left Coast Press Inc., 2007): pp. 1–11.
19. C. S. Holling. Understanding the Complexity of Economic, Ecological, and Social Systems. *Ecosystems.* 2001;4(5):390–405.

20. C. D. Oliver. Functional Restoration of Social-Forestry Systems across Spatial and Temporal Scales. *Journal of Sustainable Forestry.* 2014;33(Supplement 1):S123–S48.
21. J. Tainter. *The Collapse of Complex Societies.* (Cambridge University Press, 1990).
22. C. Oliver, M. Twery. Decision Support Systems/Models and Analyses. In *Ecological Stewardship: A Common Reference for Ecological Management.* (Elsevier, 1999): pp. 661–85.

2

A Perspective on Dynamic Systems

2.1 Systems Theory and Complexity

This chapter will give a brief, nontechnical introduction to socioenvironmental systems as complex systems. The intention is to give the reader sufficient appreciation to begin forming his/her framework that views resources, the environment, and people through the complex systems perspective. More technical and detailed information can be found in the references.

Weaver [1] distinguished different types of systems, from "organized simplicity" to "disorganized complexity" (Figure 2.1). Disorganized complexity addresses systems of a very large number of variables, each acting in its erratic way where methods of statistical mechanics apply with averages describing the behavior. He also described the in-between region of "organized complexity" that covers most of the environmental, resource, and social systems discussed in this book. They are system of a highly interrelated, large number of variables that display some degree of organization. "Organized simplicity" systems involve a few variables whose relationships are well known.

At about the same time that Weaver described different kinds of systems (Figure 2.1), the field of cybernetics was activated that addresses problems of regulatory systems involving a diverse set of components from machines to humans communicating through positive and negative feedback loops [2, 3].

This approach of integrating disciplines and seeking solutions to problems in the organized complexity region gained scientific popularity in the 1950s with the establishment of General Systems Theory as "the formulation and deduction of those principles which are valid for systems in general" [4, 5, 6]. An extensive collection of essays related to the subject are found in [7].

The Santa Fe Institute (SFI) formed a parallel field of "complexity science" that addresses the same region of problems with a different emphasis [8, 9]. The founding of SFI was followed by others.

There are disagreements on the definition of complex systems or even on the need for one, as well as on ways of measuring the degree of complexity. One definition is "a system in which large networks of components with no central control and simple rules of operation give rise to complex collective behavior and sophisticated information

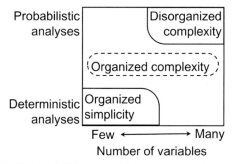

Figure 2.1 Conception of relation of different problems and analysis approaches in science. Based on [1].

processing" [8]. Adaptability is usually taken as a property and the systems are referred to as complex adaptive systems (CAS) although systems such as weather patterns are complex but not adaptive [11, 12].

The emphasis is on nonlinear, dynamic interactions among components. The interactions can be strong or weak and are subject to change. They can also disappear or appear. Another frequently studied feature is pattern formation–self-organization [13]. The complicated interactions give rise to "emergent properties" of the whole that are not explainable by studying those of the components.

Computer simulations constitute a significant area of activity. They enable many simulated components with different assigned behaviors to interact and their emergent properties to be observed. The patterns formed by self-organization can be compared among models to seek and explain commonalities and differences. Studies are pursued in diverse fields such as physics, biology, economics, and social sciences by adapting methods of one to others – at times forced rather than appropriate.

A distinction between systems theory and complexity is that systems theory is confirmatory while complexity theory is exploratory [14]. The concepts in both lead to an insightful, informative comprehension.

2.2 Some Terminology of Complex Systems

The terms below are commonly used when discussing complex systems. Some terms predate the emergence of complexity science as a discipline [15].

- *Agents*, actors (*entities*): Fundamental active units that interact and form systems. They can be atoms, organic cells, individuals, or community units.
- *Boundaries:* The demarcation of those elements (agents, actors) that are part of the system from those that are part of the system's environment.
- *Emergence*: Properties arising mainly from the interactions of a system's constituent parts that are not observed in the separate constituent parts.

- *Self-organization*: Emergent regularities, patterns, organized behavior without an internal or external controller [16].
- *Building blocks*: Interacting entities that are simple systems and form the basis of more complex systems.
- *Feedback loops*: Loops within a system where the output of a part comes back as input after some transformation. This process can have an amplifying effect or an equalizing one that reduces the deviation from the desired level. Feedback is referred to as positive/reinforcing or negative/balancing, respectively.
- *Nonlinearity*: Behavior where the output is not consistently proportional to the input.
- *Adaptability*: Change of behavior and configuration to accommodate a change in the environment through learning or other processes.
- *Uncertainty*: A condition of limited knowledge that makes it impossible to describe precisely the existing state, future outcome, or only some of the possible outcomes.
- *Vibrancy*: A property of being vigorous, often pulsating, and full of activity.
- *Vulnerability*: The probability that a system or subcomponent will experience harm when exposed to a hazard or stress.

2.3 A Conceptual Scheme of Complex Systems

A conceptual schema for the evolution of dynamic systems is suggested (Figure 2.2) using the example of life and society. Terms are discussed in this chapter. Chapter 3 will address the current studies of the details.

The development shown in Figure 2.2 follows the general pattern below:

1. Initially there are relatively isolated simple entities with very weak interactions, if any. Groups start forming as interactions get stronger but they are within close localities. With the evolution of life, the notion of resources emerges. Early on, resources are strictly limited to raw materials. Groups formed may overlap and start displaying some rather simple structure, performance, behavior patterns. Some properties of self-organization may be displayed (Figure 2.2a–b).
2. With increasingly complicated groups, specialization gets more pronounced, such as animals forming social groups. Group interactions become stronger. Structures, hierarchies or networks in the present terminology get more complex. All these formations are very sensitive to initial conditions, which typically are hypothesized but not well known (Figure 2.2c–d).
3. Once initiated, this dynamism constantly develops and changes at different rates, at times very gradual and at times rather dramatic. Interaction patterns and intensities get more complex and spatially more extended. Technological advances, authoritarian and other organizational hierarchies, craftsmen and mercantile networks form, combine, or disappear. New tools replace older ones. Systems get more elaborate and difficult to understand from the properties of the subsystems. Increasing complexities generally generate new problems that may be confronted with one or more layers of

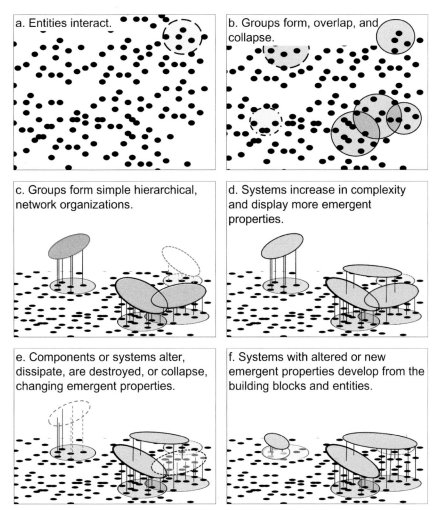

Figure 2.2 Development of complex systems from individual entities (dots) to systems with emergent properties (shaded areas).

complexity [17]. Declines or disappearances generally leave behind skills and institutions that can be viewed as building blocks for the next phase. Somewhere between order and disorder or conflicts and resolution lies a state of desirable vibrancy (Figure 2.2e–f) [18].

Another way of looking at Figure 2.2 is Figure 2.3. In Figure 2.3a, declines without replacement are depicted. Declines can be gradual or abrupt. Figure 2.3b shows a system being replaced by another one before it declines. The replacement can actually cause the decline. In Figure 2.3c the replacement can evolve after the onset of decline and retain some of the subcomponents. In all three cases there would be replacements that never mature.

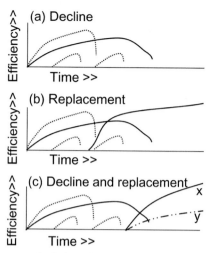

Figure 2.3 A system or subsystem declining with several interactions and consequences and another subsystem sometimes replacing it or succeeding it.

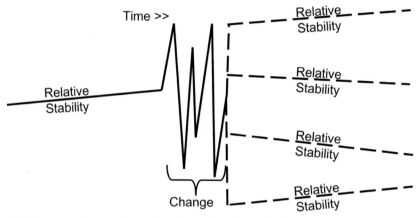

Figure 2.4 Resource change pattern: slow change, then dramatic change or collapse, then reorganization of the system components.

The declines of Figure 2.3 can occur to different subsystems or to the whole system and at different rates. If too many changes happen in temporal and spatial proximity, a temporary chaotic situation may take place that is likely to somehow stabilize (Figure 2.4). The new organizational forms may be quite different from the previous one for better or worse. Its form seems to be largely based on the conditions immediately after the change ("initial conditions"). The remaining entities, building blocks, and subsystems begin interacting to form the new systems and their behaviors.

Examples of these collapses and reorganizations can be found in biodiversity systems (Figure 2.5) [19–22] and human societies [23], among others. "Mass extinctions" of species

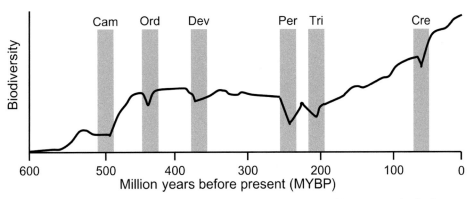

Figure 2.5 Increase in biodiversity for past 500 million years interrupted by "mass extinction" events (Chapter 13) and recovery periods [19, 21, 22]. Cam = Cambrian geologic period, Ord = Ordovician, Dev = Devonian, Per = Permian, Tri = Triassic, Cre = Cretaceous.

have occurred periodically during the Earth's existence. The preexisting plant and animal families did not necessarily recover following these mass extinctions; sometimes, completely different organisms dominated. For example, mammals have been most prevalent since the Cretaceous extinction of 65 million years ago, whereas dinosaurs were dominant previously.

2.4 Modeling

Models are simplified depictions of reality as perceived by the person or persons constructing them. They are essential tools for dealing with complexity and can be anything from an architect's mockup to a concise mathematical equation. All scientific statements are models describing observations and relationships. Models also allow the study of systems where experimentation with the real one may be impossible, too expensive, or take too long [24].

A system is typically defined by a set of variables. The choice of variables used to depict the system's behavior sets its boundaries. The range of the values that the defining variables can take defines the "state space" of the system. Some variables may be left out because they are considered insignificant in the context: changing so very slowly that they may be viewed as constant, namely a parameter, for the time scale of interest; or simply difficult to quantify. There may also be variables whose relevancy to the issue at hand are not known about at the time. Thus defined, the system has a structure and function. The state of a system at any one time is where it happens to be in its state space. When starting from an initial condition, the system tends to converge to a subspace within the state space known as a "basin of attraction" [25].

Models can be too simple or too complicated. Sophistication is not necessarily productive. Both constructing a good model and collecting the data to be used can be very time consuming. International efforts that involve people with diverse backgrounds improve ideas of the data needed and enable collection of otherwise inaccessible data, but add confusion that complicates and lengthens the processes.

Computer models enable one to compress time and space. They can provide solutions, optimizations, and capability of playing what-if games but can also provide understanding of the behavior of the system under study that may at times be rather counterintuitive. The sensitivities to initial conditions, parameter values, and responses to a variety of exogenous inputs can be studied. Creative software tools have been developed to support studies of specific problems and research in general [26].

The models relevant to this book are constructs depicting the structure of the system under study, with the interactions, dependencies, feedback loops, and time delays perceived by the modelers. Two main approaches of relevant computer modeling are "system dynamic models" and "agent-based models" [27, 28].

System dynamic modeling was developed by J. Forrester in the 1950s [29, 30] and has three basic features: feedback loops, stocks, and flows. A stock is any entity that accumulates or depletes over time. The entities in the stock are homogenous and thus do not exhibit individuality. A flow is the rate of accumulation of the stock.

Agent-based models allow heterogeneity in the stocks by assigning individuality to the entities in the stocks that are referred to as agents. The individuality allows the model to express more variation in response to the inputs.

2.5 Hierarchies and Networks

Two structural forms in the current literature address complex systems: hierarchies (Figures 1.1, 2.6) and networks (Figure 2.7). Sometimes integrative use of the two fits the context better.

The hierarchical nature of complexity gained recognition with Simon [31]. A hierarchy is an organization of entities forming levels with an ordering of above, below, and the same. In general, behavior becomes slower as one goes up the levels; and upper levels put constraints on the behavior of the lower levels. The levels are defined by common properties, the choice of which depends on the particular problem being studied and modeled. Information flows up the levels in a filtered mode [32, 33]. The term panarchy (Figure 3.1b) was introduced to avoid the authoritarian connotation of the word hierarchy.

Recently, systems are also being studied as networks, with advances made possible by massive data collections and analyses coinciding with advancing computing technologies. Networks are sets of nodes and links (Figure 2.7). They can be seen in social networks, information networks, computer networks, transportation networks (Figure 4.4), distribution networks, and biological networks (Figure 13.3). Network science is an integrative field of study involving mathematics, physics, computer sciences, biology, and sociology; it lends itself better to mathematical analyses than does the study of hierarchies.

Systems develop from a lower (Figure 2.2) or higher hierarchy level, such as a government colony or species isolated in a refugium (Chapter 13). They usually begin

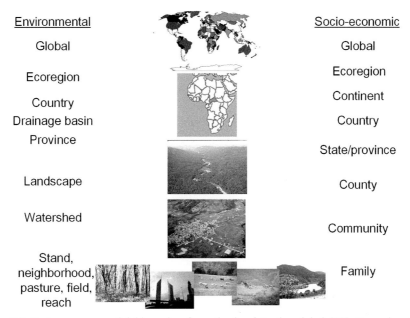

Environmental		Socio-economic
Global		Global
Ecoregion		Ecoregion
Country		Continent
Drainage basin		Country
Province		State/province
Landscape		County
Watershed		Community
Stand, neighborhood, pasture, field, reach		Family

Figure 2.6 Systems as a spatial hierarchy, from the local to the global [32]. Reproduced with permission of Taylor and Francis Group.

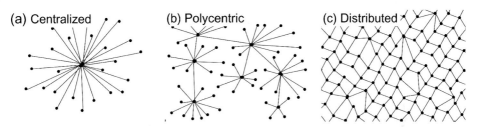

Figure 2.7 Examples of different network organizations. (dots = "nodes," lines = "links")

as a single control (Figure 2.7a) known as a "flat organization" and a "centralized network." These often expand into a well-developed hierarchical organization (Figures 1.1, 2.6), depicted as a "polycentric network" (Figure 2.7b). As interactions expand, the organization becomes increasingly "distributed" (Figure 2.7c).

Well-distributed networks have information, actions, or other influences flowing in a variety of directions, in contrast to hierarchies where the flows are primarily in vertical channels. A well-distributed network can receive information at any node and distribute it throughout the system, allowing any agent to respond as appropriate. Such networks in complex systems are more capable of responding successfully to outside influences that could destroy the system. In a hierarchy where only a few entities make the decisions, the variety of responses are limited to these entities; whereas a well-distributed network allows more entities to understand the situation and respond – and to gain the response of others [35].

Some of the measures developed in network analysis include average path length, degree of a node, modularity. Average path length is the average number of links spanned along the shortest paths for all possible pairs of nodes in the network. The degree of a node is the number of links coming in or out of a node. Modularity is the existence of communities or modules with higher connectivity within than with other modules.

Two network structures that approximate the real-world networks are small-world networks and scale-free networks. Small-world networks are those that have a small average path length relative to the number of nodes. Scale-free networks are those with few high-degree nodes and many low-degree ones [8, 35–39].

References

1. W. Weaver. Science and Complexity. *American Scientist*. 1948;36:536–44.
2. N. Wiener. *Cybernetics: Control and Communication in the Animal and Machine*. (MIT Press, 1948).
3. W. R. Ashby. *An Introduction to Cybernetics*. (Chapman and Hall, 1956).
4. L. Von Bertalanffy. *General System Theory*. (George Braziller, 1969).
5. K. E. Boulding. General Systems Theory – The Skeleton of Science. *Management Science*. 1956;2(3):197–208.
6. J. W. Forrester. System Dynamics, Systems Thinking, and Soft OR. *System Dynamics Review*. 1994;10(2–3):245–56.
7. G. Midgley. *Systems Thinking*. (Sage London, 2003).
8. M. Mitchell. *Complexity: A Guided Tour*. (Oxford University Press, 2009).
9. M. M. Waldrop. *Complexity: The Emerging Science at the Edge of Chaos*. (Simon & Schuster, 1992).
10. M. Gell-Mann. What Is Complexity? In *Complexity*. (Wiley, 1995).
11. S. A. Levin. Ecosystems and the Biosphere as Complex Adaptive Systems. *Ecosystems*. 1998;1(5):431–6.
12. R. Dodder, R. Dare. Complex Adaptive Systems and Complexity Theory: Inter-related Knowledge Domains. *ESD 83: Research Seminar in Engineering Systems, October 31, 2000*. (Massachusetts Institute of Technology, 2000): 14 pp.
13. F. Heylighen. Complexity and Self-Organization. *Encyclopedia of Library and Information Sciences*. 2008;3:1215–24.
14. S. E. Phelan. A Note on the Correspondence between Complexity and Systems Theory. *Systemic Practice and Action Research*. 1999;12(3):237–46.
15. W. Ashby. Principles of the Self-Organizing Systems. *Facets of System Science*. 1991;6(1/2):102.
16. T. De Wolf, T. Holvoet, editors. Emergence Versus Self-Organisation: Different Concepts but Promising When Combined. In *International Workshop on Engineering Self-Organising Applications*. (Springer, 2004).
17. J. Tainter. Complexity, Problem Solving, and Sustainable Societies. In *Getting Down to Earth: Practical Applications of Ecological Economics*. (Island Press, 1996): pp. 61–76.
18. J. P. Crutchfield. Between Order and Chaos. *Nature Physics*. 2011;8:17–24.
19. J. W. Kirchner, A. Weil. Delayed Biological Recovery from Extinctions Throughout the Fossil Record. *Nature*. 2000;404(6774):177–80.
20. D. B. Botkin, M. J. Sobel. Stability in Time-Varying Ecosystems. *The American Naturalist*. 1975;109(970):625–46.

21. A. Hallam. *Catastrophes and Lesser Calamities: The Causes of Mass Extinctions.* (Oxford University Press, 2005).
22. D. Huddart, T. Stott. *Earth Environments: Past, Present, and Future.* (Wiley-Blackwell, 2012).
23. M. R. Dove, P. E. Sajise, A. A. Doolittle. *Beyond the Sacred Forest: Complicating Conservation in Southeast Asia.* (Duke University Press, 2011).
24. J. D. Sterman. All Models Are Wrong: Reflections on Becoming a Systems Scientist. *System Dynamics Review.* 2002;18(4):501–31.
25. B. Walker, C. S. Holling, S. R. Carpenter, A. Kinzig. Resilience, Adaptability and Transformability in Social-Ecological Systems. *Ecology and Society.* 2004;9(2):5.
26. H. Rahmandad, J. D. Sterman. Reporting Guidelines for Simulation-Based Research in Social Sciences. *System Dynamics Review.* 2012;28(4):396–411.
27. H. Rahmandad, J. Sterman. Heterogeneity and Network Structure in the Dynamics of Diffusion: Comparing Agent-Based and Differential Equation Models. *Management Science.* 2008;54 (5):998–1014.
28. A. Borshchev, editor. From System Dynamics and Discrete Event to Practical Agent Based Modeling: Reasons, Techniques, Tools. In *Proceedings of the 22nd International Conference of the System Dynamics Society*, July 25–29. (Citeseer, 2004).
29. D. C. Lane. The Power of the Bond between Cause and Effect: Jay Wright Forrester and the Field of System Dynamics. *System Dynamics Review.* 2007;23(2/3):95–118.
30. S. Albin. Building a System Dynamics Model (MIT System Dynamics in Education Project, 1997 [Accessed August 16, 2017]). Available from: https://ocw.mit.edu/courses/sloan-school-of-management/15-988-system-dynamics-self-study-fall-1998-spring-1999/readings/building.pdf.
31. H. A. Simon. The Architecture of Complexity. In *The Science of the Artificial.* (MIT Press; 1969): pp. 192–229.
32. S. N. Salthe. Summary of the Principles of Hierarchy Theory. *General Systems Bulletin.* 2002; 31: 13–17.
33. H. H. Pattee. *Hierarchy Theory: The Challenge of Complex Systems.* (George Brasiller, 1973).
34. C. D. Oliver. Functional Restoration of Social-Forestry Systems across Spatial and Temporal Scales. *Journal of Sustainable Forestry.* 2014;33 (Supplement 1):S123–S48.
35. Y. Bar-Yam. General Features of Complex Systems. In *Encyclopedia of Life Support Systems (EOLSS), UNESCO* (EOLSS Publishers, 2002).
36. M. Newman. *Networks: An Introduction.* (Oxford University Press, 2013).
37. M. E. Newman. Modularity and Community Structure in Networks. *Proceedings of the National Academy of Sciences.* 2006;103(23):8577–82.
38. S. H. Strogatz. Exploring Complex Networks. *Nature.* 2001;410(6825):268.
39. P. Cilliers. Boundaries, Hierarchies and Networks in Complex Systems. *International Journal of Innovation Management.* 2001;5(02):135–47.

3

Change, Sustainability, and Related Concepts

3.1 Change

A societal challenge is to become aware and acknowledge that change is innate to the environment, resources and human systems. The history of human beings in almost all aspects is one of change, although people commonly presume stability. What constitutes the societal challenge now is the awareness and acknowledgment of the complex, dynamic linkages between the biophysical, economic, and social components of the environmental system.

People often regard things that change slowly as stable; however, change can occur almost imperceptibly slowly or suddenly at infrequent intervals. Changes can be reversible or irreversible; and, people can induce, avoid, and/or direct change with proper understanding.

Changes occur because of internal dynamics of systems or external forces. The impact of changes may be benign, easily adapted to, and/or more crucial. Crucial changes would require appropriate perception of the situation and actions ranging between ignoring and taking dramatic action. A wise management plan based on available information, knowledge, and underlying uncertainties can often take advantage of changes that could otherwise be harmful. Some external forces can become disasters such as floods and earthquakes depending on both the size of the disturbance and the susceptibility (or resistance) of the target [1]. Currently, major changes that are most deliberated are the demographics/population age distribution, biodiversity, water, and climate. These vary spatially and temporally, as previously mentioned.

Although people are open to discussing change, human mental models tend to be averse to them. Infrastructures, institutions, behaviors, and uncertainties about what will emerge seem to promote "wait and see" attitudes that may be counterproductive.

This chapter introduces the concepts and approaches to understanding and managing change in socioenvironmental systems.

3.2 Sustainability

Responses to changes – preventive, adaptive, reactive – are presently discussed under the label of sustainability. Although the concept can be traced to eighteenth-century German forestry (von Carlowitz, 1713) [2], "sustainability" was recently made highly visible as "sustainable development" in the UN 1987 Brundtland report as: "development that meets the needs of the present without compromising the ability of the future generations to meet their own needs" [3]. In spite of being a powerful concept, somethings remained vague – such as what and how to change and what and how to keep static.

Understanding and explaining changes, responding to changes, and channeling changes in desirable ways are the main thrusts of a relatively new field of "sustainability science." Terminologies such as "robustness," "vulnerability," "transformability," "adaptability," and "risk reduction" are often used. Sustainability science has initiated new approaches that are highly interdisciplinary and integrate concepts from complexity science. Tools and methodologies are being developed revealing unique and at times counterintuitive aspects of confronted situations. [4–15].

There is also a dispute over the compatibility of economic growth and sustainability [12, 16].

3.3 Resilience, Tipping Points, and Regime Shifts

Resilience is a major aspect of sustainability studies and is introduced here by referring to Holling's adaptive cycle framework (Figure 3.1). Resilience studies are largely concerned with the capability of individuals, communities, and large populations to absorb perturbations. Several concepts from the work of Resilience Alliance will be briefly presented.

The adaptive cycle is common in socioenvironmental cycles and contains four phases: exploitation, conservation, release, and reorganization (Figure 3.1) [17], similar to the change and reorganization in Figure 2.2. Panarchy is a hierarchy of adaptive cycles. The lower cycles are smaller, more frequent, and faster; and the higher ones, larger, less frequent, and slower [17, 18].

Keeping sustainability and resilience explorations as separate areas of study has been suggested [19].

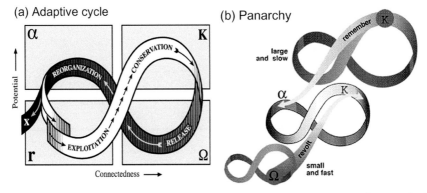

Figure 3.1 (a) Adaptive cycle and (b) panarchy behaviors of complex systems at different scales [17]. Reproduced by permission of Island Press.

3.3.1 Adaptability and Transformability

Adaptability of a complex system is the ability to change a system's resilience in a constructive manner and retain its previous behavior and/or function through learning, experience, knowledge, and evolution.

Transformability of a system is its capability to transition to a different system with a new structure, function and with an altered set of variables if the existing one becomes untenable for various reasons [20–22]. A system can transform in many ways, depending on which interactions are affected. Consequently, there can be many events that can make a system untenable and many others that can transform it and prevent its becoming untenable.

3.3.2 Tipping Points and Regime Shifts

The variables chosen for depicting a system change at different rates and can be categorized as fast- and slow-moving. Some are considered to be controlling variables (typically slow-moving ones) with tipping points also referred to as thresholds. Once over a tipping point, the system dramatically changes its structure and function, commonly shifting to a new one. A tipping point may also be reached because of feedback loop delays in the system. Undesirable tips can be avoided by monitoring the movement of the controlling variable or, if possible, changing the tipping point. The shift is referred to as a "regime shift" (critical transitions) and may or may not be reversible. Even if reversible, the effort of reversing may be too costly. The new regime may be the result of a cascading effect of the initial tipping point, once reached, triggering other shifts [23–29].

Figure 3.2 is a simplified two-dimensional depiction of changes in the state of a system responding to changes in a controlling variable or driver. (a) and (b) are reversible. (c) is reversible but does not return by the same pathway on which it initially changed.

Anticipation of critical transitions is another area of study. Increasingly slow recovery from a minor perturbation and increasingly high variance are features signaling an approaching threshold. Changes of these features signal transitions in many complex systems. The structure of the underlying network, modularity, homogeneity/heterogeneity

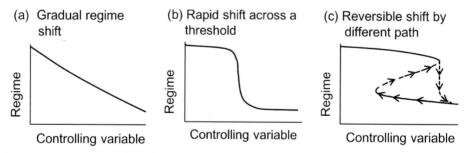

Figure 3.2 Three regime shifts discussed in text [23–27, 30–32].

of subcomponents, and connectivity are also indicators of potential susceptibility to critical transitions [23, 24, 27, 30–32].

3.3.3 Leverage Points and Appropriate Action

The functioning and structure of complex systems are commonly not well understood. A point where a small effort produces a major change is a "leverage point," and one purpose of understanding complex systems is to identify and understand leverage points. Leverage points may be identified/hypothesized with the intended outcomes of improvements. Intended improvements and similar activities need to be closely monitored for unexpected, undesirable changes elsewhere in the system.

Major characteristics of complex systems are nonlinearities through which small changes at one point may produce a big one at some other point. And, interventions at another chosen point may be ineffective and need revisions.

Even when leverage points are identified for a system, knowing how to change them for desired effects is another challenge. Forrester [11] claimed that people often successfully identify leverage points, but push them the wrong way and so do not achieve their purpose.

3.4 Case Studies

Case studies are learning tools to train one to think in terms of complex systems. Quite a few are discussed in the literature [4, 5, 33]. The following four have been chosen because they present various problems that are encountered in changing socioenvironmental systems: an ongoing issue, an issue involving a rather homogenous and educated public, involvement of spiritual values, and transboundary problems. Case studies also show how responses exacerbated or ameliorated the situation – generally not returning the system to its initial state.

1. **Goulburn–Broken Catchment**, southeastern Australia, is a well-documented, thorough study of a socioenvironmental system with many concerns. The major concern is the rise of the salty water table leading to salinization of surface soils that threatens crops (Chapter 18). European settlers removed deep-rooted vegetation that had evaporated surface water and maintained a dry, impenetrable barrier between salty ground water and surface fresh water; and so surface soils became salty. Extended droughts then lowered the water table and so reduced the salts near the surface; however, the droughts also led to more irrigation. This irrigation again eliminated the impenetrable dry barrier and allowed salts to reenter surfaces soils and destroy crops. Similarly, extended wet seasons led to the salty water table rising to critical levels and to subsequent pumping to lower it. All of these problems are costly. Biophysical, economic, social subcomponents of the system have been studied. Features with fixed or controllable thresholds, their interactions and feedbacks in between that might lead to cascading effects have been addressed [5, 34].

2. **Kristianstads Vattenrike**, Sweden is another well-studied socioenvironmental system. The area is the most productive in Sweden and includes flooded meadows historically used for haymaking and grazing. Although declared to be a wetland area of international importance, the decline in wildlife and degradation of water resources continued. This undesirable process was attributed to increasing farmed land and controlling natural water flows. With an environmentally aware public and under the scholarly leadership and communication skills of Sven-Erik Magnusson (Chapter 6), the situation was reversed. It is now a nationally and internationally highly visited reserve maintained by active management [5].

3. **Southern Androy**, Madagascar is an originally dry forested area that has been fragmented into patches by many years of intensified agriculture. An important issue is the desirable pollination of some crops by insects that live in forest patches. Local people have protected these patches because of spiritual values attached to them. Migrations of these local people to cities for a variety of economic and cultural reasons have weakened this protection (Chapter 5), the patches are diminishing, and the pollinating insects are becoming fewer.

   One suggested approach is to look at the island as a network with patches as nodes to study connectivity [24, 35].

4. **Aral Sea Basin** (rivers Amu Darya and Syr Darya), Central Asia has received much attention for water use and misuse. Crops requiring heavy irrigation, such as cotton and wheat, have led to dramatic water volume losses in the Aral Sea. Winds carrying salt from the dried lake bed have caused desertification in the basin. The transboundary nature of the river system in Central Asia has been a source of conflict rather than cooperation among the governments involved (Chapter 18). An economic and social transformation to more technology and knowledge-based industries tapping on the relatively well-educated public may be a more prudent approach, combined with a reduction in intensive agriculture use of water [36]. A similar transboundary situation that is being amicability resolved is the Rhine River Basin. It seems to be controlled by collaboration between downstream and upstream states [37].

3.5 Frameworks

A "framework" is a conceptual structure of a system that displays the interactions among subcomponents. It provides the guidelines for constructing models. Many frameworks have been developed for studying complex socioenvironmental systems, such as Figure 3.3. The choice and modification of the framework is based on the decisions of the people concerned and the specific issue. The framework chosen strongly influences the data and information needed, how the data is analyzed and interpreted, and how the results are acted on [38].

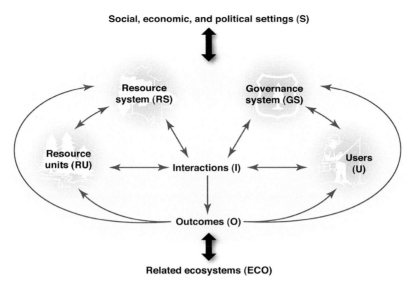

Social, economic, and political settings (S)

Resource system (RS)

Governance system (GS)

Resource units (RU) ←→ Interactions (I) ←→ Users (U)

Outcomes (O)

Related ecosystems (ECO)

Figure 3.3 One example of a framework [39]. Reprinted with permission from AAAS.

3.6 Management

Deliberate, responsible resource management is possible but a "crisis" mode typically leads to fear of taking action or to quick, shortsighted fixes that lead to more problems, as previously mentioned. The idea is to incorporate resource management for environmental, economic, and social values into our way of life. Needed are open, active, and creative minds that follow new ideas and technologies. It is difficult at first to discern good from poor new ideas.

The emphasis may seem to be quantitative but the moral and ethical aspects of management and decision-making are to be kept in mind. And decisions that do not satisfy the community are bound to fail.

Managing socioenvironmental systems is broader than recently popular financial management. Socioenvironmental systems are managed to ensure their benefits are preserved, sustained, increased, or changed. Each system has different opportunities and limitations and involves many technical issues covered in this book.

Management is coordinating the dynamics of systems over different scales of time and space to ensure relevant aspects of the system are addressed and coordinated. Higher levels of management concentrate on larger areas and longer timescales. Increasingly lower management levels are delegated to specific, smaller areas and tasks. Different levels require different skills and perspectives. A manager promoted to a higher level would need to think in a new framework than before.

Management can use different approaches depending on how certain and easily controlled the target of management is (Figure 3.4) [40–43].

Management involves decision-makers with authority; subject matter experts with both a thorough understanding of interactions among resources, people, and the environment

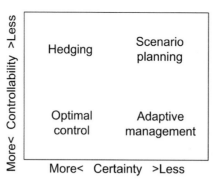

Figure 3.4 Matrix of situations when different management approaches are appropriate [41].

and a high sensitivity to local knowledge; as well as community members and other stakeholders with commonly conflicting interests. Management also involves awareness of unknowns and uncertainties, readiness to avoid or initiate changes, and willingness to modify decisions made recently or long ago. Trade-offs will always be involved.

Complex adaptive management emphasizes learning as a very significant component of the process [33]. As societies become more complex and shift to more distributed network behaviors (Figure 2.7a), the concurrence of the stakeholders becomes more vital for implementing complex decisions. An informed public becomes increasingly significant.

Some considerations apply in all management situations. The "law of requisite variety" addresses the need for management/control of complex systems to be equally or more flexible than the system being managed/controlled [44].

Unfortunately, decision-making can be a time-consuming process. Even if minimized, some unintended consequences occur. Management can best adjust to them with a dynamic decision-making approach that does not make decisions that preclude options unless necessary. The system being managed should be monitored by measuring appropriate performance indicators. Decisions made recently or long ago should be modified when appropriate. Where expedient shortcuts exist to any of the processes, managers should be aware of them and their strengths and weaknesses.

3.6.1 Data, Information, Knowledge, Understanding, Wisdom

This section introduces the relationship of data, information, knowledge, and wisdom (DIKW) [45–47]. The first step is to obtain and use data. Data are observed facts that consist of measures and descriptive observations that stand alone and have no meaning. Data collection is typically a large part of the extended management process. What data to collect depends on the questions being asked.

Information is data processed to be meaningful. Information is given in charts, graphs, tables, and other creative formats to show trends of "who," "what," "where," and "when" over time and space and among resources. Information uses a variety of methods to synthesize data, including statistical analyses and models. The data collected needs to be analyzed properly and the results conveyed in a clear way – not a simple task. The subsequent analysis may lead to new questions to ask, for which the data has not been collected.

Knowledge and understanding are applications of information to determine how and why something occurs, with knowledge being considered more deterministic. Understanding is information put into a context that allows information to be upgraded.

Within the past two or three decades a substantial amount of data has been amassed by agencies such as the United Nations Food and Agriculture Organization (UNFAO), the Energy Information Agency (EIA), the International Energy Agency (EIA), the Organization of Economically Developed Countries (OECD), the World Bank Group, the United States Central Intelligence Agency (USCIA), and others. These are tremendous helps in resource management but still have limitations.

Some confusions are the differing measurement techniques between and within regions and over time that make processing problematic. Varying measurement units are often used for the same subject; although not difficult, conversions can lead to downstream errors. Some data are reported from satellite imagery while others come from land surveys, each having a different bias. Definitions of the entity to be measured or reported vary among measuring groups. Simple errors in recording or reporting as well as forgeries or intentional obfuscations may cause serious problems. Like most data sets, the errors and confusions will increasingly be resolved as more people use the data. And, the data will allow the world's resources to be managed much more effectively.

Unfortunately, the process of generating information can be very time-consuming even with current computers and remote systems (satellite and drone) and can be at odds with the need for timeliness. Shortcuts such as using data from a different region to fill a gap can result in useless information. And, intentional or unintentional negligence in recoding or reporting data can generate worthless information.

Wisdom is the synthesis of this knowledge and understanding into an ability to make proactive decisions that incorporate not only the knowledge and understanding provided ultimately by the data, but to put it into context so that decisions handed down to be implemented transcend the immediate situation – and are capable of being implemented. The above may appear rather idealistic; however, incorrect data or synthesis can lead to misinformation, ignorance, stupidity, and folly [48].

References

1. C. D. Oliver, B. C. Larson. *Forest Stand Dynamics*, update ed. (John Wiley, 1996).
2. F. Schmithusen. Three Hundred Years of Applied Sustainability in Forestry. *Unasylva*. 2013;64 (1):3–11.
3. G. Brundtland, M. Khalid, S. Agnelli, et al. *Our Common Future*. (Oxford University Press, 1987).
4. P. Matson, W. C. Clark, K. Andersson. *Pursuing Sustainability: A Guide to the Science and Practice*. (Princeton University Press, 2016).

5. B. Walker, D. Salt. *Resilience Thinking: Sustaining Ecosystems and People in a Changing World.* (Island Press, 2006).
6. M. Mulligan. *An Introduction to Sustainability: Environmental, Social, and Personal Perceptions.* (Routledge, Taylor and Francis Group), 2015).
7. N. A. Ashford, R. P. Hall. *Technology, Globalization, and Sustainable Development: Transforming the Industrial Society.* (Yale University Press, 2011).
8. H. S. Brown. Sustainability Science Needs to Incorporate Sustainable Consumption. *Environment: Science and Policy for Sustainable Development.* 2012;1(54(1)):20–5.
9. R. W. Kates, ed. *Readings in Sustainability Science and Technology (No. 213).* (Center for International Development at Harvard University, 2010 [Accessed June 19, 2017]). Available from: http://rwkates.org/pdfs/a2010.02.pdf.
10. H. Daly. *A Steady-State Economy.* Opinion Piece for Redefining Prosperity. *Sustainable Development Commission*, UK, Vol. 24, 2008 (Accessed April 8, 2016). Available from: www.sdcommission.org.uk/publications.php?id=775.
11. D. Meadows. Places to Intervene in a System. *Whole Earth.* 1997;91:78–84.
12. J. D. Sterman. Sustaining Sustainability: Creating a Systems Science in a Fragmented Academy and Polarized World. in *Sustainability Science.* (Springer, 2012): pp. 21–58.
13. J. Fisher, K. Rucki. Re-Conceptualizing the Science of Sustainability: A Dynamical Systems Approach to Understanding the Nexus of Conflict, Development and the Environment. *Sustainability Science.* (Wiley Online Library, 2016). (10.1002/sd).
14. J. Anderies, C. Folke, B. Walker, E. Ostrom. Aligning Key Concepts for Global Change Policy: Robustness, Resilience, and Sustainability. *Ecology and Society.* 2013;18(2).
15. M. A. Beroya-Eitner. Ecological Vulnerability Indicators. *Ecological Indicators.* 2016;60:329–34.
16. G. Daily. *Nature's Services: Societal Dependence on Natural Ecosystems.* (Island Press; 1997).
17. L. Gunderson, C. S. Holling. *Panarchy: Understanding Transformations in Human and Natural Systems.* (Island Press, 2002).
18. C. R. Allen, D. G. Angeler, A. S. Garmestani, L. H. Gunderson, C. S. Holling. Panarchy: Theory and Application. *Ecosystems.* 2014;17(4):578–89.
19. C. Redman. Should Sustainability and Resilience Be Combined or Remain Distinct Pursuits? *Ecology and Society.* 2014;19(2).
20. B. Walker, C. S. Holling, S. R. Carpenter, A. Kinzig. Resilience, Adaptability and Transformability in Social-Ecological Systems. *Ecology and society.* 2004;9(2):5.
21. C. Folke, S. R. Carpenter, B. Walker, et al. Resilience Thinking: Integrating Resilience, Adaptability and Transformability. *Ecology and Society.* 2010;15(4).
22. R. W. Kates, W. R. Travis, T. J. Wilbanks. Transformational Adaptation When Incremental Adaptations to Climate Change Are Insufficient. *Proceedings of the National Academy of Sciences.* 2012;109(19):7156–61.
23. T. M. Lenton. Environmental Tipping Points. *Annual Review of Environment and Resources.* 2013;38:1–29.
24. A. P. Kinzig, P. A. Ryan, M. Etienne, et al. Resilience and Regime Shifts: Assessing Cascading Effects. *Ecology and Society.* 2006;11(1).
25. B. Walker, S. Carpenter, J. Rockstrom, A.-S. Crépin, G. Peterson. Drivers, "Slow" Variables, "Fast" Variables, Shocks, and Resilience. *Ecology and Society.* 2012;17(3).
26. A.-S. Crépin, R. Biggs, S. Polasky, M. Troell, A. de Zeeuw. Regime Shifts and Management. *Ecological Economics.* 2012;84:15–22.
27. S. J. Lade, A. Tavoni, A. Levin, M. Schluter. *Regime Shifts in a Social-Ecological System.* 2013.
28. B. Walker, J. A. Meyers. Thresholds in Ecological and Social-Ecological Systems: A Developing Database. *Ecology and Society.* 2004;9(2):3.
29. B. Walker, L. Gunderson, A. Kinzig, et al. A Handful of Heuristics and Some Propositions for Understanding Resilience in Social-Ecological Systems. *Ecology and Society.* 2006;11(1):13.
30. M. Scheffer. *Critical Transitions in Nature and Society.* (Princeton University Press, 2009).

31. M. Scheffer, J. Bascompte, W. A. Brock, et al. Early-Warning Signals for Critical Transitions. *Nature*. 2009;461(7260):53–9.
32. M. Scheffer, S. R. Carpenter, T. M. Lenton, et al. Anticipating Critical Transitions. *Science*. 2012;338(6105):344–8.
33. F. Berkes, J. Colding, C. Folke. *Navigating Socioecological Systems*. (Cambridge University Press, 2003).
34. B. H. Walker, N. Abel, J. M. Anderies, P. Ryan. Resilience, Adaptability, and Transformability in the Goulburn-Broken Catchment, Australia. *Ecology and Society*. 2009;14(1):12.
35. M. A. Janssen, Ö. Bodin, J. M. Anderies, et al. Toward a Network Perspective of the Study of Resilience in Social-Ecological Systems. *Ecology and Society*. 2006;11(1):15.
36. L. Izquierdo, M. Stangerhaugen, D. Castillo, R. Nixon, G. Jimenez. Water Crisis in Central Asia: Key Challenges and Opportunities. *Graduate Program in International Affairs, New School University*. 2010:7.
37. G. Wang, S. Mang, H. Cai, et al. Integrated Watershed Management: Evolution, Development and Emerging Trends. *Journal of Forestry Research*. 2016;27(5):967–94.
38. C. R. Binder, J. Hinkel, P. W. Bots, C. Pahl-Wostl. Comparison of Frameworks for Analyzing Social-Ecological Systems. *Ecology and Society*. 2013;18(4):26.
39. E. Ostrom. A General. Framework for Analyzing Sustainability of Social-Ecological Systems. *Science*. 2009;325(5939):419–22.
40. J. M. Anderies, A. A. Rodriguez, M. A. Janssen, O. Cifdaloz. Panaceas, Uncertainty, and the Robust Control Framework in Sustainability Science. *Proceedings of the National Academy of Sciences*. 2007;104(39):15194–9.
41. B. K. Williams, R. C. Szaro, C. D. Shapiro. *Adaptive Management: The U.S. Department of Interior Technical Guide*. (U.S. Department of Interior, 2009).
42. G. D. Peterson, G. S. Cumming, S. R. Carpenter. Scenario Planning: A Tool for Conservation in an Uncertain World. *Conservation Biology*. 2003;17(2):358–66.
43. J. Forrest, D. Novikov. Modern Trends in Control Theory: Networks, Hierarchies and Interdisciplinarity. *Advances in Systems Science and Application*. 2012;12(3):1–13.
44. W. R. Ashby. Requisite Variety and Its Implications for the Control of Complex Systems. *Cybernetics*. 1958;1(2):83–99.
45. R. L. Ackoff. From Data to Wisdom. *Journal of Applied Systems Analysis*. 1989;16(1):3–9.
46. G. Bellinger, D. Castro, A. Mills. *Data, Information, Knowledge, and Wisdom*, 2004. (Accessed November 11, 2016). Available from: www.systems-thinking.org/dikw/dikw.htm.
47. Y. Bar-Yam. From Big Data to Important Information. *Complexity*. 2016;21(S2):73–98.
48. J. H. Bernstein. The Data–Information–Knowledge–Wisdom Hierarchy and Its Antithesis. In *NASKO: North American Symposium on Knowledge Organization*. 2011;2(1):68–75.

Part II

People

4

How People Communicate and Interact

4.1 Human Settlements, Communication, Interactions

Human societies develop from the environment and resources and become complex systems based on families, communities, and increasingly larger social organizations (Figure 1.1). They create intricate physical structures, organizations, and behaviors.

People now number about seven and one half billion people on the Earth. Their distribution is not uniform (Figures 4.1, 5.3) and is the result of inertia, climates, landforms, external social forces, and internal behaviors of local societies.

About 29% of the Earth's surface is covered by land; the remaining 71% by water. Approximately 70% of this land surface is comfortable for human habitation. The other 30% is too cold, too hot, too high in elevation, too dry, or otherwise inimical for most people. Of the total land surface, 13% has been dramatically modified by cities and similar areas or by farming so that it supports little biodiversity, clean water and air, and similar ecosystem services (Figure 4.2). The remaining area is a mixture of forests, shrubs, grasslands, and other land covers from which we obtain many of the environmental values and resources (except food) necessary for survival.

People interact primarily through sight, sound, and touch. They now increasingly interact worldwide through rapidly advancing communication and transportation technologies.

4.2 Communication Technologies

Telegraph and telephone communication has been increasing for over a century. Cellular telephones, the internet, and similar technologies are accelerating the amount of communication by using modern "leapfrog technologies" that avoid the kinds of costly infrastructures needed for older technologies – such as connecting wires for non-cellular telephones. Cellular telephones and the internet began to be widely used about 1995. By 2012, there were 84 cell phones and 32 internet users per hundred people in the world [6]. Wealthy countries have more people using telephones and the internet (Figure 4.3), but people in other countries are rapidly catching up. Photovoltaic and similar "leapfrog" energy sources will probably further accelerate the distribution of telephones and the internet.

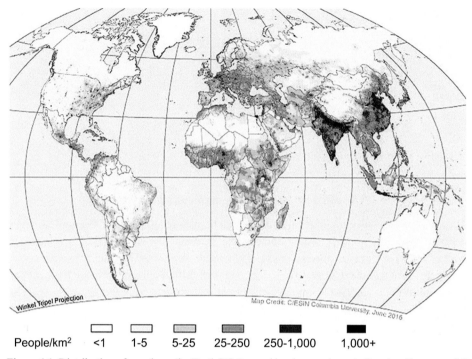

People/km²   <1   1-5   5-25   25-250   250-1,000   1,000+

Figure 4.1 Distribution of people on the Earth [1]. Legend has been enlarged. Creative Commons 3.0 Attributions License.

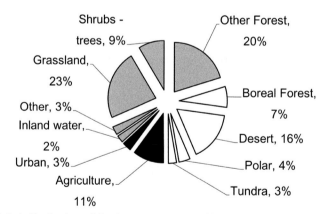

Figure 4.2 Global distribution of land covers [2–5]. White = relatively uninhabitable, black = modified by people, gray = accessible with environmental services.

Increasingly, flexible technologies that do not require installation and maintenance of a large infrastructure are being used. Some large infrastructures such as launch pads for communication satellites will always be needed; however, even these are becoming smaller and less in need of large infrastructures.

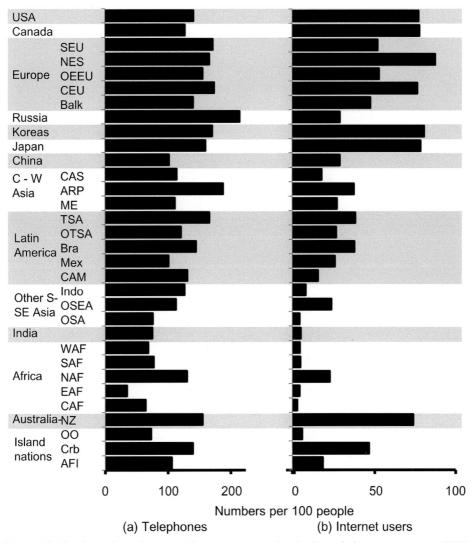

Figure 4.3 Numbers of telephones and internet users per hundred people by country groups (2013 data) [6].

4.3 Transportation

People and commodities are transported by a variety of modes (Figure 4.4). The modes are changing rapidly through new technologies, greater overall wealth, and decreasing costs. Current travel and shipping costs are largely linked to fossil fuel prices (Figure 24.11), but this link may change with new technologies and non-fossil fuel energy sources. Modes and locations of transportation are also based on historical infrastructures, climates (Chapter 7), landforms (Chapters 10–12), and social organizations and innovations (Chapters 4–5).

(a) Navigable rivers and canals

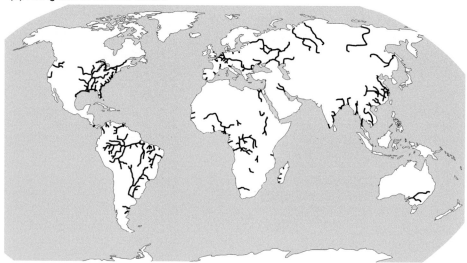

(b) Ports and shipping lanes

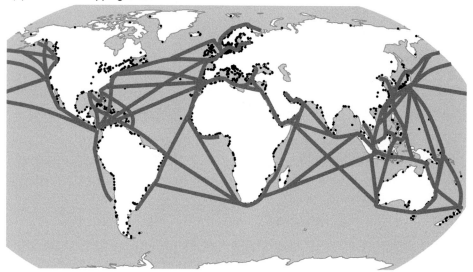

Figure 4.4 Major routes and infrastructures of different human and cargo transportation [7–9]. Major gas and water pipelines not shown. Cartography by Sabrina Szeto. 4.4a courtesy of the Rivers of the World Project. 4.4b–e made with Natural Earth; Sabrina Szeto, cartographer. 4.4f used with permission of Jeff Blossom, Harvard University.

Transportation currently produces 27% of the world's carbon dioxide emissions from burning fossil fuels (Figure 24.9). New technologies are beginning to fuel the transportation system with renewable energy, so the curtailment of transportation to reduce fossil fuels may not be needed.

(c) Railroads

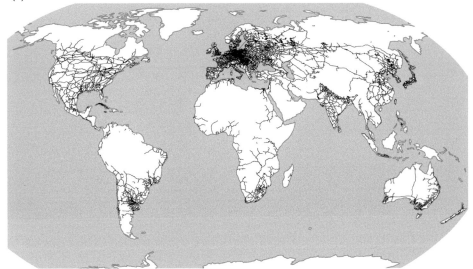

(d) Roads

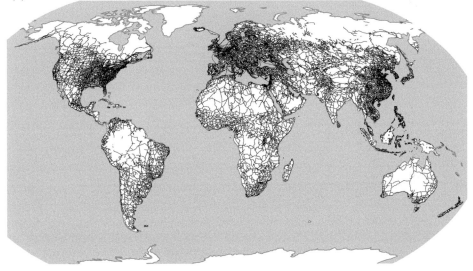

Figure 4.4 (cont.)

4.3.1 Rivers, Canals, Caravan Roads, Footpaths

Early movement of people and goods took place by water, caravans, and human walking.
Villages that grew into cities often arose at river cataracts and forks, where people lived and
help transport – and tax – goods removed from boats and carried around the cataracts.
Footpaths covered much of the world, as did caravans. Footpaths could be much steeper

(e) Airports

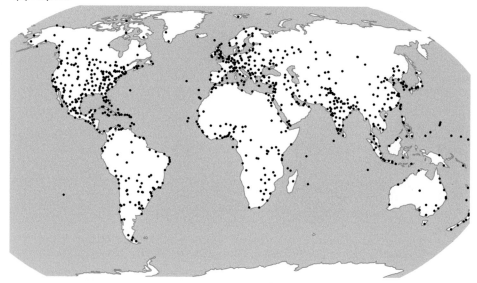

(f) Major petroleum pipelines

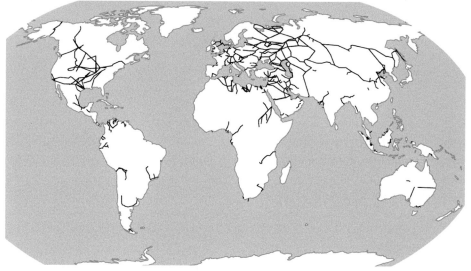

Figure 4.4 (cont.)

than caravan routes or modern highways. Caravanserais, fortified inns, commonly existed one day's travel apart along trade routes and provided caravanners and their goods safe places for the night – similar to modern truck stops.

River traffic requires relatively level areas and water but is efficient enough that canals, locks, and docks are still built and rivers dredged (Figure 4.4a). River traffic does not cross steep mountain ranges (Figure 10.3) and is limited in deserts where water is scarce (Figures 8.1, 17.4). It is predominant in areas such as the Amazon of South America because the largely alluvial and lateritic soils provide little sand and gravel for making strong road and railroad beds (Figures 10.5, 10.8).

River navigation has an interesting variation in parts of Alaska, USA. Rivers are used as highways when they freeze. Summers, when the ice has melted, are times of relative isolation.

4.3.2 Sea and Ocean Transportation

Ocean, lake, and sea transportation are also old because of the ease of letting people and goods be buoyed by water and using energy only for pushing the boats (Figure 4.4b). Major port cities flourished during times of peace, but people often retreated to hilltop residences during wars because seaside ports were easily plundered.

Port locations reflect weather, landforms, population densities, and social norms. The busiest ports are where most trading occurs and the society is seafaring (Figure 4.3b). China was heavily seafaring until the fifteenth century, but then withdrew to being a local agrarian culture until recently. Arabic merchants and seamen dominated the Indian and Red Seas slightly later. The Netherlands has also long been a seafaring society.

Seafaring ships generally move in "shipping lanes" that are efficient for travel because ocean currents and winds are used to advantage (Figures 7.6–7, 7.9). Ships commonly need to pass through narrow "gaps" such as in southeast Asia and the Mediterranean area (Figure 4.3b). Human-constructed waterways such as the Suez and Panama canals are also "gaps" that force ships through narrow passages but save much time and effort.

4.3.3 Railroads

Horse-drawn wagons moved on rails before steam engines replaced the horses in the early 1800s. These railroads first supplemented and then displaced some river and canal barge traffic. Railroads are restricted to relatively level or gently sloping terrain except where tunnels can be cut through mountains. They can move across deserts and through snow better than most other transportation forms. They have been most extensive in wealthy countries or their colonies because of relatively high initial costs of tracks.

Railroad locations are often remnants of political and economic behaviors of past decades (Figure 4.4c). Railroads in Africa are not well interconnected because they were constructed during colonial times. They were intended to move raw materials directly to ports and then to overseas home countries for further processes, keeping Africa a technology-poor "extractive economy" (Chapter 6) [10]. Railroads in Australia are not extensive because it has a small population; except in the southeast, they reflect a pattern similar to Africa of extracting and exporting raw materials. China's isolation until the 1970s can be seen in the shortage of railroads connecting it to potential trade partners to the south and west. The railroads of central Asia connect to Moscow more than to each other, reflecting Russia's control and distribution of resources and manufacturing to maintain regional dominance. The railroad across Siberia to the Pacific was to ensure Russia remained united. Railroads in the United States reflect a similar initial desire to ensure the Pacific and Atlantic areas remained united. The interconnected railroad system in India was first constructed to enable British troops to move rapidly and quell local uprisings, but has since enabled India to transport effectively within the country, do more secondary manufacturing, and export more finished products. The lack of railroads across northern North America, east-central Asia, the Amazon, and the African and Australian deserts reflect a lack of economic interest in these regions when railroad networks were expanding.

4.3.4 *Automotive Roads and Highways*

Roads using automobiles and trucks have been supplementing and/or replacing railroads since the early twentieth century (Figure 4.4d). Automobiles/trucks can travel on steeper terrain than barge and rail traffic, but good roads still require upkeep–and gravel, and sand.

Roads do not need the high initial financial investment of railroad tracks, but road transportation is generally less energy-efficient (Figure 24.12). The individuality of automobiles and trucks and many roads gives a novel combination of speed and mobility to travel and transport. The world is becoming more interconnected as formerly colonized countries that previously exported raw materials (extractive economies) [10] increasingly undertake secondary manufacture and trading with each other. Much of the new interconnection is roads, with large highways beginning to connect China and Europe (Figure 4.5).

Figure 4.5 Bridge being constructed in Eastern China to increase rapid highway traffic among Eastern Asia, Central Asia, and Europe.

4.3.5 Air Transportation

Air transportation – airplanes, helicopters, and drones – are leapfrog technologies because they allow regions and countries to bypass construction of roads, railroads, or canals for travel or transport. An airport or helicopter/drone pad is constructed instead (Figure 4.4e).

The energy requirements of air transportation are declining (Figure 24.11). Ultralight aircraft and drones may expand uses of air transportation, with drones equipped with Global Positioning Systems (GPS) enabling cheaper, safer, and quicker commodity deliveries to remote places.

4.3.6 Pipelines

Oil and gas pipelines connect much of the world for both retail use and long-distance delivery (Figure 4.4f) [11]. The oil pipeline configurations in Africa and South America suggest that much of their oil is taken to ports and exported to other continents.

If fossil fuels become less used, there is an opportunity to convert current fossil fuel pipeline routes and rights of way to ship fresh water to arid areas [12, 13]. New pipes would be needed for water, but parts of the infrastructure could be reused.

4.4 Population Concentrations and Amenities

People aggregate or disperse for safety, sharing, shortage of resources, division of labor, disagreements, and common or contrasting interests. Specialized amenities such as tools, food, entertainment, books, and teaching can be provided if a threshold number of people are close enough to create and utilize them. In that way, each user pays only a little but the specialist's aggregate income is sufficient. The critical number of people needed to provide different amenities varies with value, location, and time [14]. These values are sometimes calculated (Figure 4.6) [15, 16].

For example, if a small community and its surrounding shoppers in Missouri grew to over 12,000 people, a department store would be viable and would probably open there (Figure 4.6). Alternatively, if a large community were to reduce its population to fewer than 12,000 people, the department store would probably close because of insufficient business.

If a community declines or remains small, people often leave because the few remaining shops offer limited amenities locally. A "cascading collapse" can occur if more people leave as each shop closes. Alternatively, if a community grows, it can gain amenities that attract more people and even more amenities, creating a "cascading expansion." A small population change may affect some amenities and a large change may be needed for others.

The population forming the basis of a community is often directly or indirectly employed in resource production and secondary manufacture. These jobs are known

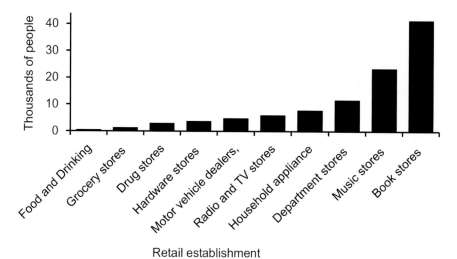

Retail establishment

Figure 4.6 Number of people needed to support different retail shops in Missouri, USA [15].

as "social drivers" [17]. They provide the initial incomes that create a turnover of money and lead to additional jobs in public service, business and finance, transportation, and amenities (Figure 5.2). The size of the "social driver" workforce changes for many reasons, such as demand of the product, labor, and trade laws, fluctuations in a country's currency, and technologies that reduce or expand the number of needed employees.

For this book, a small city with enough amenities to remain viable will be referred to as a "critical population small city." The population size that is "critical" for each amenity both varies with the society and its institutions and changes over time.

4.5 Population Densities and Dwellings

People live in a variety of dwellings, densities, and community sizes (Figure 4.7) [18]. Nearly 10% of the world's land area would be occupied by residences if everyone in the world lived in detached houses. These residences would occupy land presently providing biodiversity and other ecosystem services (Chapter 15).

Enjoyable, high-density living can be achieved through appropriate planning and architecture that create mixtures of mid-rises, townhouses, and others arranged in creative ways and intermixed with open spaces [18]. High-density communities require less area for the same number of people and so require fewer road, power, and water infrastructures [19]. Unless surrounded by communities of detached houses, a high-density community of moderate population occupies so little space that rural areas outside of the community are quite close and accessible for work, recreation, or leisure.

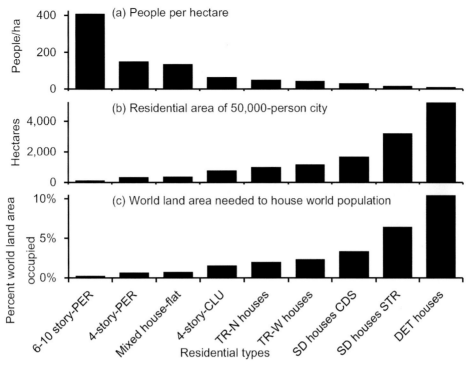

Figure 4.7 Effects of different residential dwelling types on (a) people per hectare, (b) size of residential area of a 50,000-person city, (c) world land area needed for residences if all people adopted the residential type. PER = perimeter; #-story = apartment buildings of # stories; "flat" = apartment; CLU = cluster units; TRN = narrow terrace houses (townhouses); TRW = wide terrace houses; SD = semidetached; DET = detached [18].

References

1. CIESIN. *Population Density Grid, 2015: Global Palisades* (NY Columbia University, 2016 [Accessed June 7, 2017]). Available from: http://sedac.ciesin.columbia.edu/downloads/maps/ gpw-v4/gpw-v4-population-density/gpw-v4-population-density-global-2015.jpg.
2. UN FAO. *FAOSTAT: Download Data* 2009 (United Nations Food and Agriculture Organization, 2009 [Accessed September 16, 2012]). Available from: http://faostat3.fao.org/download/.
3. UN FAO. *State of the World's Forests 2011.* (United Nations Food and Agriculture Organization; 2011).
4. Z. Liu, C. He, Y. Zhou, J. Wu. How Much of the World's Land Has Been Urbanized, Really? A Hierarchical Framework for Avoiding Confusion. *Landscape Ecology.* 2014;29(5):763–71.
5. UN FAO. *Global Forest Resources Assessment 2000: Main Report.* (United Nations Food and Agriculture Organization, 2001).
6. US Central Intelligence Agency. *Country Comparison.* (US Central Intelligence Agency, 2016 [Accessed June 24, 2016]). Available from: www.cia.gov/library/publications/the-world-factbook /index.html.
7. *Rivers of the World Atlas.* (Holland Pioneers in International Business, 2010).
8. Natural Earth. *Free Vector and Raster Map Data*, 2017 (Accessed June 4, 2017). Available from: www.naturalearthdata.com.

9. J. Blossom. *Global Oil Pipelines: Using Google Earth as a Classroom Tool*. (Harvard Map Collection, Harvard College Library, 2009 [Accessed June 4, 2017]). Available from: http://worldmap.harvard.edu/data/geonode:global_oil_pipelines_7z9.

10. D. Acemoglu, J. Robinson. *Why Nations Fail: The Origins of Power, Prosperity, and Poverty*. (Crown Business, 2012).

11. R. Kandiyoti. *Pipelines; Flowing Oil and Crude Politics*. (I. B. Tauris, 2008).

12. E. Gies. Northern Cyprus Sees Hope in Water Pipeline. *The New York Times*. 2013.

13. P. Tremblay. *Turkey's Peace Pipe to Cyprus*. (Al Monitor; 2015 [Accessed December 1, 2016]). Available from: www.al-monitor.com/pulse/originals/2015/10/turkey-cyprus-water-pipeline-delivers-fears.html.

14. L. J. King. *Central Place Theory*. (Regional Research Institute, West Virginia University, 1985 [Accessed May 8, 2017]). Available from: https://ideas.repec.org/h/rri/bkchap/06.html.

15. J. Simon, C. Braschler, J. A. Kuehn, J. Croll. *Potential for Retail Trades in Rural Communities*. (University of Missouri, 1993).

16. R. C. Coon, F. L. Leistritz. *Threshold Population Levels for Rural Retail Businesses in North Dakota, 200*. (Department of Agribusiness and Applied Economics, Agricultural Experiment Station, North Dakota State University, 2002).

17. G. Björklund, R. Connor, A. Goujon, et al. Chapter 2. Demographic, Economic and Social Drivers. In *The UN World Water Development Report 3: Water in a Changing World*. (UNESCO, Earthscan, 2009): pp. 29–40.

18. G. Towers. The Implications of Housing Density. In *Challenges and Opportunities in Housing: New Concepts, Policies and Initiatives* (CIB WC, 2002): p. 69.

19. E. Puurunen, A. Organschi. Multiplier Effect: High Performance Construction Assemblies and Urban Density in Us Housing. In *Mitigating Climate Change*. (Springer, 2013): pp. 183–206.

5

People, Societies, Populations, and Changes

5.1 Human Development and the Current Situation

The human brain's capacity has changed little over the past 75 thousand years [1], despite inclusion of minor amounts of Neanderthal and other genes in some Eurasian peoples [2, 3]. During this time, however, the ways people live have changed dramatically. Human history is often discussed in terms of distinct events such as the development of the printing press, the scientific revolution, and the industrial revolution. Others view the current global culture as emerging more gradually through many small human interactions, transactions, learning, adopting, and adapting. Both perspectives recognize that global culture changed from "hunter-gatherer" lifestyles to "agriculture" lifestyles and now is changing to "technological" lifestyle. These lifestyles are dramatically different from each other.

Modern people lived primarily in hunter-gatherer communities during the first 80% of their existence. They began to adopt agrarian lifestyles ("Neolithic Revolution") about 12 thousand years ago, possibly as glaciers retreated and sea levels rose (Chapter 9) [1, 4]. The change may prove to have begun earlier since much archeological evidence has yet to be discovered and many of the older coastal settlements would be up to 120 meters below present sea level (Figures 9.4–5).

The number of people increased dramatically during the past 5 thousand years (Figure 5.1). The transitions were probably not as abrupt and clear as previously thought by Deevey [5].

The small groups of the "hunter-gatherer lifestyles" ("toolmaking cultures") were probably preoccupied with providing food and shelter for themselves. The groups were in irregular contact, hampered by slow communication and travel as well as by geographic obstacles such as mountains, seas, and deserts. The isolation probably led to local breeding and sharing of ideas and to groups of people who appeared, thought, and behaved similarly but were different from other isolated groups.

The agriculture lifestyle led to much larger groups living in cities. With much more efficient food production and a critical mass of people, it led to more diverse activities [8]. The cities expanded into kingdoms and empires that had even more diverse activities, including expanded trade. Eurasia from the Baltic Sea to China and North Africa were vibrantly trading with rather cosmopolitan cities along the routes that shifted in significance over time [9].

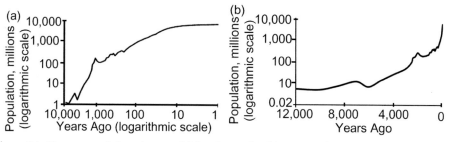

Figure 5.1 Human population changes: (a) log–log scale; (b) same projection as (a), but with non-logarithmic horizontal scale [5–7].

Contacts among previously isolated groups revealed a diversity of appearances and behaviors. Reaction to this diversity varied from mutual exchanges of behaviors and/or people to emulation, interbreeding, enslavement, or genocide. Racial appearances are inherited; and a continuing common mistake has been to assume different behaviors – religion; dress; and physical, intellectual, and artistic prowess – are also racially inherited. Recent research is showing that people are more genetically uniform than most animal species and that human behaviors are learned, not inherited. Racial differences are generally superficial [10].

A result of the agriculture lifestyle was an increasingly diverse environment where belief systems, ideologies, lifestyles, and efforts to understand the environment flourished. The human species' roles as laborer, innovator, and consumer were well underway. People often depleted their local resources and conquered and enslaved others. Migrations and wars became common when timber, clean water, specific minerals, and food became locally scarce and technologies were not sufficient to ship these from afar [11].

Many societies have been rapidly shifting in the past 1,000 to 500 years to a "technological" lifestyle in which fewer people are involved in agriculture. Primary characteristics of this new lifestyle are use of energy, information, and sophisticated tools; specialization of labor; and far-reaching communication and transportation abilities (Figures 5.2, 4.3–4). Proportionately more people are involved in production of goods from resources, finance, sales, and services including teaching [9, 12, 13].

Scientific inquiry has progressed, providing many new tools and techniques; this inquiry was systematized into methodologies and taught. Public schools were also established, although at varying rates in different countries. The concept of universities expanded rapidly and allowed instruction in recent intellectual developments and exchanges of ideas among the better educated.

The advances have generally made life easier and more secure. Deaths from common diseases such as strep throat, appendicitis, polio, breast or prostate cancer, and influenza have declined. The world's literacy rate has also improved. Meats and other foods can easily be stored without spoiling in warm weather, so less is wasted. Modern communication has given earlier warnings of windstorms, floods, and other disasters as well as better access to health care. Inequalities within and among nations will be addressed in Chapter 6.

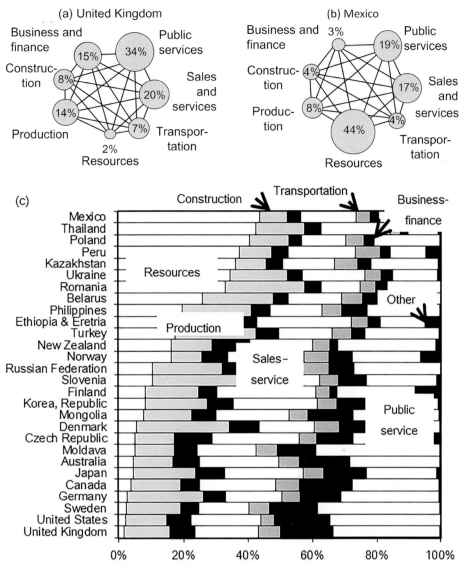

Figure 5.2 Proportion of workforce in each economic sector: (a) the UK, (b) Mexico, (c) countries with different proportions of workforce in agriculture and scientific-industrial occupations [14, 15].

The changes from agrarian to technology-focused cultures appear to occur at different rates and in diverse ways. There is no underlying "law" of social evolution that means all countries will advance toward scientific industrial societies, contrary to some theories [16]. However, there does seem to be a pattern to the changes.

The technological lifestyle employs people at a range of learning levels – from barely being able to read and write to highly knowledgeable positions that require considerable initial study and constant learning. There are generally more skills with more specialization, closer coordination and cooperation, and often fewer hours of work each week. Some jobs are no longer needed or are increasingly automated, while new ones are created.

The recent times when people are strongly affecting the Earth is referred to as the "Anthropocene" [17–19] and is being proposed as a formal geological "epoch" to follow the "Holocene" ("Recent") Epoch. Different times are being proposed for the beginning of the Anthropocene – the year 1800, the beginning of the nuclear age, and others [20].

The progression of human lifestyles has historically shown advances in the uses and modifications of raw materials and changes in values. Unless temporarily forced by external disturbances, communities have not returned to their earlier lifestyles except for some experimental communes.

Since World War II, international movement, businesses, communication technologies, scientific and educational exchanges, nongovernmental organizations (NGOs), and sports and entertainment events have brought people throughout the world together. They are increasingly aware of each other's diverse perspectives, problems, similarities, and differences. Increasingly more students spend part of their studies in foreign countries. These encounters both modify embedded, unnecessary prejudices toward racially and culturally different people and lead to celebrations of the diversity – through multicultural events, ethnic food restaurants, global sports activities, international film exchanges, and others. Increased global communication and movement is leading to more interracial marriages and mixing of social behaviors, religious beliefs, foods, and traditions.

The global integration has also led to polarization and violence between racial, religious, or otherwise "different" groups. Ironically, races and behaviors will probably become so intermixed within a few generations that such polarization will become difficult. At the same time, the diversity that makes much of the present world fascinating may be lost to uniformity.

There are still isolated groups in the hunter-gatherer and agriculture lifestyles that are little aware of modern activities, but many poor people in the less developed world are becoming increasingly aware of what is available in developed countries. This awareness often causes social tensions. There are also undesirable aspects of industrial/technological lifestyles, as will be brought up throughout the book.

5.2 Urban and Rural Changes

The shift from agricultural to technological societies is still occurring in many places. The world's rural population dropped from 66% to 47% of the total population between 1960 and 2014 and became less than half in 2009 [21]. Rural and urban population patterns vary dramatically among regions (Figure 5.3).

Many things change in rural and urban areas as countries shift to the technological lifestyle, highlighted by a dramatic movement to the cities (Figure 5.4).

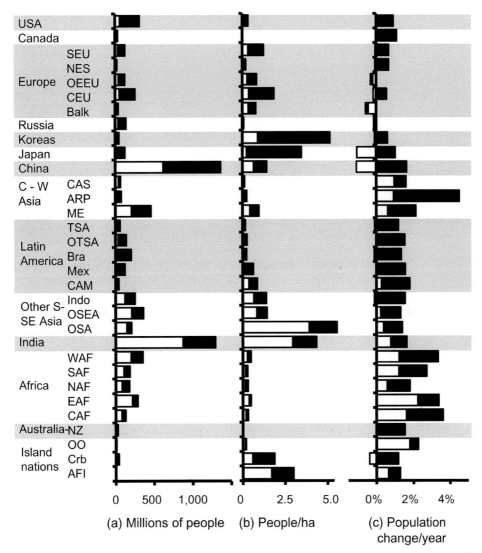

Figure 5.3 Rural (white) and urban (black) population, densities, and changes by country groups [21]. Changes based on total population.

Large farm owners on fertile soils increase their productivity using the benefits of new technologies and economies of size [23]. Modern transportation and refrigeration allows these producers to ship their crops efficiently to distant markets. They can often out-compete local farmers at these end points who are trying to raise crops on poorer soils. They also have capital to purchase more farmland and replace many farm workers with a few large machines.

(a) Rural subsistance

(b) Marginal lands
abandoned

(c) People move to cities

Figure 5.4 Common pattern of population movements during shift from agricultural to technological lifestyles [22]. Figure 5.4a courtesy of The Harvard Forest, Harvard University.

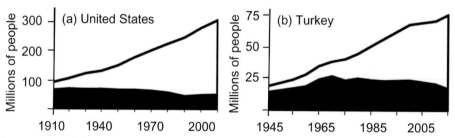

Figure 5.5 Rural (black) and urban (white) population changes in (a) the United States [33] and (b) Turkey [34–36].

More food is produced, preserved, and transported from productive agricultural land at a low price. Consequently, the least efficient "subsistence" farmers, who barely raise enough crops to feed themselves, often abandon their farms and migrate to cities (Figures 5.4–6) [22, 24–27].

The displaced former farm owners and laborers face varying degrees of hardship or benefit. The cities can become overwhelmed with the population influx and unemployment if not prepared [21, 28]. The displaced people from English "clearances" – clearing the land of subsistence farmers – often found work in the emerging factories of the industrial revolution [29], while the "clearances" in Scotland led to unemployment, poverty [30], and eventually emigration to North America and Australia [31].

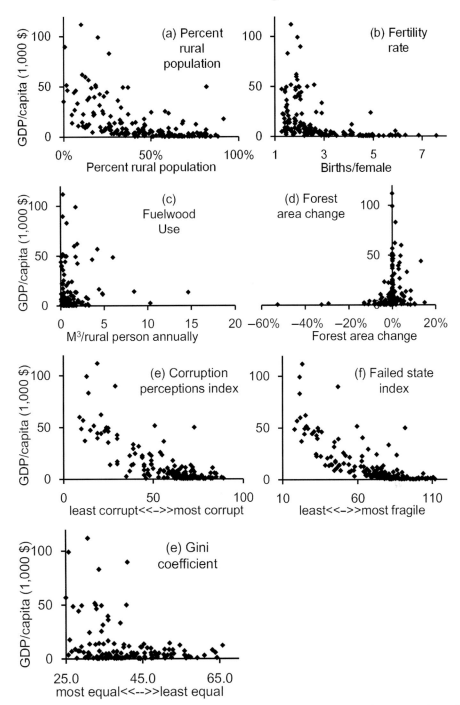

Figure 5.6 Relations of demographics and social behaviors of the world's countries to gross domestic product per capita (GDPpc) [39, 45–49]. 5.6e includes data courtesy of Transparency International; 5.6f includes data used with permission of The Fund for Peace.

5.2.1 Effects on Cities

People move to cities for several reasons. Subsistence farmers on infertile soils abandon their farms when they cannot compete with larger, increasingly productive farms. Young people often move to cities because they have little work since their parents still manage the farms. By the time the old retire, their descendants are engaged in careers elsewhere and reluctant to take charge of their parents' farms (Figure 5.7) [32]. Even on productive farms, machinery commonly displaces laborers; and they, too, migrate to cities. The lure of cities large enough to have amenities further attracts rural people (Figure 4.6), even though former farmers are often disadvantaged in cities because employment in the nonfarm sectors generally requires different skills and education.

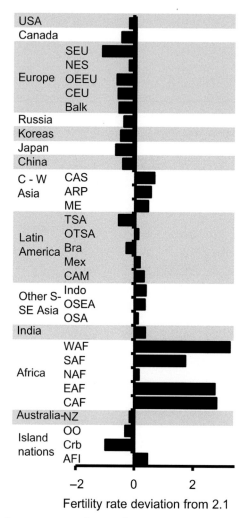

Figure 5.7 Deviations of country groups from replacement fertility rate of 2.1 [50].

Most countries with strong economies (high "gross domestic product per capita" [GDPpc], Chapter 6) have people concentrated in urban areas (Figure 5.6a). Urban areas provide cultural diversity. This concentration and diversity typically provide vibrancy by generating new ideas and innovations that lead to more complex societies. In addition, more people can be provided with food, water, energy, and medical attention efficiently in cities.

5.2.2 Effects on Rural Environments

The overall effects of urbanization have been to change people's pressures on various resources and the environment. If urban populations become wealthy enough to import rural produce cheaply from abroad, local farmers must lower their prices and thus become further disadvantaged [37].

Abandoned farms commonly regrow to forests. Forest area in the developed world increased while it declined elsewhere during the past 15 years – with a net worldwide change of −0.2% per year (Chapter 29) [38]. Total world wood volume in forests increased (Chapter 30). The developing regions' biodiversity, soils, and hydrology changed. The unprecedentedly high world population (Figure 4.2) has left much of the remaining forests, shrublands, and grasslands little utilized (Chapters 22, 28–30) [39, 40].

On the other hand, with fewer people in rural areas, there is a smaller infrastructure of merchants, professionals, repair and service businesses as well as staff and volunteers engaged in civic activities of law enforcement, fire and other rescues, hospitals, schools, and similar activities (Figure 4.6). (See "Southern Androy" case study, Chapter 3.) Consequently, dangerous and even lawless conditions can emerge – such as armed groups growing illegal drug crops.

Rural areas are also affected by urbanization in other ways. Less fuel-wood is used as people concentrate in cities (Figure 5.6c), so forests grow crowded and become susceptible to more intensive wildfires [41], losses of habitats [42], and often reductions in available water [43]. Having less contact with people, some animal populations become more aggressive [44].

5.3 Economic Development and Social Changes

A general increase in human well-being (Chapter 6) occurs as societies move from agriculture to technological lifestyles. These have effects throughout the socioenvironmental system (Figure 5.6). Gross domestic product per capita (GDPpc), corruption perception index, failed states index, and Gini coefficient (Figures 5.6e–g) are discussed in Chapter 6.

On the other hand, the concentrated urban lifestyles and concentrated food production are polluting air and water and affecting the world's climates.

5.4 Fertility Rates and Population Changes

Fertility rates are declining to nearly stable levels in most places in the world, but the world's population will continue to increase rapidly for the next few decades.

5.4.1 Fertility Rates and Changes

People in urbanized, economically developed societies often have fewer children (Figures 5.6b, 5.7), probably for several reasons. The social infrastructures that care for people in old age and sickness mean that children are not as necessary for this care. The higher survival rate because of better public health infrastructures means that fewer children are needed to ensure some grow to maturity. Children in cities provide little help to the household and are generally a cost to feed clothe, and educate – unlike on farms where children are an asset as labor. Both urban spouses commonly have jobs and many interests besides children, and contraception methods are readily available.

The world's fertility rate has been declining, with African fertility rates declining the least (Figure 5.8). The world fertility rate has halved since 1960, from 5.0 to 2.5 births/female (Figure 5.8) [21], still slightly above the replacement rate of 2.1.

5.4.2 Populations, Age Distributions, and Changes

Changing fertility rates do not immediately change the population size in the same direction because fertility rate only affects the present year's population. The changed population of all years with a new fertility rate needs to "grow" through the age classes before the population is affected (Figures 5.9–10). In contrast to the fertility rate (Figure 5.7), total population of most countries in the world has been increasing dramatically since 1960 (Figures 5.3c, 5.8) as a result of a lower mortality rate caused by better health care with

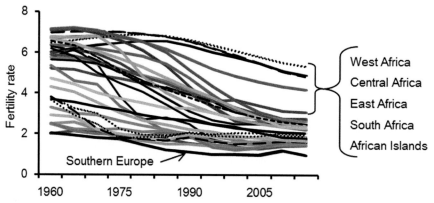

Figure 5.8 Change in fertility rates of country groups (averaged by countries) from 1960 to 2013 [6, 21].

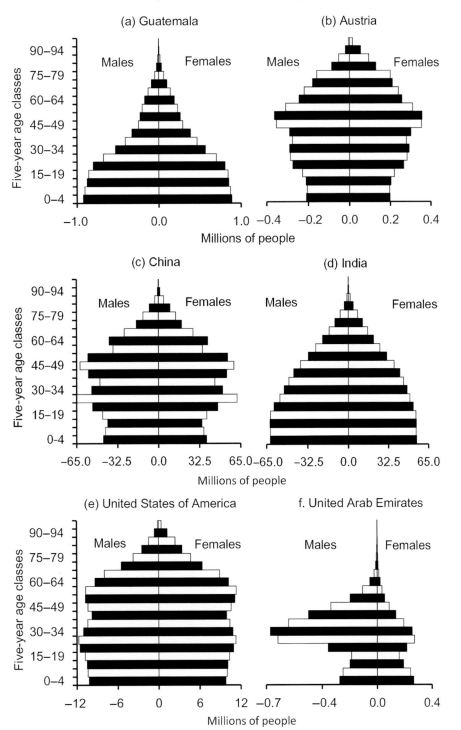

Figure 5.9 Population distributions by age and sex ("population pyramids") for selected countries, 2016. Horizontal axis scales differ [52].

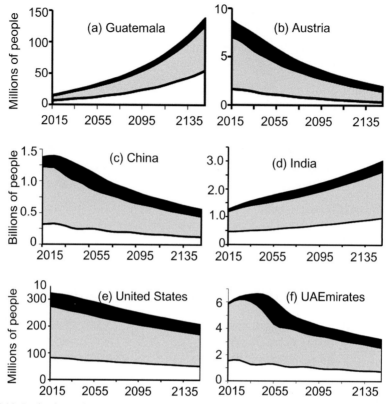

Figure 5.10 Projections of populations of countries in Figure 5.9, assuming birth rates do not change. Calculated from [52]. Horizontal bars represent age classes. White = 0–19-year age class; gray = 20–65; black = over 65.

antibiotics and other advances; the green revolution that dramatically increased the world's food supply [51]; better communication and transportation; changes in governments; and other factors.

Lags between increasing population and declining fertility rates occurs and create surges and declines in populations and irregular age distributions within countries (Figures 5.9–10).

Africa will probably experience a population surge similar to Guatemala in Figure 5.10a unless it controls its fertility rate soon. Unlike fertility rates in countries with more subsistence economies (Guatemala, for example; Figure 5.9a), fertility rates in Japan, South Korea, and most economically developed European countries are well below the replacement level of 2.1 births per female (Figure 5.9b). China's fertility rate was low during the "Cultural Revolution." China later encouraged only one child per household for many years (Figure 5.6). India and South Asia have fertility rates only slightly above replacement; however, their large total births contribute strongly to the world's population increase (Figure 5.10d).

Table 5.1 *Projected percent of population in dependent and "supporting" age group after 135 years of constant fertility rates for countries of Figures 5.9–10*

	Austria	China	Emirates	USA	India	Guatemala
Fertility rate (births/female)	1.5	1.7	1.8	1.9	2.5	3.3
Supporting age group (ages 20 to 64 years)	58%	57%	57%	57%	55%	51%
Dependent age groups						
Elderly (ages 65 and older)	23%	20%	19%	18%	13%	9%
Children (ages 19 and younger)	20%	23%	24%	25%	32%	40%

Fertility rates can change societies' behaviors and resource needs for many decades in ways that can be anticipated and planned for (Figure 5.10). For example, a sudden increase in fertility rates means many children will need tending at first; many schools will be needed within the next decades; then many more adults of working and marrying age within three to five decades; followed by many more retirees. A sudden fertility rate decrease would have the opposite effect. Ironically, the total "burden of support" of children and old people by the working age group is nearly identical whether the birth rate is high, low, or stable (Figure 5.10 and Table 5.1). The working age group supports either many young people and few old, a balance of young and old, or few young people and many old.

An imbalance in numbers of people between age classes comes with these changes; this imbalance has benefits and drawbacks. The high fertility rate in China was changed by legal mandate, while other countries have achieved it through economic development, social security, health services, and birth control promotion. The need for more housing or jobs with an expanding population has been done by stimulating certain economic sectors and promoting early retirement, while appropriate policies in a declining population would be the opposite. Unfilled jobs created by a smaller young population can be filled by importing "guest workers" (Emirates, Figures 5.9f, 5.10f), promoting liberal immigration (USA, Figures 5.9e, 5.10e), or in the long term promoting a higher fertility rate.

Xenophobic societies sometimes promote a high fertility rate to avoid the need or room for immigrants or refugees or to stimulate the economy with a constantly increasing population of consumers.

5.5 Future World Population

The projected world population may follow different trajectories (Figure 5.11), depending largely on the fertility rates of the next few years.

More people in the world will be seeking employment than will be retiring or dying for the next few decades as the older age classes "fill out," despite the fertility rate (Figure 5.12).

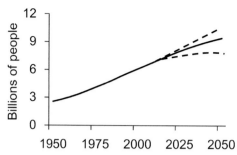

Figure 5.11 World population projection to 2050. Dashed lines show high and low possibilities. [7, 53, 54]

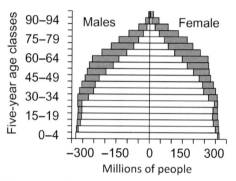

Figure 5.12 World population pyramid of 2015 (white) superimposed on the projected pyramid of 2050 (gray) [52].

Also, there will be many people purchasing residences and furniture/appliances as both the large age class matures and more countries become prosperous. The world's urban infrastructure is expected to triple by 2050 [55]. The purchase rate could then slowly decline to a replacement rate as the adult population stabilizes and the backlog is satiated.

Despite the problems of stable populations, an increasing population is a more serious threat because of potentially increasing diseases and shortages of food and water. People will be further faced with large population issues if climate changes necessitate human migrations (Chapters 9, 31).

Many of the world's environmental and social problems will be greatly alleviated if the fertility rate continues to fall. If the fertility rate could be stabilized below the replacement rate of 2.1 for 100 years, the world's population issue could be solved (Figure 5.13).

The rural exodus (Figure 5.4) has left so few people in some rural places that ecosystems can neither be tended to avoid extreme fluctuations of habitats and fire susceptibility (Chapter 28) nor policed (Chapter 28). Both rural depopulation and urban overcrowding could be countered by developing social drivers for rural areas. People could be employed to "restore" ecosystems [56–58], such as "landcare" in Australia [59–62] (Chapter 16). Additionally some agriculture operations that employed large machinery and few workers

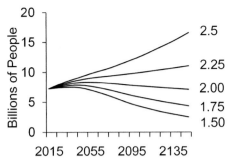

Figure 5.13 World population projections with different fertility rates. Calculated from [52].

can now be substituted with more precision production using small, precise equipment and well-trained people (Chapters 21–22). Increasing costs of big machinery, savings of irrigation water with less chemical applications, and greater production with more care could make greater technical employment in rural farming, forestry, and animal husbandry more attractive.

One choice will be whether to try to make more use of small cities in rural areas through modern communication, transportation, organization, other technologies, different organizational systems, and more diffuse extraction of food, building materials, energy, and other resources – or to continue to concentrate people and activities in large urban areas, concentrate resource extraction in small areas, and leave large parts of the world untouched. This question – conservation VS preservation – has been a major question in resources for decades, with advocates of both sides arguing that their approach will better sustain the environment and resources [63, 64].

Moderately large-sized cities (Figure 4.6) offer many more amenities than small communities. Consequently, it may be appropriate to concentrate new generations of rural people into moderately sized cities within rural regions.

References

1. W. Behringer. *A Cultural History of Climate*. (Polity Press, 2010).
2. S. Sankararaman, N. Patterson, H. Li, S. Pääbo, D. Reich. The Date of Interbreeding between Neandertals and Modern Humans. *PLOS Genetics* 2012;8(10):e1002947.
3. C. Stringer. Evolution: What Makes a Modern Human. *Nature*. 2012;485(7396):33–5.
4. A. Goudie. *The Human Impact on the Natural Environment*, sixth ed. (Blackwell, 2006).
5. E. S. Deevey. The Human Population. *Scientific American*. 1960;203(3):194–204.
6. US Census Bureau. *Historical Estimates of the World Population*. (US Dept. of Commerce; 2016 [Accessed July 1, 2016]). Available from: www.census.gov/population/international/data/world pop/table_history.php.
7. US Census Bureau. *Total Midyear Population of World: 1950–2050*: (US Dept. of Commerce, 2016 [Accessed July1, 2016]). Available from: www.census.gov/population/international/data/ worldpop/table_population.php.
8. A. R. De Souza. *A Geography of World Economy*. (Macmillan, 1990).

9. P. Frankopan. *The Silk Roads: A New History of the World*. (Bloomsbury, 2015).
10. B. A. Koenig, S. S.-J. Lee, S. S. Richardson. *Revisiting Race in a Genomic Age*. (Rutgers University Press, 2008).
11. J. Perlin. *A Forest Journey: The Story of Wood and Civilization*. (The Countryman Press, 2005).
12. I. Morris. *Why the West Rules – For Now: The Patterns of History and What They Reveal About the Future*. (Farrar, Straus, & Giroux, 2010).
13. E. R. Wolf. *Europe and the People without History*. (University of California Press, 1982).
14. US Bureau of Labor Statistics. *International Labor Comparisons*, 2016 (Accessed July 4, 2016). Available from: www.bls.gov/ILC/#laborforce.
15. United Nations. Table 4. Selected Country Occupations, 2004 (Accessed October 17, 2006). Available from: http://unstats.un.org/unsd/cdb/cdb_advanced_data_extract_fm.asp? HSrID=4670&HCrID=all&HYrID=2004&ofID=r&txtSS=New+Selection+Name.
16. W. W. Rostow. *The Stages of Economic Growth: A Non-Communist Manifesto*. (Cambridge University Press, 1990).
17. P. J. Crutzen, E. F. Stoermer. The "Anthropocene". *The Global Change Newsletter*. May 2000:17–18.
18. V. A. Sample, R. P. Bixler, C. Miller, editors. *Forest Conservation in the Anthropocene*. (University Press of Colorado, 2016).
19. C. N. Waters, J. Zalasiewicz, C. Summerhayes, et al. The Anthropocene is Functionally and Stratigraphically Distinct from the Holocene. *Science*. 2016;351(6269):aad2622.
20. Subcommission on Quaternary Stratigraphy. *What Is the "Anthropocene"? – Current Definition and Status* 2016 (Accesed September 22, 2016). Available from: http://quaternary.stratigraphy .org/workinggroups/anthropocene/.
21. The World Bank. *World Development Indicators*. (The World Bank, 2016 [Accessed May 8, 2016]). Available from: http://data.worldbank.org/indicator/EG.USE.PCAP.KG.OE.
22. The Harvard Forest. *Height of Forest Clearing and Agriculture*. (Fisher Museum, The Harvard Forest, 2015 [Accessed July 1, 2016]). Available from: http://harvardforest.fas.harvard.edu/ dioramas.
23. D. Paarlberg, P. Paarlberg. *The Agricultural Revolution of the 20th Century*. (Iowa State University Press, 2000).
24. H. M. Raup. Old Field Forests of Southeastern New England. *Journal of the Arnold Arboretum*. 1940;21:266–273.
25. C. D. Oliver. Policies and Practices: Options for Pursuing Forest Sustainability. *The Forestry Chronicle*. 2001;77(1):49–60.
26. C. D. Oliver, L. L. Irwin, W. H. Knapp. *Eastside Forest Management Practices: Historical Overview, Extent of Their Application, and Their Effects on Sustainability of Ecosystems*. General technical report PNW-GTR–324. Pacific Northwest Research Station, US Forest Service, 1994.
27. K. Ramachandran, P. Susarla. Environmental Migration from Rainfed Regions in India Forced by Poor Returns from Watershed Development Projects. In *Environment, Forced Migration and Social Vulnerability*. (Springer, 2010): pp. 117–31.
28. The World Bank. *Indicators*. (The World Bank Group, 2016 [Accessed July 2, 2016]). Available from: http://data.worldbank.org/indicator/.
29. P. K. O'Brien. Agriculture and the Industrial Revolution. *The Economic History Review*. 1977;30 (1):166–81.
30. T. Devine. Social Responses to Agrarian 'Improvement': The Highland and Lowland Clearances in Scotland. in *Scottish Society, 1500–1800*. (Cambridge University Press, 1989): pp. 148–68.
31. J. M. Bumsted. *The People's Clearance: Highland Emigration to British North America, 1770–1815*. (University of Manitoba Press, 1982).
32. H. B. Glick, C. Bettigole, D. Routh, et al. Wyoming's Aging Agricultural Landscape: Demographic Trends among Farm and Ranch Operators, 1920–2007. *Rangelands*. 2014;36 (6):7–14.

33. US Census Bureau. *2010 Census Urban and Rural Classification and Urban Area Criteria*. (US Dept. of Commerce [Accessed February 7, 2015]). Available from: www.census.gov/geo/refer ence/ua/urban-rural-2010.html.

34. Istanbul Metropolitan Municipality. *Population and Demographic Structure* (Istanbul Metropolitan Municipality, 2008 [Accessed February 1, 2014]). Available from: www .ibb.gov.tr/sites/ks/en-us/0-exploring-the-city/location/pages/populationanddemographicstruc ture.aspx.

35. TUIK. *Census of Population: Social and Economic Characteristics of the Population of Turkey*. (Turkish Statistical Institute, 2014 [Accessed February 10, 2014]). Available from: www.turk stat.gov.tr/Kitap.do?metod=KitapDetay&KT_ID=11&KITAP_ID=12.

36. C. D. Oliver, F. A. Oliver, L. Yonavjak, et al. The Transition from Timber to Multiple Value Management: A Paradigm Shift in Forestry, in Bhojvaid P. P., editor, *Sustainable Forestry Management for Multiple Values: A Paradigm Shift*. (Forest Research Institute, 2014).

37. J. E. Stiglitz. *Globalization and Its Discontents*. (Norton, 2002).

38. UN FAO. *FLUDE: The Forest Land Use Data Explorer*. (United Nations Food and Agriculture Organization, 2015 [Accessed May 25, 2016]). Available from: www.fao.org/forest-resources-assessment/explore-data/en/.

39. UN FAO. *Global Forest Resources Assessment 2015*. (United Nations, 2015).

40. C. D. Oliver, N. T. Nassar, B. R. Lippke, J. B. McCarter. Carbon, Fossil Fuel, and Biodiversity Mitigation with Wood and Forests. *Journal of Sustainable Forestry*. 2014;33(3):248–75.

41. R. N. Sampson, D. L. Adams, S. S. Hamilton, et al. Assessing Forest Ecosystem Health in the Inland West. *Journal of Sustainable Forestry*. 1994;2(1–2):3–10.

42. C. D. Oliver. Forests with Crowded Trees of Small Diameters–a Global Issue Affecting Forest Sustainability, in Baumgartner D. M., Johnson L. R., DePuit E. J., editors. *Small Diameter Timber: Resource Management, Manufacturing, and Markets*. (Washington State University Cooperative Extension, 2002): pp. 1–7.

43. P. K. Barten, J. A. Jones, G. L. Achterman, et al. *Hydrologic Effects of a Changing Forest Landscape*. (National Research Council of the National Academies of Science, USA, 2008).

44. P. Canby. The Cat Came Back: Alpha Predators and the New Wilderness. *Harper's Magazine*. 2005:95–102.

45. S. Deb. The Human Development Index and Its Methodological Refinements. *Social Change*. 2015;45(1):131–6.

46. UN FAO. *ForesSTAT* 2016 (United Nations, 2016 [Accessed July 3, 2016]). Available from: http://faostat.fao.org/site/626/DesktopDefault.aspx?PageID=626#ancor.

47. Transparency International. *Corruption Perceptions Index 2015: Downloads and Methodology*, 2015 ([Accessed June 24, 2016]). Available from: www.transparency.org/cpi2015#downloads.

48. N. Haken, J. Messner, K. Hendry, et al. *The Failed States Index Ix: 2013*. (The Fund for Peace, 2013).

49. US Central Intelligence Agency. *Country Comparison*. (US Central Intelligence Agency, 2016 [Accessed June 24, 2016]). Available from: www.cia.gov/library/publications/the-world-fact book/index.html.

50. The World Bank. *Fertility Rate, Totals (Births Per Woman)*. (The World Bank Group, 2015 [Accessed July 1, 2016]). Available from: http://data.worldbank.org/indicator/SP.DYN .TFRT.IN.

51. W. S. Gaud. *The Green Revolution: Accomplishments and Apprehensions. AgBioWorld*, 1968 (Accessed November 11, 2015). Available from: www.agbioworld.org/biotech-info/topics/bor laug/borlaug-green.html.

52. US Census Bureau. *Mid-Year Population for Five Year Age Groups*. (US Department of Commerce, 2016 [Accessed July 1, 2016]). Available from: www.census.gov/population/inter national/data/idb/region.php?N=%20Results%20&T=10&A=separate&RT=0&Y=2016&R=-1&C=AU,CH,GT,IN,AE,US.

53. UNDP. *World Population Prospects: The 2015 Revision. Key Findings and Advance Tables*. United Nations, 2015 Contract No.: ESA/P/WP 241.

54. UN Department of Economic and Social Affairs Population Division. *World Population Prospects: The 2015 Revision, Key Findings and Advance Tables* (United Nations, 2015).

55. K. C. Seto, B. Güneralp, L. R. Hutyra. Global Forecasts of Urban Expansion to 2030 and Direct Impacts on Biodiversity and Carbon Pools. *Proceedings of the National Academy of Sciences.* 2012;109(40):16083–8.

56. D. C. Dey, E. S. Gardiner, J. M. Kabrick, J. A. Stanturf, D. F. Jacobs. Innovations in Afforestation of Agricultural Bottomlands to Restore Native Forests in the Eastern USA. *Scandinavian Journal of Forest Research.* 2010;25(S8):31–42.

57. J. A. Stanturf. *Restoration of Boreal and Temperate Forests.* (CRC Press, 2015).

58. J. A. Stanturf, B. J. Palik, R. K. Dumroese. Contemporary Forest Restoration: A Review Emphasizing Function. *Forest Ecology and Management.* 2014;331:292–323.

59. J. Cary, T. Webb. Landcare in Australia: Community Participation and Land Management. *Journal of Soil and Water Conservation.* 2001;56(4):274–8.

60. E. Compton, R. B. Beeton. An Accidental Outcome: Social Capital and Its Implications for Landcare and the "Status Quo". *Journal of Rural Studies.* 2012;28(2):149–60.

61. A. Curtis, M. Lockwood. Landcare and Catchment Management in Australia: Lessons for State-Sponsored Community Participation. *Society and Natural Resources.* 2000;13(1):61–73.

62. R. Youl, S. Marriott, T. Nabben. *Landcare in Australia.* (SILC and Rob Youl Consulting Pty Ltd, 2006).

63. C. Oravec. Conservationism Vs. Preservationism: The "Public Interest" in the Hetch Hetchy Controversy. *Quarterly Journal of Speech.* 1984;70(4):444–58.

64. R. W. Righter. *The Battle over Hetch Hetchy: America's Most Controversial Dam and the Birth of Modern Environmentalism.* (Oxford University Press, 2005).

6

Influences on Human Well-Being and Cultures, Societies, and Technologies

6.1 Measures of Social Development/Well-Being

Chapter 5 described how some peoples developed complex societies, primarily unintentionally. In the process, they created both benefits and problems to themselves and the environment. Chapter 5 also discussed changing populations.

This chapter will discuss those factors that lead to complex societies, how societies can change, and how problems or social collapses can be avoided.

Social development is "a group's ability to master its physical and intellectual environment to get things done." It is considered "the bundle of technological, subsistence, organizational, and cultural accomplishments through which people feed, clothe, house, and reproduce themselves, explain the world around them, resolve disputes within their communities, extend their power at the expense of other communities, and defend themselves against others' attempts to extend power" [1].

"Development" or "well-being" need to be expressed objectively both to determine which countries are progressing and to determine what factors contribute positively and so can be promoted [2]. Many "indexes" of development and well-being have been constructed and measured for most countries [3]. They are based on subjective and objective analyses and attempt fairness through various processes. Although appearing redundant [4], the many indexes help reflect the variety of interpretations of well-being and development. They also help indicate what things can be done to promote well-being as defined in the different ways. Selected indexes are described below and compared in Table 6.1:

- Gross Domestic Product per capita (GDPpc) is a measure of the average globally valued money turnover per person in a country [5].
- (Gross National Income per capita) GNIpc is similar to GDPpc, but also includes income from investments abroad [6].
- Human Development Index (HDI) measures the degree to which each country's citizens are provided for by amalgamating such social factors as equality, life expectancy, education, poverty, wealth distribution, and others [7].
- Environmental Health (EH) is a component of the Environmental Performance Index (EPI) that examines the health quality of the environment – such as air and water pollution and human health [8, 9].

Table 6.1 *Coefficients of determination (R^2) comparing indexes of well-being for the world's countries;[A] the best fits (R^2) of linear and nonlinear analyses are shown.*

	GDPpc	GNIpc	HDI	CHIE	EH	GDPpcPPP
GDPpc	–	–	–	–	–	–
GNIpc	0.79	–	–	–	–	–
HDI	0.60	0.77	–	–	–	–
CHIE	0.34	0.63	0.77	–	–	–
EH	0.35	0.53	0.82	0.64	–	–
GDPpcPPP	0.89	0.92	0.77	0.60	0.36	–
Happy Pl	0.01	0.07	0.23	0.07	0.15	0.05

[A]Legend (indexes are described and referenced in text):
GDPpc = Gross Domestic Product per capita (current US$)
GNIpc = Gross National Income per capita
HDI = Human Development Index [16]
CHIE = Coefficient of Human Inequality [7, 16]
EH = 2016 Environmental Health [9, 8]
GDPpcPPP = GDP per capita ($PPP)
HPl = Happy Planet Index [14, 15]

- Coefficient of Human Inequality (CHIE) is a component of HDI. It averages the inequalities in health, education, and income.
- GDPpcPPP: A Purchasing Power Parity (PPP) conversion factor was developed to account for the ability for local goods and services in some countries to be purchased at low prices [10–12]. It is the GDPpc adjusted for Purchasing Power Parity (PPP), described below.
- Purchasing Power Parity Advantage (PPPA), developed for this book, is not shown in Table 6.1 and will be discussed later.
- The Happy Planet Index (HPI) considers satisfaction with life and gives high rankings to countries in the southern hemisphere in South America and Asia (Figure 6.4). It contains "Well-being," "Ecological Footprint" [13] "Governance," "Life Expectancy," and other factors [14, 15].

Table 6.1 shows that GDPpc, GNIpc, GDPpcPPP, and HDI are closely related.

There is a moderately close inverse relationship between PPPA and both GDPpc ($R^2 = 0.62$) and HDI ($R^2 = 0.64$); therefore, including PPP to form GDPpc(PPP) [14, 15] and GNIpc(PPP) [7] creates a closer relation to HDI than simply GDPpc (Table 6.1).

The CHIE is not correlated to the HPI, but negatively correlated to all others. It is closely correlated to HDI in part because it is a component of HDI [7, 16]. EH is closely correlated to HDI.

HPI is the least correlated with the other indexes.

The "National Well-being Index" is not analyzed here. It is a Gallup World Poll averaging of the subjective question of ranking one's life compared to the possible best and worst [14, 15, 17–19]. The National Well-being Index is also a subset of the Happy Planet Index.

The next few pages will discuss these and the other relationships.

6.2 Comparisons of Four Measures of Well-Being

Four indexes will be discussed in detail below: GDPpc, HDI, GDPpcPPP, and HPI.

6.2.1 Human Development Index and Gross Domestic Product Per Capita

Many social organizations have been proposed and abandoned, including enlightened despots, aristocracies, and anarchy. Two recent philosophies – socialist and free enterprise – have focused on the well-being of the individual, with different relations of society to individuals.

The socialist philosophy regards society as responsible for the welfare of the individuals and is reflected in the United Nations' HDI [7, 16, 20]. HDI is closely correlated with the country's infant mortality ($R^2 = 0.87$) and life expectancy ($R^2 = 0.81$), among other factors.

The free enterprise philosophy gives high marks to the GDPpc and its related GNIpc and GDPpcPPP [5, 21]. It regards individuals as responsible for organizing the government. GDPpc has been criticized for its insensitivity to wealth distribution and to values not readily expressed in monetary forms. GDPpc is slightly less closely correlated than HDI to infant mortality ($R^2 = 0.72$) and life expectancy ($R^2 = 0.65$).

GDPpc distinguishes between countries with medium and high economic development better than HDI (Figure 6.1). All countries achieve a similar HDI once they reach a threshold GDPpc, with little increase in HDI with increased GDPpc.

Figure 6.2 compares average GDPpc, GDP-PPP/capita, and HDI values by regions. The high degree of development and well-being of countries in the northern hemisphere

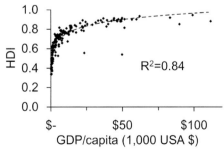

Figure 6.1 Relationship between HDI and GDPpc for countries [5, 7].

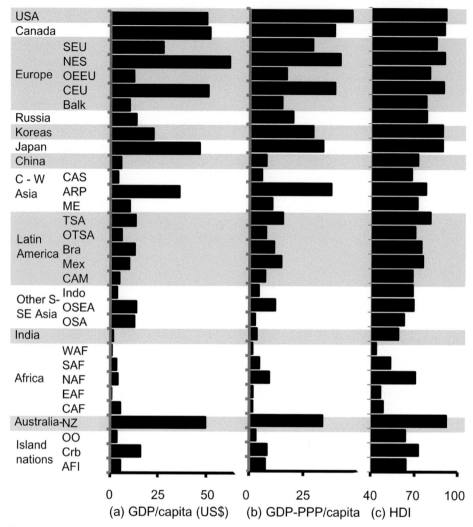

Figure 6.2 Gross domestic product (GDP) per capita, Gross domestic product-purchasing power parity (GDP-PPP) per capita, and Human Development Index (HDI) by country groups [5, 7, 12].

compared to the southern as measured by all three indexes in Figure 6.2 led to the "North–South" development awareness [22].

The organization of many countries lies somewhere between these two philosophies. The philosophical differences are explicit in the socialist government of the former Soviet Union and the more free-enterprise governments of Western Europe, North America, and elsewhere. Although these groups of countries have achieved similar HDIs, the former Soviet countries lag dramatically behind in their GDPpcs (Figure 6.2).

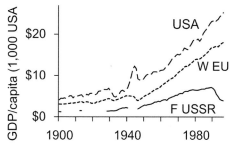

Figure 6.3 Changes in Gross domestic product per capita for the USA, Western Europe (W EU) and total former Soviet Union (F USSR) [23]. Data courtesy of the Maddison Project.

Both the Soviet Union and Western countries were developing in economically similar patterns from 1900 through about 1950 (Figure 6.3); however, the Soviet Union's GDPpc grew much slower in the following five decades. Two possibilities can explain the deviations after 1950:

- In the 1930s and 1940s, the Soviet Union was controlled by an extremely repressive government that may have hindered growth, whereas a different regime may have maintained pace with Western economies.
- The installation of large, stable infrastructures such as dams, terracing, and factories in the Soviet Union allowed rapid economic growth at first; however, because the highly centralized system did not encourage the people's input and creativity, it became too rigid compared to the more vibrant systems of the West.

6.2.2 Purchasing Power Parity and Happy Planet Index

A GDPpc in West Africa that is 4% of the United States' GDP does not mean that average West Africans can obtain only 4% of the food, clothing, and other household goods of an average United States resident. Rather, GDPpc reflects the amount of both domestic and foreign goods that can be purchased.

PPPA (Purchasing Power Parity Advantage) was developed for this book to show how many standard "baskets" of goods could be purchased with one US dollar in a region if one "basket" cost one US dollar in the United States (Figure 6.4) [10]. For example, the PPP of India in 2014 was 16.98 [24], meaning that 16.98 rupees could purchase the same standard "basket of domestic goods" in India as one US dollar in the United States. Since the international currency exchange rate of rupees to dollars was 63.47 rupees/dollar [24], a dollar could purchase nearly four times (63.47/16.98 = 3.74) as much locally in India as in the United States. The calculation (3.7, in the case of India) is referred to as the "Purchasing Power Parity Advantage" (PPPA); it is the ratio of the local currency exchange rate (to US dollars) and the PPP converted to dollars.

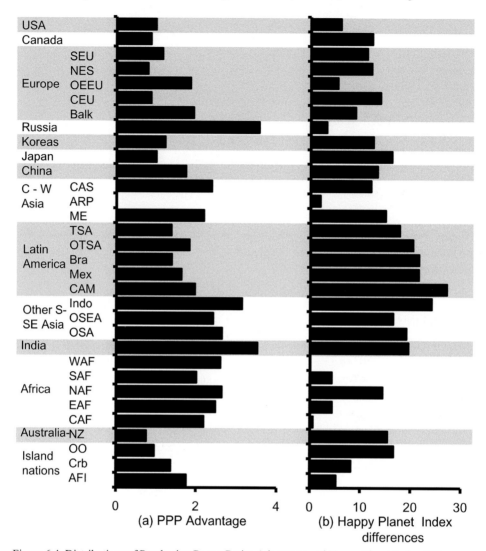

Figure 6.4 Distributions of Purchasing Power Parity Advantage and Happy Planet Index differences by country groups [12, 14]. "Differences" indicate difference from lowest-scoring country group. Some data courtesy of the New Economics Foundation.

The wide variability of well-being of countries relative to PPPA suggests that a high PPPA can have various causes. A high PPPA can occur in countries with high rural or urban populations ($R^2 = 0.35$) and with high and low fertility rates ($R^2 = 0.18$); consequently, the PPPA Index does not reflect other development indicators of countries. Only moderate correlations exist between PPPA and infant mortality ($R^2 = 0.55$) and life expectancy ($R^2 = 0.52$). As countries develop, some suggest that the disparity between GDPpc and PPP may decline [25].

A high PPPA gives poor countries a false sense of well-being – an apparent sense of well-being but limited access to international trade. People in countries with high PPPA can deplete and/or degrade local resources such as forests because they cannot afford to import wood or fossil fuels (Chapter 30); and they would be motivated to export more than sell domestically because the foreign currency exchange advantage would give them a higher price.

Figure 6.4b shows the difference between each HPI and the lowest, West Africa, to accentuate HPI differences.

6.3 Factors Related to Social Development and Well-Being

The development, stability, or decline of complex societies' well-being have been attributed to various factors or their shortages [26–33] (this list does not indicate an order of importance):

- *Resources*: Sufficient water, food, energy, minerals, and forests;
- *Government organization*: Appropriate laws;
- *Infrastructure*: An infrastructure that leads to appropriate interactions among the people;
- *Conservation of the environment*: Unpolluted water or air, protected biodiversity, and similar factors that may keep people physically and mentally healthy;
- *People's behavior/culture*: Behaviors that are not conducive to corruption, violence, uncooperativeness, or other social dysfunctions.

Each factor has multiple measurements that indicate the factor's condition for nearly every country (Table 6.2). The importance of each factor for well-being was analyzed for this book by comparing each measure of each factor to three indicators of human well-being for the world's countries: GDPpc, HDI, and HPI. The highest correlation (of both linear and nonlinear regressions) for each comparison is in Table 6.2. Those factors most closely correlated to the well-being indicator should indicate their importance relative to a country's well-being as defined by the well-being indicator. It should then be possible to prioritize attention to those most closely related factors to improve a country's well-being. (Although correlation does not necessarily mean causality, the following discussion will address possible causality in the relationships.)

Table 6.2 shows that the same factors have roughly similar correlations to HDI and GDPpc. The Happy Planet Index was quite poorly correlated with any of the factors. When promoting the Happy Planet Index, a different approach apparently must be taken. Each factor and its relation to the indexes will be discussed in the order of highest to lowest correlation with GDPpc and HDI.

6.3.1 People's Behaviors

The high correlations of many measures of people's behaviors with both HDI and GDPpc indicate that a safe, cooperative society is important to a country's well-being (Table 6.2).

Table 6.2 *Coefficients of determination (R^2) between different factors that might affect country development and selected development/well-being indexes*

	GDP per capita (current US$)	Human Development Index (HDI)	Happy Planet Index
Resources[B]			
Food produced (kcal)/(person day) [34–36]	0.47	0.62	0.07
Calories consumed/(person day) [34–36]	0.48	0.67	0.09
Government[C]			
Property rights [37, 38]	0.57	0.46	0.00
Trade freedom [37, 38]	0.34	0.50	0.02
Tax burden % of GDP [37, 38]	0.26	0.22	0.03
Infrastructure			
Unemployment rate [24]	0.05	0.02	0.04
Internet users/capita [24]	0.63	0.82	0.14
Transportation length/1000 people [24]	0.25	0.38	0.03
Transportation/million km^2 [24]	0.17	0.28	0.00
Total phone users/capita [24]	0.28	0.69	0.11
Internet users/capita [7]	0.58	0.73	0.07
Expected years of schooling [7]	0.41	0.81	0.11
Literacy rates [7, 16]	0.15	0.72	0.16
Environmental Condition			
Ecosystem vitality [8, 9]	0.31	0.45	0.03
Environmental performance [9]	0.62	0.77	0.12
People's Behaviors			
Corruption perceptions index [39]	0.68	0.53	0.00
Failed states index 2013 [40]	0.66	0.74	0.06
Control of corruption (2014) [3, 41]	0.62	0.50	0.01
Government effectiveness [3, 41]	0.65	0.70	0.04
Political stability and absence of violence [3, 41]	0.32	0.42	0.01
Regulatory quality [3, 41]	0.57	0.56	0.00
Rule of law [3, 41]	0.61	0.56	0.00
Voice and accountability [3, 41]	0.31	0.40	0.02

[B] The following resource measures had all R^2 values below 0.25: forest area [42, 43], forest area/capita [42, 43], forest volume (growing stock) [42, 43], wood volume/capita [42, 43], total population [24, 44], population/km^2 [24, 44], country area [24, 44], annual precipitation [45], renewable water/capita[46], water withdrawn/capita [46], total fossil fuel reserves [47], and fossil fuel reserves/capita [47].

[C] The following government measures had all R^2 values below 0.25: income tax rate (%) [37], corporate tax rate (%) [37], government expenditures as % of GDP [37], Gini coefficient (income inequality) [7].

An alternative explanation is that, as more wealth (high GDPpc) is distributed to more people, they have more to gain by cooperativeness and peace [48, 49]; however, a low correlation between either GDPpc or HDI indexes of well-being and the Gini coefficient, an indication of income equality, does not suggest that wealthy or developed countries are necessarily more egalitarian (Table 6.2).

Little corruption and violence are found in the highly developed countries. Countries of uniform and mixed ethnicity, different sizes, and on different continents can be found with both high and low corruption and violence measures. Some countries can be highly corrupt but not violent; but others may be the opposite. Societies also change behaviors with time – the currently virtuous Scandinavians were once the very violent "Vikings."

The cooperative behavior among farmers allowed great strides in agriculture production in the nineteenth-century United States. It also allowed those cooperative regions to avoid crop pests that devastated areas where farmers did not cooperate [50]. (See "Kristianstads Vattenrike" case study, Chapter 3.)

6.3.2 Infrastructures

Many infrastructures that provide education, communication, information, and transportation are correlated with HDI and GDPpc. It is probable that better infrastructures generate more interactions, more creation of new ideas and technologies, and thus more vibrant societies. The most recent technologies – internet and cell phones – seem most closely related to development and non-corrupt, safe societies ($R^2 = 0.65$ to 0.77). Other technologies such as transportation may be less correlated because most people in the world now have access to them.

Education is commonly a priority of both socialist and free enterprise states. Most countries with high HDIs are highly literate for several possible reasons: literacy is a measurement criterion for the HDI, most socialist countries promote education, and education in those countries is commonly free. Although countries with high GDPpc are highly literate, not all highly literate countries have high GDPpc.

Unlike literacy, all countries with relatively high internet and telephone use are relatively wealthy – they have a relatively high GDPpc. It is unclear if telephone and internet use lead to high HDI and GDPpc values, or if these countries are wealthy enough to adopt telephones and internets before less developed countries.

6.3.3 Government Organization

Much of the organization, size and economic dominance of the government showed relatively little relation to the country's development except possibly property rights (Table 6.2). Further analyses for this book showed that corruption and failed states indexes also are closely and negatively correlated to property rights ($R^2 = 0.88$, 0.74, respectively), as also suggested elsewhere [32]. Governments that own and sell resources are not dependent on their citizens' wealth and taxes to provide their financial operating base. Consequently, they tend to neglect the citizens' welfare such as private property rights. Private property avoids problems of noncooperation with common property ownership ("tragedy of the commons" [51]) but can create problems if individual property owners refuse to cooperate for a common good ("tragedy of the anti-commons" [52]).

The relation of trade to HDI and GDPpc depends on whether raw materials or more finished goods are traded. Local wealth is generated by local processing (secondary manufacture) of resources, which creates more jobs both by those engaged in secondary manufacture and by those providing goods and services to those thus engaged (inclusive economies [26]). Alternatively, communities extracting raw materials benefit little from the minimal circulation of money if the resources are extracted and sent elsewhere ("extractive economies"). Inappropriate trade can also make some segments of a society so wealthy that they import goods instead of buying them locally, reducing the prosperity of other segments ("Dutch disease" [53]).

A positive relationship has been reported between good governance and "National Well-being" – the subjective index referred to earlier [54].

6.3.4 Environmental Condition

The high correlations of the Environmental Performance Index (EPI) to social development and well-being (GDPpc and HDI) are probably because social development leads to cleaner environments (Figure 6.5).

- Less developed countries tend to be inefficient in use of such resources as agriculture land area and fuel-wood consumption and so leave little room for biodiversity and forests.
- Less developed societies are focused on their immediate needs of food and shelter and are unable to address longer-term issues of the environment and future resources.

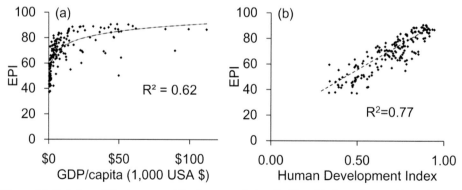

Figure 6.5 Relation of Environmental Performance Index [8, 9] to (a) Gross Domestic Product per capita [5] and (b) Human Development Index [7]. Data used with permission of the Environmental Performance Index.

- More developed societies tend to follow laws and rules, so environment and resource stewardship is more effective.
- More developed societies are more educated and linked into communication systems and so are more likely to understand environmental and resource issues.
- Developing countries sometimes use more resources such as cement and metals to develop their infrastructures (Figure 27.9), while developed countries use fewer because their infrastructures are in place (Chapter 27). On the other hand, developed countries may have a larger "environmental footprint" [13], where they use a disproportionate share of the world's fossil fuels for other reasons.

A clean environment (EPI) is more closely related to an overall well-being (HDI) than to extreme wealth (GDPpc) (Figure 6.5).

The disproportionate use of resources such as fossil fuel shows that some behaviors of developed countries are harmful to the environment and that there is no simple solution of either exploiting the environment [55] or preserving it [56]. More nuanced decisions by an educated world will probably best help both social development and the environment (Chapter 3). It is more promising – but a big challenge – to address the environment by bringing all people toward the developed social condition while also changing the undesirable aspects of development.

6.3.5 Possession of Resources

The resources that a country possesses have surprisingly little to do with its development (HDI or GDPpc; Table 6.2) except possibly food resources. The low correlation is probably because most people have learned how to use the resources they have – making behavioral adjustments and trading for things when needed.

A low correlation between property rights and food produced ($R^2 = 0.38$) but a high correlation of food produced to HDI and GDPpc (Figure 6.6) indicate that food is a priority as countries become developed – whether or not property rights are emphasized. Most countries that produce less than the minimum daily recommended Calories per capita have low HDIs and low GDPpc. Most countries that produce more than the minimum Calories have high HDIs; probably they export the excess food, feed plant foods to animals and eat meat, or produce biofuels (Chapters 22, 24).

To a large extent, societies can overcome limited water resources by storing, conserving, circulating, recycling, and desalinating (Chapter 19). Many societies that are hindered by water shortages probably lack appropriate knowledge, technology, or organization.

Other resources are also not well correlated with either HDI or GDPpc (Table 6.2). Possession of fossil fuels has often not helped a country's well-being.

Forests are not extensively utilized (Chapter 29). In countries with low HDI and GDPpc, the valuable logs have often been exported with little secondary manufacture and therefore little in-country employment (Figure 29.7) [26]. The remaining trees of lower quality are often used inefficiently for fuelwood. Forests are very important for biodiversity, but few are willing to compensate countries to protect it (Chapter 16).

Countries with concentrations of minerals that are valuable in small volumes can export these; however, like fossil fuel and valuable trees, this will best help a country's HDI or GDPpc if secondary manufacture adds value to the minerals before export and if the benefits are distributed fairly to the country's citizens.

6.3.6 Wars

The above analyses suggest that any society can create or lose a strong influence on social development – peaceful behaviors that avoid corruption and violence. Violent conflicts not only harm people and reduce people's living condition, they greatly harm the environment and resources:

From a personal observation, civil unrest is a big cause of environmental degradation. The lawlessness that exists during wartime allows illegal and unsustainable exploitation of natural resources, and displacements of people fleeing conflict zones lead to fast and unplanned settlements in areas where forests and other biodiversity are then unsustainably exploited for food and shelter. The collapse of institutions means protected areas are no longer protected; and the poverty that results from the war means governments don't have enough resources to protect the environment; and the people's livelihoods only depend on exploiting natural resources.

(Erik Ndayishimiye,
Citizen of Rwanda [57].)

6.4 Cultures, Groupthink, Mental Models, Societies, and Institutions

Social development and well-being can be promoted by people changing how they think and interact. Five overlapping concepts are discussed below: cultures, groupthink, mental models, societies, and institutions.

6.4.1 Cultures and Groupthink

Cultures have been defined with different nuances, but generally refer to "that complex whole which includes knowledge, belief, morals, law, custom and any other capabilities and habits acquired by [a person] as a member of society" [58]. They strongly influence individual people's behaviors, choices, and decisions to cooperate or compete with another, to accept a technology or particular job, and to study and experiment. Cultures are dynamic; they are learned, not inherited.

Some authorities group cultures by geographic and ethnic origin; however, many people of such groupings can have less in common with each other than with people of distant origins but similar occupations, wealth, or interests. A wealthy business person probably has more in common with another from a distant land than with a poor person of his/her own village. Engineers, scientists, farmers, fishermen, or sports enthusiasts from around the world similarly form cultures. Individuals can associate with many cultures.

Cultures have the benefits and shortcomings of "groupthink" [59]. Groupthink is the numerous unstated assumptions, actions, and relationships that a culture holds in common and accepts as fundamental. They can include marriage; family; caste/class systems; segregation; privileged groups; courteous gestures; burning forests; protecting biodiversity; not eating some foods; fighting or befriending strangers; accepting or rejecting corruption and bribes; and others. People engaged in groupthink can cooperate very efficiently and synergistically while creating very good or very bad results, depending the validity of their commonly held assumptions and behaviors.

6.4.2 Mental Models

Each person's image of the world is a model that he/she has mentally constructed. Nobody has a complete mental construction of the world, the environment, or the resources. Each person has only selected concepts and relationships among them that are used to represent the real system [60].

A person first constructs a mental model from his/her family, culture, and community. These models change and/or are reinforced with one's experiences, learning, and needs to make decisions. A young individual's mental model is relatively local. The mental ability to group enables people to conceive, form, and manage complex societies; however, it can also lead to "stereotyping," the negative aspects of groupthink described above, and xenophobia – fear or hatred of strange or foreign people, things, or behaviors.

Differences in education or in specific talents (such as mathematics, music, athleticism) can further lead to different mental models. The way people use their mental models to approximate reality is based partly on their ability to process more intricate relationships and feedback loops. Some people are more endowed with this capability than others, but like other endowments it needs nurturing.

People within a culture behave similarly because of common mental models that are implicit, inferred, or explicit. An academic culture may find physical confrontation inacceptable, but a sports culture may not.

6.4.3 Societies

Societies are groups of people who live together in a geographic region. Societies can be delimited at different scales, from a neighborhood to a country or larger. Most governing is done within and among geographic areas; and so, the collective attitudes of resident societies fundamentally affect the governing. Members of a society have some similar mental models but differences in prosperity, occupations, education, religions, and family upbringing can lead to different cultures within a society. These differences can enrich or weaken a society, depending on how society treats the differences.

The mental models of some societies have enabled them to make tremendous advances quickly and overcome hardships. For example, Japan, Germany, and England have developed high living standards; and despite devastating wars and other setbacks, they rebounded quite quickly.

6.4.4 Institutions

In time, "institutional behaviors" emerge. They are "durable systems of established and embedded social rules (conventions, norms, and legal rules) that structure social interactions" [61], but are not unexpectedly imposed by governments. Institutional behaviors regulate the relationship among people and "between social and ecological systems" [62]. They may include not throwing trash onto streets, not smoking inside public buildings, and driving on a specified side of the road.

Many, but not all, institutional behaviors are codified as laws – from a country's constitution to local regulations. "Institutional organizations" include governments and other organizations that provide stable arrangements for interacting; however, societies must embrace a behavior before it can become a successful law. Consequently, laws need to be aligned with institutional behaviors to be obeyed in complex societies that have strong social networks and not simply a hierarchy of leadership (Chapter 3). The law then safeguards against nonconformist behaviors. The importance of this local acceptance is increasingly becoming appreciated [63].

In contrast to cultural behaviors, "institutional behaviors" are explicitly stated and agreed upon.

6.5 Social and Technological Changes

Mental models, cultures, and behaviors change with generations, technologies, unanticipated situations, and encounters of people with other behaviors. The changes can be fruitful or disastrous, leading to greater well-being or the collapse of a society. People are beginning to examine how changes can be managed effectively to society's benefit.

6.5.1 Avoiding Rogue Societies with Social Change

When confronting a new situation, people can follow different behaviors depending on which behavior establishes initial dominance (Chapter 2) [64, 65]. Consequently, different groups may squabble, cooperate, flee, work hard, complain, or laugh when confronted with the same, new situation and thus create different patterns of reorganization (Figure 2.4). Rogue societies ascend by establishing a dominant, cruel mental model during times of initial reorganization. Modern technologies make these rogue societies extremely dangerous.

Different words and gestures can create different emotions and reactions in others, which can strongly influence their mental model, behavior, and resulting culture [65, 66]. For example, words and gestures of fear can stimulate an attitude of "do not change, for fear matters will become worse." Fear often reinforces rigidity where flexibility in behaviors is needed [66, 67]. Words and expressions that stimulate people to seek changes and opportunities may more effectively maintain flexibility in complex societies during times of change.

Some cultures and societies have institutionalized responses to changes that promote cooperation and continuity during chaotic times, and so help avoid adoption of a rogue mental model. Mountaineering guides, soldiers, police, and similar groups accustomed to changing conditions commonly have cultural behaviors already prepared for new situations. For example, Sherpa guides in the Himalayas react calmly and cooperatively to avalanches and other crises in mountainous regions. Soldiers define a chaotic situation as "friction" and cooperatively seek solutions, rather than initially seek blame and become divisive. Some military groups fault a leader more harshly for indecision than for a poor decision when confronted with a new situation. Governments have tried to institutionalize change with orderly transitions of power, such as elections or monarchy successions as well as constitutions that both outlive individual leaders and can be amended.

For better or worse, a new collective mental model can become institutionalized by having it adopted by many people within a culture or society. A "master narrative" can be established that is a seemingly internally consistent mixture of accurate and inaccurate analyses, emotional appeals, charismatic information or misinformation, and other messages that form a story attracting the positive or negative emotions of people [68]. Increasingly educated cultures and societies will demand more accurate information and analyses.

The arising of rogue societies can probably best be avoided by preventing countries from becoming economically or socially destabilized, by institutionalizing positive and cooperative initial reactions to change, and by educating the population. Conversely, a society can be changed by destabilizing it, creating panic when change occurs, and denigrating the importance of education. Once destabilized, the emergent behavior of a society is very difficult to predict or control.

6.5.2 Opportunities and Challenges of Emerging Technological Societies

Technological changes typically lead to different uses of resources as well as new effects on the environment that can both affect and be affected by other changes.

Human ingenuity is sometimes referred to as "the ultimate resource" [55]. It can solve some negative aspects of these changes – and some expect it to resolve all. Greater agricultural efficiency, sophisticated communication, abundant energy, rapid transportation, and institutional behaviors are examples of human ingenuity. Not all ingenuity has been productive, such as the development of more "creative/clever" destructive weapons.

Ingenuity can enable developing countries to avoid many large expenses in infrastructures incurred in the developed countries by adopting leapfrog technologies.

The nature of complex, technological societies will differ among various environments, countries, and rural/urban lifestyles. The transition to technological societies is based on a high rate of turnover of goods and services; each sector contributes something other people are willing to pay for, and so wealth essentially "increases" by its circulation through society [26, 69].

Modern communication and technologies can lead many people to feel entitled to more wealth and privileges, to feel disenfranchised and frustrated if these expectations are not realized, and to take violent actions as a result. Many perceived inequities are real and need to be corrected; others occur when people innocently form higher expectations than can possibly be realized; and still others are falsely instilled by those seeking personal advantage. The challenge is to identify and correct the true inequities; to manage people's expectations so they are not unrealistic; and to prevent the intentional spreading of false information.

References

1. I. Morris. *Why the West Rules – For Now: The Patterns of History and What They Reveal About the Future.* (Farrar, Straus, & Giroux, 2010).
2. N. Hicks, P. Streeten. Indicators of Development: The Search for a Basic Needs Yardstick. *World Development.* 1979;7(6):567–80.
3. The World Bank. *World Development Indicators.* (The World Bank Group, 2016 [Accessed July 5, 2016]). Available from: http://data.worldbank.org/indicator/NY.GDP.PCAP.CD.
4. M. McGillivray. The Human Development Index: Yet Another Redundant Composite Development Indicator? *World Development.* 1991;19(10):1461–8.
5. The World Bank. *GDP Per Capita (Current US$).* (The World Bank Group, 2016 [Accessed June 24, 2016]). Available from: http://data.worldbank.org/indicator/NY.GDP.PCAP.CD.
6. The World Bank. *World Development Indicators.* (The World Bank, 2016 [Accessed May 8, 2016]). Available from: http://data.worldbank.org/indicator/EG.USE.PCAP.KG.OE.
7. UNDP. *Human Development Index (HDI).* (United Nations Development Programme, 2015 [Accessed June 24, 2016]). Available from: http://hdr.undp.org/en/content/human-development-index-hdi.
8. Yale University Data-Driven Environmental Group and CIESIN Columbia University. *2016 Environmental Performance Index (2016 Epi).* (Yale University, 2016 [Accessed June 24, 2016]). Available from: http://epi.yale.edu.
9. A. Hsu, et al. *Environmental Performance Index.* (Yale University, 2016).
10. R. Dornbusch. *Purchasing Power Parity.* (National Bureau of Economic Research, 1985).
11. L. H. Officer. The Purchasing-Power-Parity Theory of Exchange Rates: A Review Article. *Staff Papers.* 1976;23(1):1–60.
12. The World Bank. *PPP Conversion Factor, GDP.* (The World Bank Group, 2017 [Accessed January 31, 2017]). Available from: http://data.worldbank.org/indicator/PA.NUS.PPP.
13. E. Lazarus, G. Zokai, M. Borucke, et al. *Working Guidebook to the National Footprint Accounts: 2014 Edition.* (Global Footprint Network, 2014).
14. Centre for Well-being. *Happy Planet Index: The Data.* (New Economics Foundation; 2016 [Accessed June 24, 2016]). Available from: www.happyplanetindex.org/.

15. S. Abdallah, J. Michaelson, S. Shah, L. Stoll, N. Marks. *The Happy Planet Index: 2012 Report. A Global Index of Sustainable Well-Being*. (New Economics Foundation, 2012): 26 pp.
16. S. Deb. The Human Development Index and Its Methodological Refinements. *Social Change*. 2015;45(1):131–6.
17. J. F. Helliwell, C. P. Barrington-Leigh, A. Harris, H. Huang. *International Evidence on the Social Context of Well-Being*. National Bureau of Economic Research Working Paper 14720, 2009.
18. R. Veenhoven. *World Database of Happiness*. (Erasmus University Rotterdam [Accessed July 12, 2017]). Available from: http://worlddatabaseofhappiness.eur.nl.
19. Global Footprint Network. Our Work. (Footprint Network [Accessed July 12, 2017]). Available from: www.footprintnetwork.org/our-work/.
20. J. Selim, et al. *Human Development Report 2015: Work for Human Development.* (United Nations Development Programme, 2015).
21. R. M. Auty. *Resource Abundance and Economic Development*. (Oxford University Press, 2001).
22. W. Brandt. *North–South: A Programme for Survival: Report of the Independent Commission on International Development Issues*. (MIT Press, 1980).
23. The Maddison Project. *Historical Statistics of the World Economy* 2013 (Accessed October 17, 2016). Available from: www.ggdc.net/maddison/maddison-project/home.htm.
24. US Central Intelligence Agency. *Country Comparison* (US Central Intelligence Agency; 2016 [Accessed June 24, 2016]). Available from: www.cia.gov/library/publications/the-world-factbook/index.html.
25. T. Piketty, L. J. Ganser. *Capital in the Twenty-First Century*. (Harvard University Press, 2014).
26. D. Acemoglu, J. Robinson. *Why Nations Fail: The Origins of Power, Prosperity, and Poverty.* (Crown Business, 2012).
27. W. W. Rostow. *The Stages of Economic Growth: A Non-Communist Manifesto*. (Cambridge University Press, 1990).
28. A. J. Toynbee, D. C. Somervell. *A Study of History*. (Oxford University Press, 1987).
29. J. Tainter. *The Collapse of Complex Societies*. (Cambridge University Press, 1990).
30. J. Diamond. *Collapse: How Societies Choose to Fail or Succeed*. (Penguin, 2005).
31. P. R. Ehrlich, A. H. Ehrlich, editors. Can a Collapse of Global Civilization Be Avoided? *Proceedings of the Royal Society B*. (The Royal Society, 2013).
32. T. Burgis. *The Looting Machine: Warlords, Oligarchs, Corporations, Smugglers, and the Theft of Africa's Wealth*. (PublicAffairs, 2015).
33. T. Homer-Dixon. *The Upside of Down: Catastrophe, Creativity, and the Renewal of Civilization*. (Island Press, 2010).
34. UN FAO. *FAO Statistical Yearbook*. (United Nations Food and Agriculture Organization, 2009 [Accessed October 8, 2014]). Available from: www.fao.org/docrep/014/am079m/PDF/am079m00a.pdf.
35. UN FAO. *FAO–Food Security Indicators*. (United Nations Food and Agriculture Organization, 2013 [Accessed June 25, 2016]). Available from: Food-Security-Statistics@FAO.org.
36. UN FAO. *Food Security Indicators: Statistics*. (United Nations Food and Agriculture Organization, 2016 [Accessed June 25, 2016]). Available from: www.fao.org/economic/ess/ess-fs/ess-fadata/en/#.V24xg7grLIU.
37. The Heritage Foundation. *2015 Index of Economic Freedom: Download Raw Data*: (The Heritage Foundation and The Wall Street Journal, 2015 [Accessed November 26, 2016]). Available from: www.heritage.org/index/download.
38. T. Miller, A. B. Kim, K. Holmes, cartographers. *Index of Economic Freedom: Promoting Economic Opportunity and Prosperity*. (The Heritage Foundation and Dow Jones and Company, 2015).
39. Transparency International. *Corruption Perceptions Index 2015: Downloads and Methodology*. 2015 (Accessed June 24, 2016). Available from: www.transparency.org/cpi2015#downloads.
40. N. Haken, J. Messner, K. Hendry, et al. *The Failed States Index IX: 2013*. (The Fund for Peace, 2013).

41. D. Kaufmann, A. Kraay, M. Mastruzzi. *The Worldwide Governance Indicators: Methodology and Analytical Issues*. (The World Bank, 2010).
42. UN FAO. *ForesSTAT2016* [Accessed July 3, 2016]. Available from: http://faostat.fao.org/site/626/DesktopDefault.aspx?PageID=626#ancor.
43. UN FAO. *Global Forest Resources Assessments*. (United Nations; 2016 [Accessed July 3, 2016]). Available from: www.fao.org/forest-resources-assessment/explore-data/flude/en/.
44. US Central Intelligence Agency. *The World Factbook 2014–15*. Agency UCI, editor: (US Government Publishing Office, 2015).
45. The World Bank. Table 3.2. *World Development Indicators: Agriculture Inputs* 2015 (Accessed November 26, 2015). Available from: http://wdi.worldbank.org/table/3.2.
46. UN FAO. *AQUASTAT Database*. (United Nations Food and Agriculture Organization, 2016 [Accessed June 25, 2016]). Available from: www.fao.org/nr/water/aquastat/data/query/index.html?lang=en.
47. British Petroleum. *BP Statistical Reviews of World Energy, June 2013–2016* (Accessed August 30, 2017). Available from: www.bp.com/statisticalreview.
48. P. Collier. *Breaking the Conflict Trap: Civil War and Development Policy*. (World Bank Publications, 2003).
49. P. Collier, A. Hoeffler. Greed and Grievance in Civil War. *Oxford Economic Papers*. 2004;56(4):563–95.
50. A. L. Olmstead, P. W. Rhode. *Creating Abundance*. (Cambridge University Press, 2008).
51. G. Hardin. The Tragedy of the Commons. *Science*. 1968;162(3859):1243–8.
52. M. Coelho, J. Filipe, M. Ferreira. *Tragedies on Natural Resources a Commons and Anticommons Approach*. School of Economics and Management, University of Lisbon Report 0874–4548, 2009.
53. J. E. Stiglitz. *Globalization and Its Discontents*. (Norton, 2002).
54. J. F. Helliwell, H. Huang, S. Grover, S. Wang. *Empirical Linkages between Good Government and National Well-Being*. National Bureau of Economic Research Working Paper 20686, 2014.
55. J. L. Simon. *The Ultimate Resource 2*. (Princeton University Press, 1998).
56. R. F. Noss. Sustainability and Wilderness. *Conservation Biology*. 1991;5(1):120–2.
57. E. Ndayishimiye. Personal Communication, in: Oliver C, editor, *Mr. Erik Ndayishimmiye, Citizen of Rwanda*. When Master of Environmental Management Student at the Yale Forestry School, he wrote this quote in answer to a question on a final examination (December 2015) in the "Global Resources and the Environment" course taught by Chad Oliver, ed. (2015).
58. S. E. B. Tyler. *Primitive Culture: Researches into the Development of Mythology, Philosophy, Religion, Language, Art and Custom*. (John Murray, 1891).
59. I. L. Janis. *Victims of Groupthink: Psychological Studies of Policy Decisions and Fiascoes*. (Houghton Mifflin, 1972).
60. P. N. Johnson-Laird. *Mental Models: Towards a Cognitive Science of Language, Inference, and Consciousness*. (Harvard University Press, 1983).
61. G. M. Hodgson. The Evolution of Institutions: An Agenda for Future Theoretical Research. *Constitutional Political Economy*. 2002; 13(2):111–27.
62. S. Anastasopoulou, V. Chobotova, T. Dawson, T. Kluvankova-Oravska, M. Rounsevell. Identifying and Assessing Socio-Economic and Environmental Drivers that Affect Ecosystems and their Services. In *The RUBICODE Project: Rationalizing Biodiversity Conservation in Dynamic Ecosystems*. (European Commission Sixth Framework Programme [Contract Number: 036890]): 86 pp.
63. E. Ostrom. A General Framework for Analyzing Sustainability of Social-Ecological Systems. *Science*. 2009;325(5939):419–22.
64. N. A. Christakis, J. H. Fowler. *Connected: The Surprising Power of Our Social Networks and How They Shape Our Lives*. (Little, Brown, and Company, 2009).
65. J. Jay, G. Grant. *Breaking through Gridlock: The Power of Conversation in a Polarized World*. (Barrett-Koehler, 2017).
66. B. L. Fredrickson. The Value of Positive Emotions. *American Scientist*. 2003;91(4):330–5.
67. D. Kahan. Fixing the Communications Failure. *Nature*. 2010;463(7279):296–7.

68. J.-F. Lyotard. *La Condition Postmoderne: Rapport Sur Le Savoir [the Postmodern Condition: A Report on Knowledge]*. Bennington G, Massumi B, editors: (Les Editions de Minuit [French]; University of Minnesota Press [English], 1979).

69. R. W. Dimand. *The Origins of the Keynesian Revolution: The Development of Keynes' Theory of Employment and Output*. (Stanford University Press, 1988).

Part III

Climates

7

Global Distribution of Climates

7.1 Distribution of Climates and Vegetation

Climate is the prevalent atmospheric conditions "near the Earth's surface over a long time and includes temperature, precipitation, humidity, wind, barometric pressure, and other phenomena" [1]. Weather is similarly the atmospheric conditions near the Earth's surface over a short time. Climate and weather patterns affect people, landforms, biodiversity, and every resource.

Climate and weather are complex systems (Chapter 2), with weather changing rapidly and climates changing slowly. Climates exist in patterns that can be somewhat predicted from the latitude and position relative to oceans and mountains [2]; and similar climates can be found in geographically different places. Weather is predicted less accurately from hot and cold fronts, high- and low-pressure systems, humidity, storms, clouds, and similar phenomena.

Figure 7.1 shows an interpretation of the world's ecological zones based largely on areas of similar climates [3]; other classifications of vegetation also exist [2, 4–8].

Global ecological zones are named by the kind of cover they "potentially" support and probably reflect a bias in favor of forests (Chapter 14). The classification is valuable if this bias is recognized. Global vegetation cover types are groups of similar ecological zones (Chapter 14) and can roughly be classified within a two-dimensional state space (matrix) along the climatic gradients of moisture and temperature (Figure 7.2) [10]. Mountains do not fit well into this matrix because they contain dramatic variations in moisture and precipitation within a short distance. The cover types have variations, gradations, and overlaps (Figure 14.6).

7.2 Energy and Heating of the Earth

The Earth's energy comes from several sources, but the Sun contributes over one thousand times as much as the others (Chapter 23). Only solar energy will be discussed here.

Most energy that the Earth receives from the Sun is initially in short wavelengths [2, 11, 12]. Much of it passes through the atmosphere and is converted to longwave energy (heat) when it reaches land and water. It is then moved by radiation, conduction, and

Global Ecological Zones

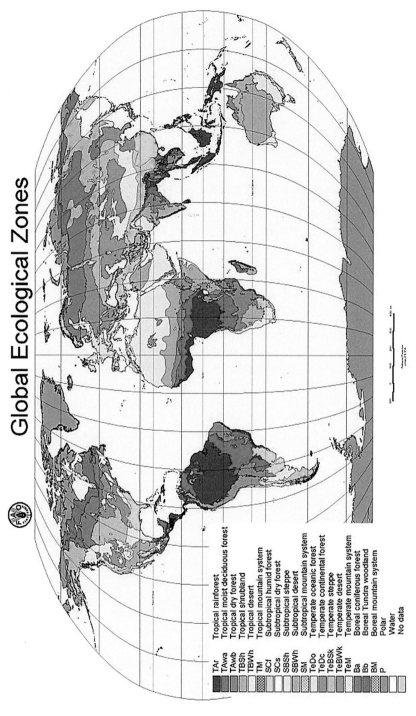

TAr	Tropical rainforest
TAwa	Tropical moist deciduous forest
TAwb	Tropical dry forest
TBSh	Tropical shrubland
TBWh	Tropical desert
TM	Tropical mountain system
SCf	Subtropical humid forest
SCs	Subtropical dry forest
SBSh	Subtropical steppe
SBWh	Subtropical desert
SM	Subtropical mountain system
TeDo	Temperate oceanic forest
TeDc	Temperate continental forest
TeBSk	Temperate steppe
TeBWk	Temperate desert
TeM	Temperate mountain system
Ba	Boreal coniferous forest
Bb	Boreal Tundra woodland
BM	Boreal mountain system
P	Polar
	Water
	No data

Figure 7.1 Locations of "Global Ecological Zones." Reproduced with permission, United Nations, Food and Agriculture Organization. (A black and white version of this figure will appear in some formats. For the color version, please refer to the plate section. Color plate 3.)

	Humid <<-->> Dry				
Hot ^	Tropical-	Rain forest	Moist deciduous forest	Dry forest	Shrubland	Desert
	Subtropical-	Humid forest	Dry forest		Steppe	Desert
	Temperate-	Oceanic forest	Continental forest		Steppe	Desert
	Boreal-	Coniferous forest			Tundra woodland	
Cold <<	Polar-			Polar		

Figure 7.2 Global vegetation cover types classified within a two-dimensional matrix of moisture and temperature. Names based on Figure 7.1.

convection into the water and terrestrial Earth surface, the atmosphere, or beyond into space (Figure 23.2).

Heat energy circulates among different substances at different rates through absorption, conduction, convection (circulation), and radiation, discussed in Chapter 23. Rocks and soil radiate more heat in the day and cool at night more quickly than water.

Water plays a large role in moderating and distributing heat throughout the Earth. Water has both high thermal conductivity and turbulence that moves heat away from its surface and high "specific heat capacity," so it needs to absorb or release a lot of heat in order to change its temperature a small amount [13]. Compared to many liquids, water also absorbs a lot of energy ("latent energy") without changing temperature when it melts (313 British Thermal Units per kilogram [BTUs/kg]) or evaporates (2,142 BTUs/kg) – changes states between solid (ice), liquid (water), and gas (dissolved in the atmosphere). It releases the same energy when it freezes or condenses [13].

Rocks, soils, water, and other solid materials convert shortwave energy from the Sun into heat energy that then heats the adjacent air. Heat builds up in still dry air and is conducted very slowly; lethal air temperatures (above 70 degrees Centigrade [°C]) can be found about two millimeters above the soil in direct sunlight with no turbulence. On the other hand, heat is circulated quickly by turbulent air as hot air moves and mixes with other air [2, 14].

As air is heated through contact with a hot part of the Earth, the air molecules move more rapidly and increase air pressure. If possible, the heated air expands to relieve the pressure, releasing heat in the process. The expanded air is lighter than air immediately above it and so floats upward, allowing other air to move into the vacated space beneath in a turbulent manner and to be heated in its turn. This rising of warm, light air creates low atmospheric pressures at the equator and other places where air rises above warm land or water.

As air expands and rises when at the equator or other warm places or when going over mountains, it comes under less pressure, expands, and cools at a rate of 10°C/km of altitude

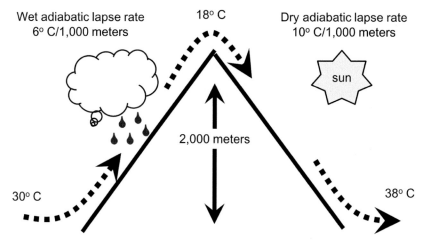

Figure 7.3 Effects of Wet and Dry Adiabatic Lapse rates as initially moist air (dotted lines) rises over mountain, drops its moisture, and returns to original elevation on other side.

providing there is no water condensation. This value is the Dry Adiabatic Lapse Rate (Figure 7.3). Dry air increases in temperature at the same rate when it warms while descending. Commonly, the rising, cooling air can no longer hold all of its water vapor; and the water condenses and forms clouds and rain and/or snow. The condensation of air – conversion from a gaseous to liquid state – releases so much latent energy that the net change in temperature is only 6°C/km of altitude (Wet Adiabatic Lapse Rate) [15]. Air would increase in temperature at this rate as it descended if there were moisture to absorb; however, it usually descends in dry areas and so descends at the Dry Adiabatic Lapse rate [2]. Consequently, land on leeward sides of mountains are commonly drier and warmer than on windward sides.

When heated air remains in the vicinity but moves vertically and changes local weather, the weather pattern is known as "convective." Warm, upward-moving turbulent air can rise high enough for a cloud to form as water vapor condenses (Figure 7.4a). These "cumulus" clouds can lead to rain and/or thunderstorms.

Cold air becomes compact and heavy, causing it to descend, and often creates very cold and even subfreezing draining or stationary air immediately above the ground – known as "frost drains" or "frost pockets" [14].

When the warm, cold, moist, or dry air moves to another place by the wind, it creates a weather system known as "advective" or "orographic." For example, moist air blown by tropical winds is forced to rise when it reaches a tropical island (Figure 7.4b), causing it to cool, condense, and release water as rain. Once the excess moisture is released and the air drops and warms as it returns to sea level on the other side of the island, the air begins absorbing moisture and the clouds dissipate. The leeward sides of islands and mountains are known as "rain shadows." They are commonly warmer and drier than windward sides because the air rises at the wet adiabatic lapse rate and falls at the dry adiabatic lapse rate (Figure 7.3).

Figure 7.4 (a) convective and (b) advective (moving right to left over island) weather patterns. (a in Ukraine; b in St. Croix Island.)

7.3 Air Movement and the Atmosphere

Sunlight does not reach the Earth's surface in uniform concentrations. The Sun is so much larger than the Earth that sunlight reaching the Earth can be considered coming in parallel rays (Figure 7.8). The Earth's surface is perpendicular to the Sun's rays at the equator and receives a high concentration of sunlight there. Its surface is increasingly at an angle to the Sun's rays as one moves toward the poles, and the concentration of sunlight is less. The result is greater heating of the Earth by the sun at the equator than at higher latitudes [2, 16, 17]. The tilting of the Earth changes the location of greatest concentration of sunlight and so causes seasons, as will be discussed later.

From a global perspective, the Earth is a smooth ball covered by an extremely thin layer of atmosphere. This atmosphere partially behaves with laminar properties as it adheres to the underlying land and water through gravity, friction, and the land's roughness. Unless otherwise disturbed, the air tends to accompany the underlying Earth as it rotates.

The air also partially behaves with non-laminar properties. Air can move from areas of high pressure to low pressure even if they are at variance to the movement of the underlying Earth. This movement of air across the Earth's surface creates winds, some of which are highly predictable.

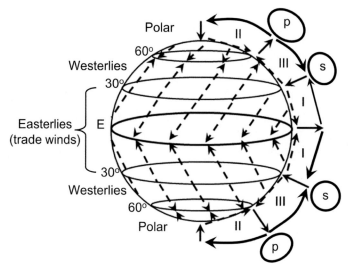

Figure 7.5 Generalized global circulation of air (s = subtropical jet streams; p = polar jet streams; I = Hadley cells; II = Polar cells; III = Ferrel cells).

The greatest concentration of sunlight at the equator means that the ground and water, and then the air, become hotter there than at progressively higher latitudes [2, 16, 17]. As a result, the air begins to rise above the equator, displacing high elevation air and drawing low elevation air toward the equator from both north and south (Figure 7.5). The place where air moving in from the north and south meet is the "convergence zone." The convergence zone is at the equator in the spring and autumn, but shifts slightly with the seasons, as will be discussed.

Most air movement occurs in the troposphere, the lowest 20 kilometers or less of the atmosphere (Chapter 8). At the convergence zone, air in the upper troposphere is displaced by rising warm air and is pushed toward the poles. One might expect the displaced air to move all of the way to the poles, by which time it would have cooled. There, one would expect it to return to low elevations and replace the air being pulled to the convergence zone by the air rising from there. In fact, there is a modification of this expected cycle. The high elevation air being displaced from the equator becomes sufficiently cool and heavy at about 30 degrees north and south latitude so that it drops, displaces some low elevation air which is pushed toward the equator, and creates an air circulation pattern [18] (Figure 7.5, I, Hadley cells). This downward movement of air at about 30 degrees latitude creates semi-permanent "belts" of high pressure over the oceans in the northern and southern hemispheres. Over land, these same latitudes can contain high pressure in winter and low pressure in summer, as will be discussed. Different amounts of land in the northern and southern hemispheres create slightly different air movements and climates between the hemispheres.

The ground, water, and air are heated least at the poles, and so this air remains heavy. Another rising air pattern occurs at about 60 degrees north and south latitude, with the rising air drawing in the heavy, low elevation polar air. These create a second circulation pattern of high elevation air being displaced toward the poles and dropping to replace the low elevation polar air (Figure 7.5, II, Polar cells).

Between these two circulation patterns is a third one (Figure 7.5, III, Ferrel cells) that cycles in the opposite direction, probably in part because of friction from the other two cells. This one moves low elevation air toward the poles between 30 and 60 degrees latitude, where it meets the polar circulation pattern and rises (II). It moves toward the equator at high elevations until it meets the equatorial circulation pattern (I) at 30 degrees latitude and descends with it.

7.4 Prevailing Winds, Evaporation, and Precipitation

The Earth's rotation causes its surface to move constantly to the East. The rotation causes nearly all of the Earth to receive some solar radiation every twenty-four hours, helping keep temperatures relatively moderate.

Another effect of this rotation is to deflect the winds (Figure 7.5) from moving vertically (north–south between the equator and poles) to moving east–west or diagonally in predictable patterns. This deflection is known as the "Coriolis effect" [2, 16–18]. The equator is 40,000 km in circumference, so the land at the equator is moving at nearly 1,700 km/hour (40,000 km/24 hours). At 60 degrees latitude, the Earth's circumference is only 20,000 km, so the land is moving slower, and a different rate occurs at each latitude.

The Earth's rotation drags the air and water along with it through the laminar flow properties and so the air and water can first be assumed to move at the same speed as the land. However, if air near the Earth's surface is displaced to higher or lower latitudes, it retains some of its speed from its former position. If moving poleward, the air moves eastward faster than the underlying land. But, it moves westward faster than the underlying land if moving toward the equator. The result (Figures 7.6, 7.9) is that the most common winds ("prevailing winds") in the tropics are from the east ("Easterly Winds," or "Trade Winds"). Prevailing winds in mid-latitudes are from the west ("Westerly Winds") [16].

Mountain peaks in the land surface can be as much as half the height of the troposphere. They cause waves, deflections, and eddies in the air as it passes over and around them.

Circulating at the upper limit of the troposphere are the "polar jet streams" and "subtropical jet streams" (Figure 7.5) that move rapidly eastward. They undulate north and south and influence cold/high and warm/low pressure systems and other atmospheric circulations. A westward-moving "tropical easterly jet stream" emerges at the convergence zone in parts of the world during northern hemisphere summers [2].

Moisture and clouds are common along the equatorial convergence zone because the rising air cools. Progressively cooler air can hold proportionately less gaseous water vapor. As the air cools, the excess water vapor condenses and forms clouds and often precipitation. A similar rising and condensation occurs at 60 degrees latitude, although the effect is less strong because the Earth is not as hot there.

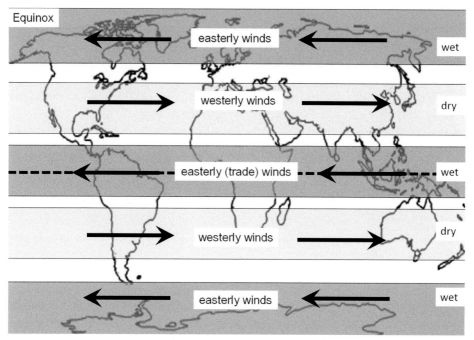

Figure 7.6 Easterly and Westerly Winds when the Sun's path is over the equator (during equinoxes).

The 30-degree latitude area is where much of the terrestrial moisture is absorbed. Deserts and other arid areas are created as the dry air descends and warms [16]. Since the air rises at the equator at the Wet Adiabatic Lapse Rate and descends on these deserts at the Dry Adiabatic Lapse Rate (Figure 7.3), it can actually become hotter in these latitudes than at the equator [2, 11, 16]. Many of the large deserts of the world are at these 30-degree latitudes (Figure 7.1).

7.5 Oceans, Currents, and Climates

Water is important for circulating the Earth's energy [2]. Its high specific heat, high energy absorption/release with changes of states, conductivity, and turbulence can carry its heat (or cold) elsewhere and release it. Ice reflects very much solar radiation, being white (or nearly white). Much of the Earth's surface is covered with water. The water circulates in predictable patterns that help determine the Earth's climates.

Winds generated around the equator (Trade Winds) push surface ocean waters westward. When they reach the eastern sides of continents, the waters build up in elevation. The accumulated water flows poleward, the path of least resistance, where it moves faster than the underlying solid Earth at the higher latitudes (Coriolis effect) and so flows eastward (Figure 7.7). Westerly Winds at these latitudes then push the surface water eastward

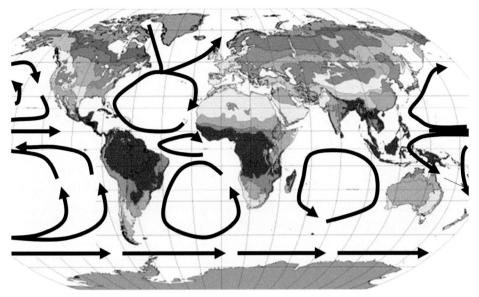

Figure 7.7 Ocean surface currents showing mirroring patterns in the northern and southern hemispheres [3, 12, 19]. Base map same as Figure 7.1.

until it reaches the western sides of continents at high latitudes [2, 16, 17]. These currents are distinct, relatively narrow flows of water that are noticeable to sailors by their temperature, color, and direction of water flow. The North Atlantic current is known as the "Gulf Stream," and the North Pacific one is known as the "Kuroshio" ("Japanese Current") [16].

Their momentums then carry the water in these currents toward the equator, completing circular flows of the ocean surface waters that are mirror images in the northern and southern hemispheres (Figure 7.7). Waters moving toward the equator on the western sides of continents move slower than the Earth at the decreasing latitudes (Coriolis effect), creating a turbulent, westward movement of the surface waters. The turbulence causes deeper, cold water to come to the surface to their east, near the continents. The turbulent water brings much phytoplankton to the surface and thus provides abundant food for fish; the cold waters also contain much dissolved oxygen that supports fish. Consequently, waters just to the west of continents are some of the world's richest fishing areas [16], but cold for swimming.

As ocean currents carry warm water from near the equator up the east sides of continents, the warm water saturates the air and creates "subtropical humid forests" along east coasts (Figure 7.1). When the same currents cross the oceans to high latitudes on the west sides of neighboring continents, the Westerly Winds blow across them, gain heat and humidity, and deposit these as rain and snow on the relatively cool land. These western, upper latitude rainy areas have moderate temperatures because of the oceans' effects. They support "temperate rain forests" (temperate oceanic forests of Figure 7.1), which can be found in similar continental positions and similar latitudes in much of the world. The climates of these regions are so similar that Douglas fir (*Pseudotsuga menziesii*) and Sitka spruce (*Picea sitchensis*) can grow vigorously in all of them – northwestern Europe, Chile, and New Zealand – although they are only native to the Pacific Northwestern North America.

A deep-sea ocean current, the "Thermohaline Circulation System," takes about 1 thousand years to complete and buffers short-term climatic fluctuations. It receives cold, shallow water from the North Atlantic, circulates it deep into the ocean throughout much of the world, and releases it to the surface in the southern hemisphere and the North Pacific Ocean [2, 12].

7.6 Earth's Tilt and Seasons

The Earth rotates on an axis that is not perpendicular to the Sun. Consequently, the place perpendicular to the Sun's rays (and receiving the highest concentration of sunlight) changes throughout the year from about 23 degrees North at the summer solstice, to the equator during the spring and fall equinoxes, to about 23 degrees South at the winter solstice (Figure 7.8).

The convergence zone shifts in latitude with this change of heat concentration, although it only moves about 10 degrees North and South because of a lag in the Earth's heating and cooling (Figure 7.8) [16]. The other circulation patterns contract and expand, moving poleward in summer and away from the poles in winter (Figure 7.9).

Climates can be understood from the wind patterns, ocean currents, and changes of the convergence zones (Figures 7.6–7, 7.9). Climate names discussed here do not necessarily correspond to Figure 7.1 (Chapter 14). "Mediterranean" climates are typified by hot, dry summers and cool, rainy winters. They are common on the non-eastern parts of continents, just poleward from deserts and abutting temperate rain-forests at higher latitudes. Mediterranean climates are actually a combination of the desert climate in summer when the convergence zone moves toward its hemisphere's pole and the desert climate moves over the area, and a temperate rain forest climate in winter when the convergence zone has moved to the other hemisphere, shifting the temperate rain forest climate to lower latitudes.

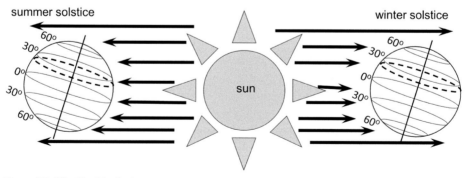

Figure 7.8 The Earth's tilt that causes seasons.

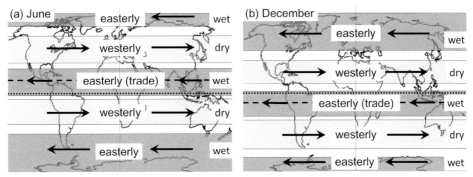

Figure 7.9 Global distribution of winds, convergence zone (dashed line), and wet and dry zones during (a) June and (b) December solstices.

"Seasonal tropical forests" – with wet and dry seasons – are common slightly poleward of the continuously humid tropical forests along the equator. Seasonal tropical forests take on dry weather patterns when the convergence zone moves to the far hemisphere, shifting more desert-like weather to these areas. They take on wet weather patterns similar to their lower latitude, neighboring humid tropical forests when the convergence zone has shifted into their hemisphere and forced the humid tropical weather over their location.

Oceans or seas, as well as nearby terrestrial areas, generally have "maritime climates" characterized by less extreme temperatures than inland areas. Ocean waters circulate and mix, thus moderating coastal temperatures. Large water bodies change temperature much more slowly than land surfaces, often being warmer in the winter and cooler in the summer. Air moving from oceans or seas to nearby lands commonly contain much water as vapor and clouds, which also moderate temperatures [16]. In temperate and polar climates, much energy is also absorbed during spring melting of seasonal snows and ice and released during autumn freezing.

Land areas far from oceans or seas are not moderated by moist winds. Their temperatures fluctuate more between seasons, usually creating hot summers and cold winters – known as a "continental climate." In winter, there is little ocean wind or humidity; and the heat radiates outward with little atmospheric water to absorb and radiate it back. Consequently, the lands become cold. They cool the overlying air until it becomes heavy, its weight creating high pressure areas. In summers, the same lands can become very hot, heating the air, causing it to rise, and thus creating low pressure areas. Much of the precipitation occurs during the hot season, when the rising air creates thunderstorms and rain – convective patterns (Figure 7.4a).

Monsoon rains occur when convectively rising air over hot lands in summer creates low pressure areas that "suck" moist air from a nearby or distant ocean or moist area. The most dramatic monsoon is in Southern Asia, where the Tibetan plateau draws moist air off the Indian Ocean in summer. As the moist air rises, the moisture condenses and drops extremely large amounts of rain. A similar behavior also occurs in Arizona, USA, where the dry, rising air brings moist air that causes late summer rains (Figure 18.3d). The "dog days" of the southeastern United States – humid, rainy days in late summer – are caused by similar phenomena.

7.7 Other Climate Behaviors

Other climatic patterns and unusual variations can also be observed. Because of wet and dry adiabatic lapse rates, cooler climates can be found by traveling either up mountains or poleward at low elevations. Plant and animal populations often become separated when they move in different directions or speeds with climate changes [2]. A general rule, "Hopkins Bioclimatic Law," is that a change of 190 meters in elevation has similar climatic effects on living creatures as a change of 100 kilometers in latitude [20].

Mediterranean climates sometimes produce "imbats" or "land–sea" breezes. These blow from land to sea in the morning when the relatively warmer sea causes air to rise above it and suck in cooler air from the land. By afternoon the land has become warmer than the sea, the air rises more there, and the breezes blow back toward the land. These imbats are believed to have stimulated seafaring cultures before boats were designed to sail against the wind.

Unexpected climates can be found at unusual juxtapositions of seas and mountains. For example, Batumi, Georgia (Europe) is at nearly the same latitude as Chicago and New York (North America) and Beijing and Vladivostok (Asia); however, it has a much warmer climate because it is protected by mountains to the north and south, and has warm, Westerly Winds from the Black Sea to the west and from the Caspian Sea to the east (when the wind shifts). Just southwest of Batumi, the mountainous Black Sea coast has a temperate rain forest climate, caused by Westerly Winds blowing across the Black Sea.

7.7.1 Mountain Climates

Mountains create different temperature and precipitation patterns on their slopes (Figure 7.3), thus providing a diversity of climates within a small area that often harbors a diversity of species and people [21]. Mountains also affect climates and species quite far away. They can block species migrations with climate change, especially if they are oriented east–west or have high ridges and steep valleys oriented that way. Blockages probably account for the few plant species north of the Alps in Europe; species had difficulty returning across the Alps after the glaciers receded. Deep, east–west valleys along the west coast of North America can also block mountain species from migrating.

Some mountains are oriented north–south such as the North American Rocky, Cascade, and (older) Appalachian Mountains; the South American Andes; and those surrounding the East African Rift. They divert tropical Easterly Winds or temperate Westerly Winds and so create cool, rainy windward climates and hot, dry leeward climates. Their north–south orientations allow easier north–south flow of tropical and polar air, creating more weather extremes and rapid changes in North and South America than in Eurasia [22]. These north–south orientations also allow species to migrate more readily toward cooler and warmer latitudes with changing climates.

Other mountains developed an east–west orientation, such as those separating tropical and Mediterranean Eurasia from temperate Eurasia. They inhibit the north–south flow of tropical and polar air, reducing the extreme weather changes. Several effects of the east–west orientation are (Figure 7.1):

- Central Europe has a large area of the "temperate oceanic forest" ecoregion that extends hundreds of miles inland; however, both North and South America have very small areas of similar climate because the Cascade and Andes Mountains block and alter the Westerly Winds shortly after they reach land. The cooler summers in Europe are also partly because, unlike North America, the east–west Pyrenees, Alps, and Carpathians limit warm winds from the south.
- The "temperate continental forest" extends much farther inland in North America than in eastern Asia, because high mountains near the Asian coast block the maritime winds from the Pacific.
- The arctic circle in Alaska has much colder winters and warmer summers than in Finland. Moderating maritime air from the Gulf Stream flows relatively unimpeded to Finland; however, the east–west Chugach Mountains in southern Alaska block moderating effects that the Kuroshio current might otherwise have on northern Alaska.
- The monsoon rains are much stronger in southern Asia than in southern North America because the Himalaya Mountains create a much higher and steeper gradient on which the oceanic winds drop their moisture.
- The rise of the Himalaya Mountains and other ranges in southern Eurasia about 30 million years before present (MYBP) also contributed to extirpation, and probable extinction, of species. Then, when the recent glacial advances/retreats began about 2.6 MYBP, climates became cooler and the species were trapped between the colder climate coming from the north and the colder climate on the mountains to the south. This trap probably reduced the range of Douglas fir (*Pseudotsuga*) and hemlock (*Tsuga*) tree species, which eventually disappeared from Europe 100,000 years ago. *Sequoia* tree species, once common in many places in Eurasia, was last known to exist there 2 MYBP, near Istanbul [23].
- The rise of the Himalaya Mountains is also believed to have stimulated a decline of atmospheric carbon dioxide by its reacting rapidly with silicate and carbonate minerals uncovered in a warm climate (Figures 8.3, 8.4), thus reducing atmospheric greenhouse gases and enabling the recent ice age [24, 25] – in conjunction with the Milankovitch Cycles (Chapter 9).

7.7.2 Erratic Weather

Climates can create interesting, sometimes violent weather patterns, usually with their winds, extreme temperatures, rains, or lightning. The occasionally extreme variation is

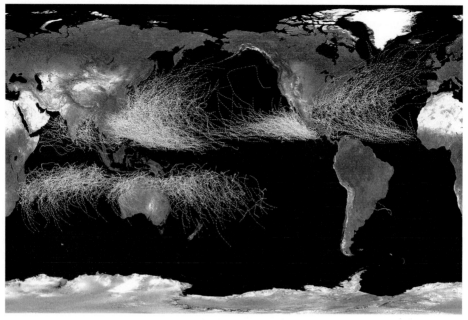

Figure 7.10 Paths of major windstorms of tropical origin [27]. Reproduced courtesy of Wikimedia Commons.

often as important for resource management as the average condition. Several of these patterns are worth noting.

A moist air mass moving over a mountain and drying and warming a frigid area in the process (Figure 7.2) can create a sudden warming of over 40 degrees Centigrade within a few hours or days. These winds are known as "Chinooks" or "Santa Anna's" in North America and "fohn" winds in Germany [26]. When a Chinook stops, the temperature can again drop [2, 16].

During summer solstices when the convergence zone is nearest a hemisphere's pole, low-pressure systems generate violent storms as air rushes in and rises from both the north and south. The Coriolis influence then creates a whirling effect that creates winds over 150 kilometers per hour, which radiate outward for tens to hundreds of kilometers. These storms move along the Trade Winds and can move poleward, often as they reach land (Figure 7.10). They are common in southeastern North America, Central America, Southeast Asia, the Bay of Bengal in the Indian Ocean, and the southwest Pacific Islands. In different places, they are known as hurricanes, typhoons, cyclones, or "willy-willies" [2, 16, 17]. Winds and rain from these storms commonly reach land in temperate oceanic forest ecoregions (Figure 7.1). They can be quite strong and sometimes very destructive.

More local convergences of high and low pressures can create tornados, which can be extremely destructive. They occur in many parts of the world.

The ocean circulation system in the South Pacific Ocean oscillates between the east and west at irregular intervals of 2 to 7 years (average 3 to 4 years) – sometimes not reaching the western coast of South America and at other times not reaching the western Pacific Islands, southeastern Asia, and Australia (Figure 7.7). The oscillation is known as "El Niño" or "ENSO" (El Niño Southern Oscillation) and contains "El Niño" and "La Niña" phases. The period of rainfall known as El Niño lasts 12 to 18 months and occurs in western equatorial South America (Ecuador and Peru) when the circulating cold waters do not reach the South American coast [2]. The changing ocean currents can cause fish populations to move over 100 kilometers from their usual places. The warmer waters in the eastern Pacific Ocean also reduce the air pressure and Trade Wind force, causing precipitation to fall in the central Pacific that otherwise falls in the western Pacific Islands, southeastern Asia, and Australia, creating periodic droughts in these areas. Strong ENSOs have created droughts that catalyzed forest fires in the southwest Pacific Islands in 1982–3 and again in 1997. The changes in overlying atmospheric pressures are transmitted to other parts of the globe. For example, the Pacific atmospheric pressure changes affect the weather in Central America, southern North America, and elsewhere.

References

1. The Columbia Electronic Encyclopedia. *Climate.* (Columbia University Press, 2013 [Accessed March 14, 2016]). Available from: http://encyclopedia2.thefreedictionary.com/climate.
2. D. Huddart, T. Stott. *Earth Environments: Past, Present, and Future.* (Wiley-Blackwell, 2012).
3. UN FAO. *Global Forest Resource Assessment.* 2000.
4. D. M. Olson, E. Dinerstein, E. D. Wikramanayake, et al. Terrestrial Ecoregions of the World: A New Map of Life on Earth a New Global Map of Terrestrial Ecoregions Provides an Innovative Tool for Conserving Biodiversity. *BioScience.* 2001;51(11):933–8.
5. A. L. Takhtadzhian. *The Floristic Regions of the World.* (Nauka, 1978).
6. M. D. Udvardy. *A Classification of the Biogeographical Provinces of the World.* (IUCN and Natural Resources Morges, 1975).
7. O. W. Archibold. *Ecology of World Vegetation.* (Springer Science & Business Media, 2012).
8. R. G. Bailey, H. C. Hogg. A World Ecoregions Map for Resource Reporting. *Environmental Conservation.* 1986;13(03):195–202.
9. UN FAO. *Global Ecological Zones Map.* (United Nations, Food and Agriculture Organization, 2000 [Accessed September 10, 2017]). Available from: www.fao.org/geonetwork/srv/en/main .home.
10. UN FAO. *State of the World's Forests 2011.* (United Nations Food and Agriculture Organization, 2011).
11. Z. Merali, B. J. Skinner. The Atmosphere: Composition, Structure, and Clouds. in*Visualizing Earth Science.* (Wiley, 2009): pp. 424–55.
12. B. J. Skinner, S. C. Porter, J. Park. *Dynamic Earth.* (Wiley, 2004).
13. S. E. Manahan. Green Chemistry, Biology, and Biochemistry. In *Environmental Science and Technology: A Sustainable Approach to Green Science and Technology.* (CRC Press, 2006): pp. 61–98.
14. R. Geiger, R. H. Aron, P. Todhunter. *The Climate near the Ground.* (Rowman & Littlefield, 2009).
15. S. E. Manahan. The Atmosphere: A Protective Blanket around Us. In *Environmental Science and Technology: A Sustainable Approach to Green Science and Technology.* (CRC Press, 2006): pp. 181–204.

16. R. A. MacArthur. Climates on a Rotating Earth, in MacArthur R. A., editor, *Geographical Ecology: Patterns in the Distribution of Species*. (Harper & Row, 1972). pp. 5–19.

17. Z. Merali, B. J. Skinner. Global Circulation and Weather Systems. In *Visualizing Earth Science*. (Wiley, 2009). pp. 456–87.

18. W. K. Hamblin, E. H. Christiansen. *The Earth's Dynamic Systems*, tenth edition. (Prentice Hall, 2001).

19. A. N. Strahler, A. H. Strahler. *Physical Geography*. (Wiley, 1997).

20. A. D. Hopkins. The Bioclimatic Law 1. *Monthly Weather Review*. 1920;48:355–7.

21. V. A. Sample, R. P. Bixler, C. Miller, editors. *Forest Conservation in the Anthropocene*. (University Press of Colorado, 2016).

22. T. Flannery. *The Eternal Frontier: An Ecological History of North America and Its Peoples*. (Atlantic Monthly Press, 2001).

23. H. Kayacik, F. Aytug, F. Yaltirik, et al. Sequoiadendron Giganteum Trees Lived near Istanbul in Late Tertiary. *Review of the Faculty of Forestry, University of Istanbul*. 1995;45(1):22.

24. T. E. Cerling, J. R. Ehleringer, J. M. Harris. Carbon Dioxide Starvation, the Development of C4 Ecosystems, and Mammalian Evolution. *Philosophical Transactions of the Royal Society B: Biological Sciences*. 1998;353(1365):159–71.

25. W. F. Ruddiman. *Tectonic Uplift and Climate Change*. (Springer Science and Business Media, 2013).

26. R. Castro, A. Mascarenhas, A. Martinez-Diaz-de-Leon, R. Durazo, E. Gil-Silva. Spatial Influence and Oceanic Thermal Response to Santa Ana Events Along the Baja California Peninsula. *Atmósfera*. 2006;19(3):195–211.

27. Nilfanion, cartographer. *Worldwide Tropical Cyclones from 1985 to 2005* [Image on the Internet]. (Wikimedia Commons, 2006).

8

Greenhouse Gases, Atmosphere, and Climates

8.1 Layers and Composition of the Atmosphere

To understand climates, it is necessary to understand the chemical composition of the atmosphere and its influence.

The Earth's atmosphere is classified as layers in ascending order from the Earth's surface [1–3]:

- *Troposphere:* Clouds, rain, snow, and most winds affecting climates and weather occur here. It varies from about 17 km thick at the equator to 8 km at the poles but can be thicker or thinner in high- and low-pressure areas. Particulate matter and some other substances injected into this layer are often washed out by rains within a few years.
- *Stratosphere:* It has a different atmospheric circulation than the troposphere and contains the ozone layer, among other things. The upper limit of the stratosphere is about 50 km above the Earth's surface. Particulates and some other substances that reach this layer can remain for many years [1].
- *Mesosphere, Ionosphere, Thermosphere,* and *Exosphere:* These layers are at increasing heights above the Stratosphere.

Compared to the Earth's diameter of about 12,750 km, the troposphere and stratosphere are extremely thin.

The Earth's atmosphere has changed since the Earth began, generally slowly enough to enable species to evolve with the changes. Many of the very short wavelengths released by the Sun are currently absorbed in the upper atmosphere (Chapter 23). The ultraviolet wavelengths, also very short, are largely blocked by the mesosphere and especially the ozone layer in the stratosphere. These short wavelengths can harm living systems. Visible light, which is also the wavelengths in which photosynthesis occurs, is largely unblocked except by clouds, smoke, and volcanic and other dust. Longer, infrared waves are largely blocked in the stratosphere and troposphere by water vapor and other greenhouse gases (Figure 23.2), while the very long radio waves are largely unblocked.

8.1.1 Chemical Composition of the Atmosphere

The Earth's atmosphere of the past hundreds of millions of years has consisted largely of relatively inert diatomic nitrogen molecules, potentially more reactive diatomic oxygen

Table 8.1 *Chemical composition of the atmosphere [3–6]*

Chemical	Symbol	Proportion (%)	Chemical	Symbol	Proportion (%)
Nitrogen	N_2	78.10%	Krypton	Kr	trace
Oxygen	O_2	21.00%	Hydrogen	H_2	trace
Water vapor	H_2O	trace to 4%	Nitrous oxide	N_2O	trace
Argon	Ar	0.90%	Carbon monoxide	CO_2	trace
Carbon dioxide	CO_2	0.04% (current)	Xenon	Xe	trace
Neon	Ne	trace	Ozone	O_3	trace
Helium	He	trace	Nitrous dioxide	NO_2	trace
Methane	CH_4	trace			

Table 8.2a *Energy and emission characteristics of various energy sources [8–12]*

	Energy intensity		Efficiency	Residence
	$BTU (10_3)$ /kg	$BTU (10_6)$ /m^3	Kg CO_2/BTU	CO_2 atmosphere residence (years)
Wood fuel	12	3	116	100
Construction wood	−36	−9	−119	−4,000
Forest wood				−100 to −500+
Natural gas	221	28	53	4,000
Gasoline	44	33	71	4,000
Anthracite	34	37	104	4,000
Bituminous	24	20	93	4,000
Lignite	22	13	136	4,000
Peat	10	4	98	2,000

molecules, and inert argon atoms (Table 8.1). Water vapor varies in concentration, and other elements and molecules exist in very small amounts. These other elements vary from highly reactive methane to unreactive helium.

Other substances such as pollen, volcanic ash, dust, smoke particles, plant aerosols, and sea spray have always been emitted to the atmosphere. More recently, people's activities have added more carbon dioxide, methane, nitrites, and sulfur compounds by burning fossil fuels (Table 8.2). Other chemicals include fluorine and chlorine, mercury, radioactive elements, plus artificial aerosols and other substances. These come from many sources including agricultural and industrial processes such as chemical, concrete, pesticide, and fertilizer manufacture and use [1, 7].

Reactions among the many chemicals, often aided by sunlight, can generate new chemicals in the atmosphere. Some, such as those containing nitrogen and sulfur, can create corrosive and otherwise harmful chemicals in the troposphere. Local and regional problems from human-generated atmospheric pollutants have long been recognized

Table 8.2b *Continuation of Table 8.2a*

	Density	CO$_2$ intensity		Sulfur	Nitrogen
	kg substance /m^3	kg CO$_2$/kg	kg CO$_2$/m^3	content (%)	content (%)
Wood fuel	240	2	456	2.0	5.0
Construction wood	240	−4	−1,032		
Forest wood	240	−2	−432		
Natural gas	128	12	1,504	0.2	0.0
Gasoline	749	3	2,352	0.2	0.1–0.5
Anthracite	1,105	4	3,883	1.0	0.7
Bituminous	833	2	1,863	1.0	1.4
Lignite	600	3	1,800	1.7	0.6
Peat	400	1	402	0	1.1

[13–16]. Realization that small concentrations can cause global harm was accentuated when DDT (dichlorodiphenyltrichloroethane), an effective insecticide, was identified as harmful to human health and threatening to the reproduction of charismatic birds [15, 17, 18].

Chlorofluorocarbons (CFCs) and related chemicals will be discussed later.

Polychlorinated biphenyls (PCBs) and dioxins have also been identified as carcinogenic to people and threatening to charismatic animals high on the food web (Figure 13.3) [19, 20]. Harmful properties of other chemicals are increasingly being discovered [21]. "Acid rain" is a general term given to abnormally high amounts of atmospheric nitric and sulfuric acids that fall to earth with rainfall. The acids are mostly generated by people [22]; acid rain harms building and automobile surfaces, vegetation, and soils. The nitrogen pollutants will be discussed below [23]. Sulfur dioxide (SO$_2$) emissions from coal-fired power plants have caused worldwide issues of acid rain. A combination of government regulations and technologies have been instituted to curb sulfur emission in some countries [24].

8.1.2 Greenhouse Gases (GHG)

Water, carbon dioxide, methane, nitrates, CFCs, HCFCs, HFCs, and other compounds to lesser extents are known as "greenhouse gases." They absorb heat energy as it radiates from the Earth and reradiate it in all directions (Figure 23.2). Since some is reradiated back toward the Earth, the net effect is a warming of the atmosphere, known as "radiative forcing" (Table 8.3). A greenhouse gas' ability to heat is numerically expressed as its "global warming potential" (GWP) compared to carbon dioxide, which is given a global warming potential of "one." Except for water vapor, greenhouse gases are in low concentrations that belie their influence (Tables 8.1, 8.3). These greenhouse gases have been increasing in the atmosphere. Atmospheric water has been increasing because of general warming by the other gases; and, the other gases have increased because of human actions.

Table 8.3 *Major greenhouse gases (except water), their temperature, lifespan, and strength of warming impact ("global warming potential") [7]*

Chemical	Atmospheric concentration	Radiative forcing		Lifespan	Global warming potential	
	(ppm)	(BTU/ (m²hr))	(% of total forcing)	(years)	(100 years)	(20 years)
Carbon dioxide	391	6.21	64%	1,000–10,000	1	1
Methane	1.8	1.64	17%	12.4	28	67
Nitrogen oxides	0.3	0.58	6%	121	265	277
Chlorofluorocarbons	0.8	0.9	9%	45	4,660	6,900
Hydrofluorocarbons	–	0.03	0.3%	13.4	1,300	3,710
Hydrochlorofluoroc- arbons	–	0.18	2%	>45	–	–
Other	–	–	1%	–	–	–
Total		9.66				

If not replenished, these gases disappear from the atmosphere after different lengths of time ("life spans") by decomposition or chemical reactions with minerals or plants.

8.2 Carbon Dioxide

Carbon exists in many places (Figure 8.1). It is usually converted to carbon dioxide (CO_2) when released to the atmosphere; and, carbon dioxide is usually converted to other carbon-containing compounds when removed from the atmosphere. Discussions of carbon emissions can be confusing because both elemental carbon and the carbon dioxide molecule are used in analyses. A carbon dioxide molecule weighs about 3.67 times as much as a carbon atom. Elemental carbon (C) is used when analyzing the stocks (Figure 8.1) or flows (Figure 8.4) of carbon among carbon dioxide and other carbon compounds. Carbon dioxide is used when discussing carbon dioxide in the atmosphere or "carbon dioxide equivalents" of other carbon-containing compounds.

Carbon dioxide is the most dominant greenhouse gas, accounting for 64% of the global warming by greenhouse gases since 1750 [7]. The amount of carbon (as carbon dioxide) in the atmosphere is small relative to the carbon in other stocks (Figure 8.1). Consequently, converting a small proportion of the carbon in vegetation, fossil fuels (gas, oil, coal), or the Earth's crust to atmospheric carbon dioxide can greatly increase the atmospheric carbon dioxide concentration.

Over the past 450 thousand years, atmospheric CO_2 has declined to about 170 parts per million (ppm) during each of the past four times of glacial maxima (Figure 9.2). It has increased to 300 ppm during the intervening times of minimum glaciation [27]. The Earth's present glacial minimum has lasted over 10 thousand years (Chapter 9); and, in the year

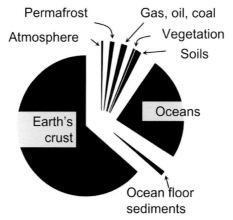

Figure 8.1 Distribution of the Earth's carbon stocks [7, 25, 26].

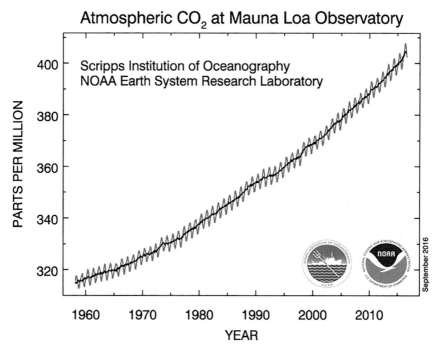

Figure 8.2 Changes in global atmospheric CO_2 from years 1959 to 2016 [28].

1750, the atmospheric CO_2 concentration was about 280 ppm) [7]. Since then, it has rapidly risen to over 400 ppm, with much of the increase occurring since 1960 (Figure 8.2).

Carbon dioxide is moved out of the atmosphere and into other stocks (Figure 8.1) by two general processes, known as "domains": the slow domain and the fast domain (Figure 8.3) [7]. The circulation of carbon within the fast domain is the subject of ongoing studies [29, 30].

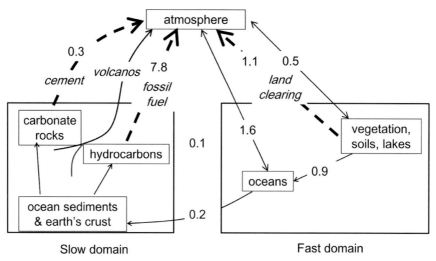

Figure 8.3 Global carbon domains, showing human-induced changes (dashed arrows) and background changes (solid arrows). Numbers = 10^{12} kg carbon/year [7].

Carbon moves to the slow domain in two ways:

- CO_2 reacts with silicates or carbonates from weathering rocks to form bicarbonates that wash into the ocean and eventually become calcium carbonate in shells, coral, and similar solid materials (Figure 8.4) [31, 32];
- Buried organic carbon in plants that remove CO_2 from the atmosphere through photosynthesis sometimes eventually becomes fossil fuels.

Movement of carbon from atmospheric CO_2 to carbon rocks can take 10 thousand years, and movement of organic carbon to fossil fuels can take millions of years.

Atmospheric CO_2 moves within a few decades into the fast domain through both absorption of carbon dioxide by lakes and oceans and photosynthetic decomposition of carbon dioxide followed by storage of the organic carbon in living and dead vegetation and soils (Figure 8.4). Some vegetation rots or burns, returning CO_2 to the atmosphere; and, oceans release CO_2 to the atmosphere when the atmospheric CO_2 concentration declines.

Carbon dioxide can move rapidly into the atmosphere from both the slow and fast domains. This movement has increased since 1750, with 88% of the increase from burning fossil fuels and producing cement. The other 12% increase of atmospheric CO_2 has been from movement from the fast domain caused by agriculture activities and converting forests to other lands uses [7]. Forest harvesting and shifting cultivation did not make a net contribution of atmospheric CO_2 or other greenhouse gases [34] despite being grouped with land use changes; however, forest harvesting could reduce more CO_2 emissions if managed differently (Chapter 28). Fossil fuel and cement contributions have been increasing recently, while land use emissions have declined dramatically.

Table 8.4 *Chemical formulas for materials in Figure 8.4*

Formula	Name	Formula	Name
Organic matter		*Other carbons*	
$C_6H_{12}O_6$	Glucose (or other sugar)	CO_2	Carbon dioxide
Variable	Coal	H_2CO_3	Carbonic acid
Variable	Petroleum	HCO_3-	Hydrogen carbonate ion
CH_4	Methane	$CaCO_3$	Calcium carbonate
C_x	Graphite	*Other*	
Silicates		H_2O	Water
SiO_2	Silicon dioxide	O_2	Oxygen
$CaSiO_3$	Calcium metasilicate	CA_2+	Calcium cation

Fast Domain	CO_2 fate
Respiration and burning	Emitted
$\quad C_6H_{12}O_6 + 6O_2 \rightarrow 6CO_2 + 6H_2O$ + energy	
Photosynthesis	Sequestered
$\quad 6CO_2 + 6H_2O$ + sun energy $\rightarrow C_6H_{12}O_6 + 6O_2$	
Carbonic acid formation (in soil and seawater)	Sequestered
$\quad CO_2 + H_2O \dashrightarrow H_2CO_3$	
Slow Domain	
Burning of fossil fuels	Emitted
$\quad$ coal or petroleum or methane $\dashrightarrow CO_2 + H_20$ + energy	
Silicate metamorphosis and CO_2 (from volcanos)	Emitted
$\quad CaCO_3 + SiO_2 \dashrightarrow CO_2 + CaSiO_3$	
Rock dissolution	Sequestered
$\quad H_2O + H_2CO_3+$ silicate minerals- $\dashrightarrow$ cations + clays	
Shell and limestone formation	Sequestered
$\quad CO_2 + CaCO_3 + H_2O \rightarrow Ca_2+ + 2HCO_3-$	
Fossil fuel formation	Sequestered
$\quad C_6H_{12}O_6 \dashrightarrow$ COAL + petroleum $\dashrightarrow CH_4 + C_x$	

Figure 8.4 Chemical reactions involved in moving carbon dioxide into and out of the atmosphere [33] in Figure 8.3. (Table 8.4 shows chemical formulas. Appendix II shows element symbols and names.)

Strategies for removing CO_2 from the atmosphere include eliminating CO_2 emissions from the slow and/or fast domains as well as storing carbon in the fast domain. All three have merit; however, given the large amount of carbon in the slow domain (rocks and fossil fuels) compared to the fast domain (vegetation and soils) (Figure 8.1) and the relative irreversibility of emitting CO_2 from the slow domain, reducing fossil fuel consumption would seem to have highest priority. Sequestering CO_2 in the fast domain (plants and soil) could effectively reduce atmospheric CO_2 for a while – until the ecosystems became crowded with carbon, burned, eliminated habitats, and/or simply did not sequester more carbon (Figure 9.8) [12, 25, 35–38].

8.3 Other Greenhouse Gases

Although carbon dioxide is the most dominant greenhouse gas, there are others.

8.3.1 Methane

Carbon in organic substances is also converted to methane (CH_4) and released to the atmosphere, although much more rarely than to CO_2 (Table 8.3). Methane concentration in the atmosphere has increased 2.5 times since 1750 and is believed to cause 17% of the global warming generated by greenhouse gases. There is much uncertainty about CH_4 origins, with new sources such as old trees being recently recognized [39, 40]. Most emissions not caused by people come from wetlands, with lesser amounts coming from lakes and oceans, volcanic eruptions, and wild animal flatulence. Most CH_4 generated by people is from agriculture – domestic ruminant flatulence, landfills and waste, and rice paddies. Fossil fuel production and use contribute some methane, as does biomass burning.

The atmospheric lifetime of CH_4 is about 7 to 11 years – much shorter than that of CO_2 (Table 8.3). About 7% of the emitted CH_4 is removed each year (Figure 8.5), and some years in the early twenty-first century had a net decline in atmospheric CH_4. Most CH_4 removal is by chemical conversion in the atmosphere to CO_2, a much weaker greenhouse gas. Smaller amounts of CH_4 are removed by the soil.

8.3.2 Nitrogen Oxides

Nitrogen (N_2) constitutes 78% of the atmosphere and is relatively inert. Small proportions of it are converted to nitrogen oxides (NO_X) and ammonia (NH_3) through microorganism and lightning. Recently an equal amount is also converted by fossil fuel combustion, nitrogen fertilizer, and feedstock manufacture (Chapter 21). Some NO_X and NH_3 move to the atmosphere, largely through human activities such as burning fossil fuels, biofuels, and biomass [1, 7].

Most atmospheric NO_X and NH_3 returns to land and oceans in precipitation as acid rain. They can beneficially increase plant growth in some places but can create problems of "eutrophication" if the nitrogen excessively increases plant growth in lakes and rivers (Chapter 18) [41]. Much of the returning nitrogen is converted back to atmospheric nitrogen (N_2) by soil microorganisms.

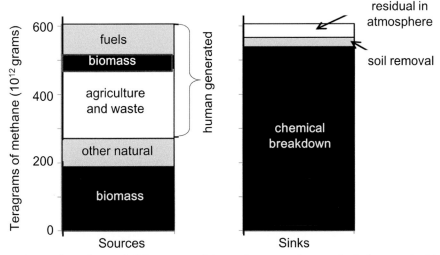

Figure 8.5 Estimated annual CH_4 sources, sinks, and net annual atmospheric increase (residual; 1980–2009) [7].

A small portion of the atmospheric NO_X and NH_3 is converted to nitrous oxide (N_2O), a very strong greenhouse gas (Table 8.3) and destroyer of stratospheric ozone, discussed below. Nitrous oxide has a long lifespan [21]; its concentration is now 16% to 20% higher than in preindustrial times; and it contributes 6% to global warming [1, 7].

8.3.3 Chlorofluorocarbons and Substitutes

Chlorofluorocarbons (CFCs) and their substitute hydrochlorofluorocarbons (HCFCs) were used as refrigerants, propellants, and cleaning solvents. They were found to deplete ozone in the stratosphere and upper troposphere – creating an "ozone hole" over parts of Australia [42] with concomitant increases in cancers and injuries to animals [1, 7, 43]. Ozone is harmful at low altitudes; but high-altitude ozone reduces the ultraviolet light passing through the troposphere. CFCs and HCFCs were banned and their atmospheric concentrations have since been declining, stratospheric ozone is building up again, and the "ozone hole" is expected to disappear within several decades. Hydrofluorocarbons (HFCs) are less harmful and have been used as substitutes [21].

8.3.4 Aerosols

Aerosols are small particles injected into the lower atmosphere by volcanic eruptions, windblown soil, tree terpenes, forest and grassland fires, and fossil fuel combustion. Aerosols can promote clouds by serving as nuclei for condensation of water vapor. The increased cloudiness and haze, "global dimming," cools the Earth in the daytime by both reflecting and reradiating incoming sunlight but warms at night by acting as a greenhouse gas [1, 7, 44, 45].

8.4 Heating the Atmosphere

The amount of energy reaching the Earth's outer atmosphere varies because the Sun's radiation fluctuates and because the Earth periodically changes its position relative to the Sun (Chapter 9). The Sun's radiation fluctuates as much as two-tenths of 1% of its radiation, roughly in proportion to its sunspot activity [46]. These fluctuations occur on an 11-year cycle and at longer, irregular intervals. The Little Ice Age (1650–1700; Maunder Minimum) may have been influenced by low sunspot activity. The Intergovernmental Panel on Climate Change (IPCC) considers the Sun's radiation fluctuation much less significant than human greenhouse gas emissions in causing current global warming [7], in contrast to others (for example, Dr. W. Soon) [47, 48]. Dr. Soon [47] has become unpopular among scientists because of his subsequent pro-fossil fuel industry stance; however, examining the veracity of his research was beyond the scope of this book.

As the Earth moves nearer the Sun in its elliptical orbit or changes its orbital shape, the amount of energy reaching the Earth changes (Chapter 9) [1].

The reflection, absorption, and reradiation of solar X-rays, ultraviolet wavelengths, and infrared wavelengths by different layers of the atmosphere have already been discussed (Chapter 7). When absorbed, the energy is reradiated as heat in all directions – both toward and away from the Earth [1, 3, 49].

Clouds, being nearly white, also reflect much energy away from the Earth, lowering its total temperature [1]. Aerojet vapor trails may be increasing the atmosphere's cloudlike formations, thus reducing the energy reaching the Earth.

Smoke, volcanic dust, loess, aerosols, and other substances can partially reflect the Sun's radiation [1, 3, 50]. They also absorb solar energy and reradiate it in all directions as heat energy. These substances can have either warming or cooling effects. Global temperatures decline following major volcanic eruptions because of this particulate matter. About 2 million years ago, fires set or promoted by early hominids for protection, game management, and cooking may have greatly increased wood smoke, and hence atmospheric reflectance and reradiance. Smoke and haze have declined in the past few decades in regions where forest fires and wood fuel use have declined [51].

The condition of the Earth's surface also affects solar energy behavior. Changing land uses can alter the reflectivity (albedo) of the Earth's surface, reflecting different amounts of shortwave radiation back into space. White snowfields and glaciers reflect sunlight [3]. The Earth's reflectance is greater during glacial periods and winters in the northern hemisphere. On the other hand, conifer forests at upper latitudes absorb more radiation because of their dark color, thus increasing the Earth's heat [52]. Permanent removal of vegetation can also reduce the water released through evapotranspiration (Chapter 17).

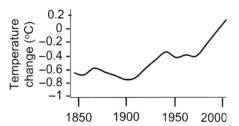

Figure 8.6 Recent changes in the Earth's temperature compared to 1986–2005 average [7, 38].

High fossil fuel use (Figures 24.2, 25.5) and other lesser factors have increased the relative carbon dioxide content of the atmosphere dramatically in the past hundred years (Figure 8.2). The high probability that this increased atmospheric CO_2 is a major cause of the disturbing increase in global temperatures (Figure 8.6) [7] merits actions to reverse the atmospheric CO_2 and other greenhouse gas concentrations.

References

1. D. Huddart, T. Stott. *Earth Environments: Past, Present, and Future.* (Wiley-Blackwell, 2012).
2. V. Smil. *The Earth's Biosphere: Evolution, Dynamics, and Change.* (MIT Press, 2002).
3. Z. Merali, B. J. Skinner. The Atmosphere: Composition, Structure, and Clouds. *Visualizing Earth Science.* (Wiley, 2009), pp. 424–55.
4. M. Pidwirny. Atmospheric Composition. *Fundamentals of Physical Geography*, second edition. (University of British Columbia, 2006).
5. H. Schlager, V. Grewe, A. Roiger. Chemical Composition of the Atmosphere. In *Atmospheric Physics.* (Springer, 2012): pp. 17–35.
6. S. E. Manahan. *Environmental Science and Technology: A Sustainable Approach to Green Science and Technology.* (CRC Press, 2006).
7. T. Stocker, D. Qin, G. Plattner, et al. Climate Change 2013: The Physical Science Basis. In *Intergovernmental Panel on Climate Change.* (Cambridge University Press, 2013).
8. J. R. Craig, D. J. Vaughan, B. J. Skinner. *Resources of the Earth: Origin, Use, and Environmental Impacts*, fourth edition. (Prentice Hall, 2011).
9. V. Quaschning. *Statistics: Specific Carbon Dioxide Emissions of Various Fuels.* (Volker-Quaschning, 2015 [Accessed April 1, 2016]). Available from: www.volker-quaschning.de/dat serv/CO2-spez/index_e.php.
10. A. Burnham, J. Han, C. E. Clark, et al. Life-Cycle Greenhouse Gas Emissions of Shale Gas, Natural Gas, Coal, and Petroleum. *Environmental Science & Technology.* 2011;46(2): 619–27.
11. US Energy Information Administration. *Carbon Dioxide Emissions Coefficients.* (US Energy Information Administration, 2015 [Accessed March 29, 2016]). Available from: www.eia.gov /environment/emissions/co2_vol_mass.cfm.
12. C. D. Oliver, N. T. Nassar, B. R. Lippke, J. B. McCarter. Carbon, Fossil Fuel, and Biodiversity Mitigation with Wood and Forests. *Journal of Sustainable Forestry.* 2014;33 (3):248–75.
13. G. G. Hedgcock. Injury by Smelter Smoke in Southeastern Tennessee. *Journal of the Washington Academy of Sciences.* 1914;4.
14. J. R. Barrett. An Uneven Path Forward: The History of Methylmercury Toxicity Research. *Environmental Health Perspectives.* 2010;118(8):A352.

15. F. L. McEwen, G. R. Stephenson. *The Use and Significance of Pesticides in the Environment.* (Wiley, 1979).
16. R. Carson. *Silent Spring.* (Houghton Mifflin, 1962).
17. J. J. Hickey, D. W. Anderson. Chlorinated Hydrocarbons and Eggshell Changes in Raptorial and Fish-Eating Birds. *Science.* 1968;162(3850):271–3.
18. US Environmental Protection Agency. *DDT – A Brief History and Status.* (EPA United States Environmental Protection Agency, 2016 [Accessed March 14, 2016]). Available from: www.epa.gov/ingredients-used-pesticide-products/ddt-brief-history-and-status.
19. J. P. Giesy, K. Kannan. Dioxin-Like and Non-Dioxin-Like Toxic Effects of Polychlorinated Biphenyls (PCBs): Implications for Risk Assessment. *Critical reviews in toxicology.* 1998;28(6): 511–69.
20. D. Hoffman, C. Rice, T. Kubiak. PCBs and Dioxins in Birds, in Beyer W. N., Heinz G. H., Redmon-Norwood A. W., editors, *Environmental Contaminants in Wildlife: Interpreting Tissue Concentrations.* (Lewis Publishers, 1996).
21. A. Ravishankara, J. S. Daniel, R. W. Portmann. Nitrous Oxide (N_2O): The Dominant Ozone-Depleting Substance Emitted in the 21st Century. *Science.* 2009;326(5949):123–5.
22. US Environmental Protection Agency. *Acid Rain.* (Environmental Protection Agency, USA, 2016 [Accessed April 1, 2016]). Available from: www.epa.gov/acidrain/index.html.
23. P. M. Vitousek, J. D. Aber, R. W. Howarth, et al. Human Alteration of the Global Nitrogen Cycle: Sources and Consequences. *Ecological Applications.* 1997;7(3):737–50.
24. M. R. Taylor, E. S. Rubin, D. A. Hounshell. Effect of Government Actions on Technological Innovation for SO_2 Control. *Environmental Science and Technology.* 2003;37(20):4527–34.
25. University of New Hampshire. *An Introduction to the Global Carbon Cycle.* (University of New Hampshire, 2016 [Accessed May 21, 2017]). Available from: http://globecarboncycle.unh.edu/CarbonCycleBackground.pdf.
26. Soil Carbon Center. *What Is the Carbon Cycle?* (Kansas State University, 2016 [Accessed March 14, 2016]). Available from: http://soilcarboncenter.k-state.edu/carbcycle.html.
27. D. M. Sigman, E. A. Boyle. Glacial/Interglacial Variations in Atmospheric Carbon Dioxide. *Nature.* 2000;407(6806):859.
28. US National Oceanic and Atmospheric Administration. *Full Mauna Loa CO_2 Record.* (U.S. Department of Commerce, National Oceanic & Atmospheric Admin., Earth System Research Laboratory, Global Monitoring Division, 2016 [Accessed July 6, 2016]). Available from: www .esrl.noaa.gov/gmd/ccgg/trends/full.html.
29. O. J. Schmitz, P. A. Raymond, J. A. Estes, et al. Animating the Carbon Cycle. *Ecosystems.* 2014;17(2):344–59.
30. P. Ciais, A. Dolman, A. Bombelli, et al. Current Systematic Carbon Cycle Observations and Needs for Implementing a Policy-Relevant Carbon Observing System, *Biogeosciences Discussions,* 10. 2013: 11447–581.
31. Z. Liu, W. Dreybrodt, H. Liu. Atmospheric CO_2 Sink: Silicate Weathering or Carbonate Weathering? *Applied Geochemistry.* 2011;26:S292–S4.
32. P. V. Brady. The Effect of Silicate Weathering on Global Temperature and Atmospheric CO_2. *Journal of Geophysical Research: Solid Earth.* 1991;96(B11):18101–6.
33. Columbia University. *The Carbon Cycle and the Earth's Climate* (Accessed December 29, 2016). Available from: www.columbia.edu/~vjd1/carbon.htm.
34. R. A. Houghton, J. House, J. Pongratz, G. Van der Werf, R. DeFries, M. Hansen, et al. Carbon Emissions from Land Use and Land-Cover Change. *Biogeosciences.* 2012;9(12):5125–42.
35. UN FAO. *FLUDE: The Forest Land Use Data Explorer.* (United Nations Food and Agriculture Organization, 2015 [Accessed May 25, 2016]). Available from: www.fao.org/forest-resources-assessment/explore-data/en/.
36. UN FAO. *Global Forest Resources Assessment 2015.* (United Nations, 2015).
37. IPCC. *Terrestrial Carbon Processes: Background.* (Intergovernmental Panel on Climate Change, 2013 [Accessed September 6, 2017]). Available from: www.ipcc.ch/ipccreports/tar/wg1/099 .htm.

38. IPCC. *Climate Change 2014: Synthesis Report.* (Intergovernmental Panel on Climate Change, 2014).
39. K. Covey, C. B. de Mesquita, B. Oberle, et al. Greenhouse Trace Gases in Deadwood. *Biogeochemistry.* 2016;130(3):215–26.
40. K. R. Covey, S. A. Wood, R. J. Warren, X. Lee, M. A. Bradford. Elevated Methane Concentrations in Trees of an Upland Forest. *Geophysical Research Letters.* 2012;39(15).
41. S. R. Carpenter, D. Ludwig, W. A. Brock. Management of Eutrophication for Lakes Subject to Potentially Irreversible Change. *Ecological Applications.* 1999;9(3):751–71.
42. E. A. Parson. *Protecting the Ozone Layer: Science and Strategy.* (Oxford University Press, 2003).
43. J. Wettestad. The Vienna Convention and Montreal Protocol on Ozone-Layer Depletion, in Miles E. L., Underdal A., et al., editors, *Explaining Regime Effectiveness: Confronting Theory with Evidence.* (The Vienna Convention and Montreal Protocol, 2002): pp. 149–70.
44. S. Strada, N. Unger. Potential Sensitivity of Photosynthesis and Isoprene Emission to Direct Radiative Effects of Atmospheric Aerosol Pollution. *Atmospheric Chemistry and Physics.* 2016;16(7):4213–34.
45. X. Yue, N. Unger, K. Harper, et al. Ozone and Haze Pollution Weakens Net Primary Productivity in China. *Atmospheric Chemistry and Physics.* 2017;17(9):6073–89.
46. J. Weier, R. Cahalan. *Solar Radiation and Climate Experiment (Sorce).* (U.S. National Aeronautics and Space Administration, 2003 [Accessed March 19, 2016]). Available from: www.earthobservatory.nasa.gov/Features/SORCE/.
47. W. W. H. Soon. Variable Solar Irradiance as a Plausible Agent for Multidecadal Variations in the Arctic-Wide Surface Air Temperature Record of the Past 130 Years. *Geophysical Research Letters.* 2005;32(16).
48. N. Scafetta, B. J. West. Phenomenological Solar Signature in 400 Years of Reconstructed Northern Hemisphere Temperature Record. *Geophysical Research Letters.* 2006;33(17).
49. US National Aeronautics and Space Administration. *Climate and Earth's Energy Budget.* (United States National Aeronautics and Space Administration, Earth Observatory, 2016 [Accessed March 14, 2016]). Available from: http://earthobservatory.nasa.gov/Features/EnergyBalance/.
50. A. E. Carlson. What Caused the Younger Dryas Cold Event? *Geology.* 2010;38(4):383–4.
51. A. D. Richardson, E. G. Denny, T. G. Siccama, X. Lee. Evidence for a Rising Cloud Ceiling in Eastern North America. *Journal of Climate.* 2003;16(12):2093–8.
52. D. B. South, X. Lee, M. G. Messina. Will Afforestation in Temperate Zones Warm the Earth? *Journal of Horticulture and Forestry.* 2011;3:195–9.

9

Past and Future Climate Changes

9.1 Long-Term Climate History

Dramatic changes have occurred in the Earth's climate over thousands to billions of years caused by combinations of factors discussed in this chapter. The climate will continue to change in ways that people can now partly understand, anticipate, and possibly mitigate and/or prepare for.

Floating on top of the Earth's hot interior mantle is a crust of rock 50 km or less thick that is broken into plates of various sizes – from larger than continents to quite small (Figure 9.1). Lighter crustal rocks composed largely of granite float higher, and commonly form dry land; and heavier crustal rocks composed largely of basalt form the submerged ocean floors. The plates slowly move at speeds up to 20 cm per year. They converge, break apart, expand, and contract based on geologic processes of plate tectonics. These processes influence all aspects of the environment and resources.

The positions of continents and islands were much different in the past (Figure 9.1). As they changed to their present configuration, ocean currents shifted and created different climate patterns, eventually leading to the present ones [1–3].

The continents had been clustered as a single continent around the South Pole 300 to 400 million years before present (MYBP). By 250 MYBP, the continent Pangea straddled the equator and was beginning to break apart, stretching nearly from pole to pole (Figure 9.1a) [4].

During the ice age of 350 to 300 MYBP, glaciers covered what are now parts of Antarctica and southern South America, Africa, India, and Australia (A, Figure 9.1a). Sometime between 200 and 130 MYBP, Laurasia – North America, Europe, and Asia stuck together – separated from the other land masses known collectively as Gondwanaland. This separation allowed the Trade Winds to circulate freely around the Earth at the tropics, moving with them the warm ocean currents that were not hindered or diverted by continents as they are today. The lack of a diversion that would have caused the present Gulf Stream in the north Atlantic Ocean means that northern Europe probably did not receive as much warmth or precipitation as it does now. Australia and South America were still attached to Antarctica, so the ocean circulation in the far south between warm and cold waters made Antarctica less frigid than it is today. North America and Eurasia were attached to each

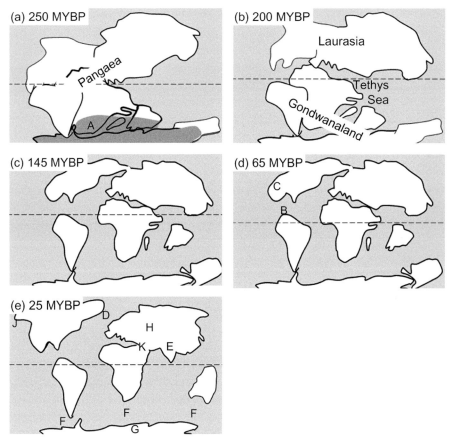

Figure 9.1 Changes in positions of continents during past 250 million years; Figure 9.5 shows present positions. (Dashed line = equator; Gray = Glaciated area of 350–300 MYBP. Letters are referred to in text.) Based on [1, 4–7].

other, so there was less cold water from the north pole moving downward into the Atlantic. About 65 MYBP, a large meteorite landed at what is now the northern Yucatan Peninsula (B) with such intensity that most dinosaur species were destroyed, allowing mammals to evolve and replace them (Figure 2.5 and Chapter 13). About this time, the Rocky Mountains emerged paralleling the Pacific Ocean in North America, probably blocking the moderating effects of the Kuroshio Current on western North America (C).

Europe and North America separated about 60 MYBP (D, Figure 9.1e). North America and Asia met at the Bering Strait, created an intermittent land bridge 1,600 km wide (north–south) during most of the last 25 million years (I), and slowed the Arctic waters moving into the North Pacific. When Australia separated from Antarctica 40 MYBP and South America separated from Antarctica shortly afterward, a cold circumpolar current began circulating around Antarctica (F), causing an ice cap to form on Antarctica and cooling many places in the world. Europe and Asia connected (H), India attached to Eurasia (E), Australia moved near Eurasia, and Africa attached to Eurasia 30 MYBP (K). Africa's attachment was a critical impact because it blocked water from circulating between the Indian and mid-Atlantic Oceans. The push of Africa and India into Eurasia caused the uplift of the Pyrenees, Alps, Carpathian, Pontus, Taurus, Zagros, Suleiman, Karakorum, Hindukush, and Himalaya Mountains parallel to the equator, dramatically changing the weather to their north and south. Finally, North and South America joined 12 MYBP [8], forcing the tropical waters in the Atlantic to flow north and augment the Gulf Stream, as well as south.

Some mountains, such as the Laurentian and Appalachian Mountains of North America, already existed during the ice age of 300 MYBP, although they were in different positions relative to the equator and poles. New mountains have formed since, associated with tectonic plate boundaries. The influence of these mountains on the Earth's climates varies with the mountains' locations, sizes, and orientations (Chapter 8).

9.2 Ice Ages and Glaciers

"Ice ages" are long, intermittent times when large glaciers are present on the Earth, advancing and retreating within these periods. "Glacial periods" are times during an Ice age when glaciers are well advanced. Ice ages occurred between 2.4 and 2.1 billion years ago as well as 850–630, 460–430, and 350–260 MYBP [1, 9–12]. A cold period occurred 200–180 MYBP but was not accompanied by large glaciation. The Earth is currently in another ice age that began 2.6 million years ago. Ice ages seem to occur when continents and mountains are positioned at high latitudes so that continental glaciers can form. During ice ages, ice sheets periodically expand during "glacial periods" of low atmospheric greenhouse gas concentrations and appropriate cyclic positions of the Earth relative to the Sun (Milankovitch Cycles, discussed later). The glaciers then contract dramatically during "interglacial periods" as greenhouse gas and Sun–Earth positions change (Figure 9.2). There are periodic extreme expansions and contractions separated by minor ones. The Earth left an extreme glaciation about 18 thousand years ago and arrived at an extreme non-glaciated period about 8 thousand years ago [6].

Glaciers begin on mountains at high latitudes where snow accumulates more than it melts, known as the "accumulation zone." The accumulated snow becomes packed, compresses, and forms ice (Figure 9.3). Its weight then forces it downward. In the process, it scrapes off and pushes ahead, carries, or rides over the vegetation, soil, and rocks in its path (Chapter 12). It moves forward at a slow speed or in quite rapid surges of 10 or more meters per day. The glacier pushes itself down to warmer elevations until it reaches places and times with a net snow melt during the year; however, even there the push from behind can sometimes maintain its presence or cause it to advance.

Glaciers expand during cool times, when the border (the "equilibrium line") between the accumulation zone and the ablation (melting) zone is at a low altitude. These occur both

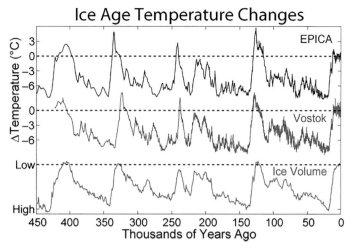

Figure 9.2 Antarctic temperature changes at two stations (upper two lines) during the last several glacial/interglacial cycles compared to global ice volume changes [6, 13–16]. EPICA and Vostok = Antarctic field stations. Reproduced with permission from the Free Software Foundation, Inc.

Figure 9.3 Glacier on higher elevation in Iceland, with tongues extending to low elevations. Farmhouses and road at bottom center show scale. Used with permission of Professor Roger Mesznik. (A black and white version of this figure will appear in some formats. For the color version, please refer to the plate section. Color plate 6.)

during small climatic coolings such as the Little Ice Age (700–150 years ago) [17] and during parts of the Milankovitch Cycles when less direct sunlight at upper latitudes lowers the temperature there while greater sunlight at the equator increases high latitude precipitation.

Continental glaciers accumulated so much water as ice that the sea level dropped by approximately 120 meters at the extreme glacial maximum about 18 thousand years ago (Figure 9.4) [12, 18].

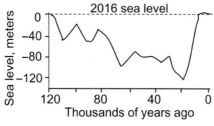

Figure 9.4 Change in sea level for past 120 thousand years. Data from [19–21].

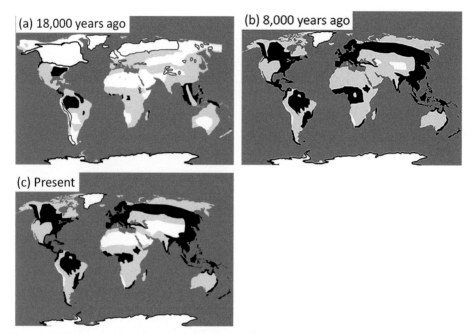

Figure 9.5 The Earth and land covers (a) 18 thousand years ago, (b) 8 thousand years ago, and (c) at present. (Light gray = shrub, grassland, tundra, open woodlands; black = closed forests; white, no borders = extreme deserts; white with thick black borders = glaciers). Based on [1, 4–6, 11, 18, 22–27].

Beginning 2.7 MYBP, the first continental glaciation of the new Pleistocene Ice Age began. The primary continental glaciers began in three places. One began in what is now northern Quebec and spread into the United States (Figure 9.5a). It merged with a second primary continental glacier that began in the Canadian Rocky Mountains and spread over western Canada and into the United States. Interior Alaska was not glaciated. A third primary continental glacier began over northern Nordic countries and Russia and expanded to cover much of Western Europe.

The glacial pattern of eastern Asia is less clear. Glaciers had existed in much of China, but their timing is uncertain [28]. Smaller glaciers covered parts of the southern Andes

Mountains of South America; Rocky Mountains in North America; and Himalayas, Alps, and Pyrenees Mountains of Eurasia.

During the glaciations, lower latitudes became hotter and drier as sunlight concentrated there and the Hadley cells contracted (Figure 7.5). The dryness is evidenced by dust in the atmosphere and drought-adapted vegetation – even in the usually moist tropics [18]. The Amazon forests apparently shrunk to refugia surrounded by grasslands [26, 29, 30], creating a species pump and greater biodiversity (Chapter 13) [29]. Southwestern corners of the northern hemisphere continents apparently had quite moist, temperate climates (presently Spain and southern California) as the jet streams were diverted around the glaciers.

The last period of melting glaciers occurred between 18 and 12 thousand years ago. The global sea level rose rapidly (Figure 9.4); and Canada and other lands that were under heavy ice are still locally rising (rebounding). The climate changed and became moist and quite mild in temperature by 8 thousand years ago (Figure 9.5b). Since then, it has gradually become drier in tropical latitudes. This dryness and other evidence suggest the Earth may be slowly returning to a period of continental glaciations [31, 32]; but the high level of greenhouse gases may counteract this return [33].

9.3 Milankovitch Cycles and Glaciation

The Earth's movement around the Sun varies in cycles of between about 20- and 100 thousand years. These cycles lead to cyclic changes in sunlight hitting the Earth, which affect glacial cycles and other weather phenomena. Three major cycle variations have been elucidated, known as the Milankovitch Cycles (Figure 9.6) [12, 18, 26]. Together they explain much of the pattern of Figure 9.2.

Cycle #1: Equinox Precession. The Earth has an elongated elliptical orbit around the Sun, so the Earth is closer to the Sun and receives more sunlight at some times of the year than at others. Times of closeness are slightly out of synchrony with the seasons, so seasons of maximum closeness slowly shift from the solstices (June and December) to the equinoxes (September and March) approximately every 10- to 12 thousand years. Seasons of maximum distance are opposite to these; a complete cycle occurs about every 23 thousand years.

When the Sun is closest to the Earth (and so maximum sunlight) during solstices, the heat is most intense at temperate latitudes because these areas are directly perpendicular to the Sun (Figure 7.8). At these times, the equator faces the Sun during equinoxes when the Earth is farthest away and so receives less intense heat. During the next 10- to 12 thousand years, the time of maximum closeness gradually shifts to the equinoxes and the equator receives intense heat, while the Earth becomes farthest away during solstices and so temperate and boreal latitudes receive less heat.

The Earth is currently approaching the time when the Sun is closest to the Earth as the equinoxes occur – and the Sun is farthest from the Earth as summer and winter

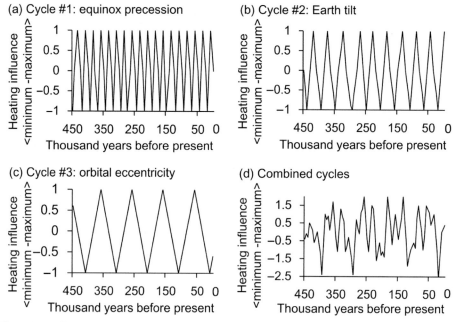

Figure 9.6 Milankovitch Cycles and their combined effects [6, 12, 26].

solstices occur. Consequently, low latitudes and the equator are increasingly receiving more sunlight; but temperate parts of the hemispheres are receiving 8% less solar radiation than they did 10 thousand years ago, when the Sun was closest to the Earth as the solstices occurred [12, 34].

Cycle #2: Earth tilt. The "Earth tilt cycle" occurs because the Earth's rotational axis "wobbles" back and forth between about 21.5 and 24.5 degrees from perpendicular to the Sun. The Earth is currently at an angle of about 23.5 degrees, and the angle is still decreasing. Less sunlight reaches the poles and more reaches the equator when the axis is at a low angle. The cycle moves from one extreme to the other in about 41 thousand years [12].

Cycle #3: Orbital eccentricity. The "orbital eccentricity" cycle occurs because the Earth's orbit around the Sun varies from nearly round to more elongated. When it is more elongated, the Earth is closer to the Sun as it moves along the narrow part of the orbit; however, the Earth's distance from the Sun is not greater when the orbit is at the long part of its elongated orbit than when the orbit is round. Consequently, the Earth receives more sunlight and becomes warmer when its orbit is more elongated. The orbit eccentricity changes from its most elongated to the most circular to most elongated in about 100 thousand years.

Figure 9.6d was created by adding Figures 9.6a through c. Note the remarkable similarity to Figure 9.2, and the periodic, extreme events when the cycles are in synchrony.

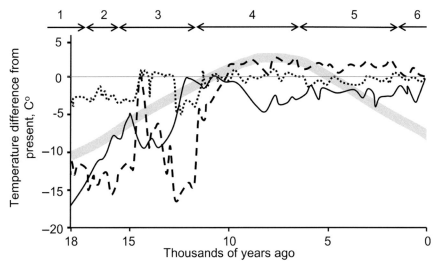

Figure 9.7 Reconstructed global temperature patterns from poles (solid and large dash line), low latitudes (dotted line) and progression of Milankovitch Cycles (thick gray line) for past 18 thousand years [11, 31, 35–38]. (Solid line = "Dome C" [Antarctica]; large dashes = "GISP2" [Greenland]; dotted line = "Cariaco" [Caribbean].)

For reasons given above, glacier cycles of the past several million years are closely related to the amount of summer sunlight at 65 degrees North latitude in summer, rather than to the total radiation reaching the Earth. This latitude is where the Ferrel and Polar Cells (Figure 7.5) converge and generate precipitation. When summer sunlight there is too weak to melt the winter's accumulation of snow, the snowfields expand into glaciers. When the sunlight there becomes strong enough during the approximately 23-thousand-year cycle for melting to exceed accumulation, the glaciers recede.

The change from hot periods of few glaciers to cold periods generally follows a long, uneven progression to a glacial maximum (Figure 9.2). During this time, glaciers expand, sea levels fall, and the atmosphere dries. Figures 9.2, 9.4 suggest the transition could occur slowly or rapidly. A general pattern of the Earth's temperature, sea level, glaciers, and atmospheric greenhouse gas concentration that occurred during the past 18 thousand years is described below, based largely on the equinox precession cycle (Figure 9.7).

Phase 1. For the 100 thousand years leading up to the extreme glaciation prior to about 18 thousand years ago, the glaciers had been slowly advancing, with intermittent, minor retreats (Figure 9.2). The sunlight concentration at the equator made the tropics dry. The glaciers' "equilibrium lines" moved farther down in elevation creating more snow in the glacier accumulation zones. Consequently, glaciers spread farther, sea levels lowered more as the glaciers took the water; lands in many places became uncovered; sea beds rose farther under the reduced weight of water; and land beneath the glaciers sank farther under the greater weight of ice. As glaciers expanded and sea levels continued to lower, the Earth's climate became drier and cooler.

Phase 2. Phase 2 occurred at full glaciation when the Milankovitch Cycles were in enough synchrony to create one of the few, extreme glaciations of the past 450 thousand years (Figure 9.2). The sea levels were low, the lands currently beneath sea level were exposed, the weather was cool and dry, and the glaciers were at a maximum.

Phase 3. Because of the lag time in heating, cooling, and snow/ice accumulation and melting, the pattern from Phase 2 continued even as the Milankovitch Cycle shifted toward less sunlight on the tropics and more on the upper latitudes. As increasingly more sunlight reached the upper latitudes, the glaciers began melting, the sea levels began rising, and the Earth's atmosphere became moister [12].

Phase 4. By about 8 thousand years ago, the solstices were in the part of the orbit closest to the Sun; and the solar radiation was maximum in the temperate zones. The remaining glaciers melted quite rapidly – with most of the continental glaciers disappearing over a period of a few thousand years (Figure 9.2). Sea levels rose to slightly above their present levels with the melting glaciers (Figure 9.4); and the Earth became moister, with both temperate and tropical zones having moderately warm weather [12].

Phase 5. As the Milankovitch Cycles continued, the solar radiation began cycling back to concentrating on the tropics, the Earth's climates seem to have become hotter and drier near the equator and are still warming near the poles. It is unclear what will happen next. Sea levels are high, but some evidence suggests the temperate zones have slowly become cooler [31, 33]. Human emissions of carbon dioxide and other greenhouse gases to the atmosphere probably have prevented the temperature reduction in temperate zones that may otherwise have been expected at this phase.

Phase 6. Phase #6 will be discussed later in this chapter (9.5 Future Climate Scenarios).

9.4 Shorter Climate Fluctuations

Superimposed on these long-term patterns are fluctuations that last several hundred years, may be regional, and have various causes (Figure 9.7). One of these, the "hypsithermal" (also known as the "Holocene maximum"), coincided with the expected warm, moist period based on the Milankovitch Cycles soon after the glaciers melted.

The steep dip in temperatures, especially in the northern hemisphere, approximately 12 thousand years ago (Figure 9.7) was the "Younger Dryas" (also known as the "Loch Lamond Stadial" and "Nahanagan Stadial" [36]). It is believed to have resulted from much of North America's frigid glacial meltwaters shifting from flowing down the Mississippi River into the Gulf of Mexico to flowing out the St. Lawrence River into the Gulf Stream and to Europe [6, 35, 36]. Other processes have been proposed [39]. Similar temperature patterns following glacial lake drainages occurred in northern Canada and Eurasia [37, 40, 41].

Eastern North America and Europe endured a warm period between about AD 900 and 1400 and then a cool period, the "Little Ice Age" [42–44]. The climate became warmer into the 1930s, followed by a plateau until about 1975 and another rise since [26].

9.5 Future Climate Scenarios

In the near future, climates will probably be warmer than in the recent past. The average temperature increase has been 0.24 degrees C (°C) per decade between 1980 and 2005 and 0.1 °C per decade between 1905 and 2005 (based on Figure 8.6). These rates are much faster than the average estimated rise of 0.01 °C per decade between 20- and 10 thousand years ago (Figure 9.2) or a rise of 0.06 °C per decade during the Younger Dryas (Figure 9.7). There probably were shorter periods within the earlier periods when temperatures did rise much faster; and it is difficult to conclude that warming will continue at recent rates. However, if the recent temperature changes were to continue, species would need to migrate and adjust at unprecedentedly rapid rates for them to survive.

Many models and predictions have been made of climate trends for the next one or several centuries (for example, [45, 46]). This book will make generalized forecasts of possible future climates for the next 1 to 5 thousand years. These long-term forecasts put short-term forecasts into context.

Long-term future climates can be inferred from climate change processes such as the Milankovitch Cycles and greenhouse gas effects; geologic reconstructions of past climates; experiences with recent climate changes; and current trends in weather, biota, and glacier behaviors. Three scenarios will be proposed:

1. The glaciers' return based on past glaciations and the Milankovitch Cycles as if the greenhouse gases were not factors influencing climates;
2. The changes expected based on past events and the Milankovitch Cycles, but assuming the elevated greenhouse gases affect climates and preclude glaciers from returning.
3. The changes expected if the greenhouse gases precluded the glaciers' returning for 1,000 years, after which the glaciers return.

The scenarios are presented in the sequence above to facilitate understanding. The Intergovernmental Panel on Climate Change (IPCC) asserts that it is "virtually certain" that recent warming from greenhouse gases will prevent "widespread glaciation" based on the Milankovitch Cycles during the next 1,000 years [33]. Consequently, the third scenario is the most likely, and the first is least.

9.5.1 Scenario #1: Greenhouse Gases Are Not a Factor

Were it not for anthropogenic increases in greenhouse gases, at some time between now and several thousand years hence, glaciers would begin returning where they had been before. Glaciers would move rapidly for about 5 thousand years (Figure 9.2). By then, they would probably contain about one-third of the ice volume of 18–20 thousand years ago. The sea level would probably be about 30 to 40 meters lower than at present (Figure 9.4). After that, the glaciers would slowly advance, with occasional short retreats, for about another 80 thousand years. It would then be at its maximum with about the same conditions as 18 thousand years ago.

For the first 5 thousand years, sea levels would lower about 0.7 meters per hundred years, the glaciers would advance slowly until they covered much of the Scandinavian countries, parts of Russia, and about half of Canada, with mountain glaciers covering upper parts of the Alps, Rocky, Himalayas, and Andes Mountains. Upper latitudes would grow cold and moister, and lower latitude temperatures would become slightly cooler but increasingly drier.

The slow changes seem distant and slight. They could be easily accommodated if people were expecting them and acted appropriately. However, the changes accumulate and could become surprising and catastrophic if ignored until they became impediments. Cities on coastal plains could find themselves far inland within a few centuries, similar to Ephesus (Figure 10.11) if sea levels fell. Harbors and some sea passages could become too shallow for large ships; and mountain passes could become much colder or even blocked by glaciers or winter snows. New transportation routes and mechanisms would need to be developed and are best planned in advance.

People would certainly migrate southward from where glaciers expanded; immediate glacial peripheries would be too cold to inhabit. Slightly farther from the poles the weather would be colder, which would require different crops that probably provide lower yields [42]. Buildings would require more heating. Near the equator, the drying and cooling would stress people and agriculture. As sea levels dropped, people would populate the newly exposed coastal plain landforms in southeastern North and South America and Asia, eastern Asia, northern Australia, and elsewhere (Figure 9.5a). Lakes such as the Caspian and Aral Seas and Great Salt Lake may expand again, perhaps creating more stable climates where people could live. People survived colder and drier periods before, but with lower populations, dramatically different lifestyles, different skills, and less technology.

9.5.2 Scenario #2: Greenhouse Gases Prevent Glaciers

If greenhouse gases permanently overrode the glaciers' returns, tropical areas would become very hot both because of greenhouse gases (Figure 8.6) and because the Sun will be close to the Earth during equinoxes for the next few thousand years. Without glaciers, however, sea levels would not recede, more water would be in the atmosphere as a greenhouse gas, and low latitudes would be hotter than had the glaciers advanced. Low latitudes would continue to dry but probably not become as dry as with glaciation. Without glaciers, the upper latitudes may remain at present temperatures or become cooler and moister, but not as cold as had glaciers arrived. Sea levels could rise if arctic and Antarctic ice continued to melt. If sea levels rose very much or if violent storms increased, people in vulnerable places such as the Nile Delta, Netherlands, Bangladesh, Thailand, Vietnam, China, southeastern United States and western equatorial Africa would need other places to live and grow crops. And, many coastal cities would need to be relocated. The differences between cool polar regions and warmer equatorial regions would probably create unusual weather patterns – possible sudden and extreme ones. There would probably be a general shift in people and biota toward middle latitudes, where most food would also be grown. Unlike Scenario 1, more land would not be exposed by a falling sea level; and all of the currently high human population may not be able to survive.

9.5.3 Scenario #3: Greenhouse Gases Delay Glaciers

If greenhouse gases overrode the Milankovitch Cycles' influences for only 1,000 years, the trend described in Scenario #2 would probably occur for the first thousand years. Then, the transition to glaciation, upper latitude cooling, lower latitude drying, and sea level receding of Scenario #1 would probably accelerate until it came into equilibrium with the climate. This transition would probably be a time of rapid and possibly violent changes in climate, weather, sea level, and glaciers.

9.6 Managing Climate Changes

Climate changes will be dramatic, but they need not be catastrophic. People cannot control the Milankovitch Cycles but can predict them. They can control greenhouse gas emissions and slowly reverse the trend. And people can plan for orderly transitions even in the face of current uncertainties. Managing entails forecasting, developing scenarios to adjust (Chapter 3), and mitigating adverse situations.

Three future climate scenarios have been forecast, above. Development of scenarios for adjusting to them will be discussed next, followed by ways to mitigate global warming.

9.6.1 Adjusting to Climate Change

Whichever scenario forecast above occurred – or if a different one unfolded – dramatic adjustments would be needed in where and how people live. Past, small climate changes such as the Little Ice Age led to violence brought on by famines, migrations, and breakdowns of social order [42–44, 47]. Very recent problems of resettling refugees in Europe and elsewhere from the Middle East shows the unnecessary strife that large-scale migrations created by climate changes could bring (Chapter 6). Such strife could be avoided using appropriate protocols such as have been developed for transboundary water issues (Chapter 18) [48].

Whether the glaciers return or not, experiences from northern Europe during the relatively small cooling of the Little Ice Age of about 700 to 150 years ago suggest that local weather would become quite harsh if temperatures cool at upper latitudes [42–44, 49]. Villages would be abandoned, warmer shelters created, and different food crops grown. With drying climates at lower latitudes, deserts would continue to expand. Recent experiences suggest the weather may become less predictable.

Management could focus on learning to predict and respond to changing climates and dealing with their uncertainties. Equally important would be improved communication; technical knowledge; and technically skilled people to manage agriculture, water, energy, minerals, forests, and biodiversity increasingly effectively.

People's adjusting would also require changing mental models – and learning to embrace the change and work cooperatively. Where and how people live and with whom would change dramatically with climate changes. The changes could be peaceful and even stimulating if they are recognized as social drivers and stimulate creative lifestyles and innovation. By not ignoring the changes until they reached a harmful threshold, last-minute decisions often involving triage could be avoided – such as decisions to allow species extinctions so people could be saved.

9.6.2 Mitigating Atmospheric Carbon Dioxide

The expected global warming could also be mitigated by either rapidly reducing greenhouse gases or putting things in the atmosphere that counteract greenhouse gases. Both of these will be discussed.

Several methods could either slow the increase of atmospheric CO_2 or reduce the total atmospheric CO_2. Some methods reduce CO_2 release from the slow domain by reducing fossil fuel and cement use (Figure 8.3). Others reduce CO_2 release from the fast domain by limiting land conversion and other biosphere activities. Still others reduce atmospheric CO_2 by removing it and storing more in the fast domain, thus increasing vegetation and soil carbon. These are compared below and in Figure 9.8 by examining the effect of several methods when done aggressively and efficiently. The reduction in fossil fuel CO_2 emissions does not create a one-to-one decrease in atmospheric CO_2 because some CO_2 is continuously being sequestered in the biosphere and other CO_2 is being released from the biosphere by human activities.

As a baseline ("#1 2014," Figure 9.8), fossil fuel burning emitted about 36.7 billion tons of CO_2 in 2014, and land use change added about 6.2 billion more tons [33, 50, 51]. Of this, about 11 billion tons of CO_2 were returned to the biosphere (fast domain) [33, 50]. Some other CO_2 went into the ocean, but CO_2 in the ocean returns to the atmosphere as other atmospheric CO_2 is removed and so is considered part of atmospheric CO_2. The net CO_2 emissions to the atmosphere were 31.9 billion tons in 2014.

Six methods for reducing atmospheric CO_2 are examined below and compared in Figure 9.8. The analysis assumes that all methods could be implemented fully and immediately.

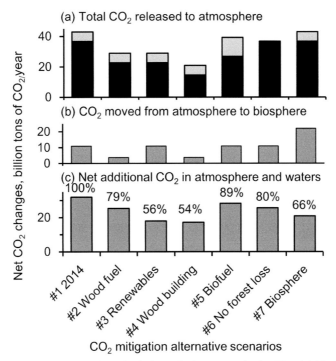

(a) Total CO_2 released to atmosphere

(b) CO_2 moved from atmosphere to biosphere

(c) Net additional CO_2 in atmosphere and waters

100%   79%   56%   54%   89%   80%   66%

Net CO_2 changes, billion tons of CO_2/year

#1 2014   #2 Wood fuel   #3 Renewables   #4 Wood building   #5 Biofuel   #6 No forest loss   #7 Biosphere

CO_2 mitigation alternative scenarios

Figure 9.8 Current conditions and methods for reducing atmospheric CO_2. (a) black = fossil fuel emissions (slow domain); gray = emissions from vegetation and soil (fast domain). (b) net changes in biosphere CO_2 (fast domain). (c) net CO_2 increase in atmosphere; percentages show net CO_2 emissions compared to #1.

#1. **2014 (baseline).** The 2014 condition.

#2. **Wood fuel.** Using all new wood that grows each year (not standing volume) as wood fuel to substitute for fossil fuel would reduce fossil fuel use and emissions, but also reduce CO_2 biosphere sequestration by a lesser amount (Figure 28.5). Net atmospheric emissions would be reduced by 21%/year [52–54]. No additional CO_2 would be sequestered in the forest each year, but less would be lost from fires and rotting. Forest biodiversity could be harmed or enhanced depending on how the forests were managed (Chapters 28–30).

#3. **Renewables.** Using renewable energy such as wind, solar, and hydroelectricity and conservation such as better insulation would reduce fossil fuel use. If the fossil fuel reduction from renewables and conservation is assumed to equal fossil fuel reduction from using all wood growth for fuel (#2, above), renewable energy would reduce net CO_2 emissions by 44%.

#4. **Wood building.** Using all wood growth for construction and the wood waste for fuel would reduce fossil fuel use [54] and reduce net atmospheric CO_2 emissions by 46%/year (Chapters 24, 28). Other considerations would be similar to using all wood growth for fuel, #1, above.

#5. **Biofuel.** Growing non-cellulosic biofuels by clearing forest land. Nearly 10% of the world's forest land would be needed to grow enough sugarcane biofuel to reduce net fossil fuel CO_2 emissions by 11% (Table 24.2). The cleared forest would reduce carbon uptake as well as biodiversity, and may increase water scarcity if irrigated. If land-clearing CO_2 emissions are averaged over 50 years, this biofuel substitution for fossil fuels would result in a net decline of 11%/year in CO_2 emissions. The same CO_2 emission decline would result from palm oil biofuel and about 5.5% decline from corn biofuel (Table 24.2). Converting cellulosic biomass to biofuels is technically feasible and useful (Chapter 24) [55–57]; however, intensive forest plantations grown for cellulosic biofuels are probably not.

#6. **No forest loss.** Carbon dioxide emissions could also be mitigated by reducing emissions from the biosphere (fast domain) by stopping conversion of forests to other uses [58]. Net CO_2 emissions would decline by 20%/year if all such conversions ceased. Others use a different emissions rate [59]. The increased demand for meat by developing countries may make this method difficult (Chapter 22).

#7. **Biosphere.** More CO_2 could be stored in the biosphere (fast domain). This analysis assumes the amount stored in the biosphere is doubled from 11 to 22 CO_2 tons/year (Chapter 7). Doubling this sequestration rate by active management would only increase the biosphere carbon by 0.6% per year, but would reduce net atmospheric carbon dioxide emissions by 34% per year. It should be possible to increase the fast domain sequestration by this amount through better management of forests (Chapter 29) and rangelands (Chapter 22) for at least a few decades [60, 61]. This storage would not reduce fossil fuel consumption.

The effects of these methods on CO_2 emissions and fossil fuel use vary considerably. Wood building (#4) and using renewable energy (and conservation, #3) reduce net CO_2 emissions the most (Figure 9.8c). Carbon sequestration in the biosphere (#7) is moderately successful in reducing CO_2 emissions. Biofuels (#5) do little to reduce emissions. No forest loss (#6) and wood fuel use (#2) do only slightly better than biofuels.

Wood building (#4) reduces fossil fuel use the most (Figure 9.8a), with renewable energy (and conservation, #3) being next in reduction. No forest loss (#6) and sequestration in the biosphere (#7) do not reduce fossil fuels; they recapture the fossil fuel emissions in the fast domain – the biosphere. Biofuel use (#5) is slightly more effective in reducing fossil fuel emission than #6 and #7, but reduces forest land and water.

Other methods not analyzed include converting non-forest/nonfood biomass to biofuels [62] that would otherwise rot and emit CO_2 anyway. An estimated 4.4 billion tons of CO_2 from fossil fuels could be saved annually from this biomass. Forest wildfires in the United States are emitting 4% to 6% as much CO_2 as all fossil fuels burned there [63]; this forest biomass could be converted to biofuel or wood chips that save fossil fuel before being emitted to the atmosphere (currently as wildfires).

"Carbon Capture and Storage" (CCS) technology is intended to remove CO_2 emissions from fossil fuels by chemical-mechanical means [33, 50]. The world's accessible fossil fuel reserves are finite (Chapter 24), and CCS as well as no forest loss and biosphere sequestration would neither prolong them nor provide substitutes.

9.6.3 Intentionally Changing Climates

People may be able to reduce the Earth's temperature through "geoengineering" [64], such as by creating more clouds – and thus shade – using atmospheric emissions that partially block sunlight. People have been moderately effective at locally controlling frost by several means(Chapter 21); controlling wind with trees ("shelterbelts," Chapter 12); or controlling local heat and cold with deciduous shade trees, architecture, stoves, or air conditioners. Large-scale attempts to change climates through cloud seeding have sometimes been effective at generating precipitation; but sometimes the precipitation has landed where not intended.

Further tinkering with the atmosphere to solve past problems would probably have many unintended consequences [33], just as DDT and PCBs did (Chapter 8). Unlike some other environmental and resource systems, the atmospheric system is probably safer to people if it remains restored and preserved in its pre-Anthropocene condition rather than being modified.

References

1. P. Cattermole. *Building Planet Earth*. (Cambridge University Press, 2000).
2. P. V. Brady. The Effect of Silicate Weathering on Global Temperature and Atmospheric CO_2. *Journal of Geophysical Research: Solid Earth*. 1991;96(B11):18101–6.
3. D. A. Evans. A Fundamental Precambrian – Phanerozoic Shift in Earth's Glacial Style? *Tectonophysics*. 2003;375(1):353–85.
4. C. C. Plummer, D. McGeary, D. H. Carlson. *Physical Geology*, ninth edition. (McGraw-Hill, 2003).
5. J. S. Monroe, R. Wicander. *Physical Geology*. (Brooks/Cole, 2001).
6. B. J. Skinner, S. C. Porter, J. Park. *Dynamic Earth*. (Wiley, 2004).
7. US Geological Survey. *Historical Perspective*. (U.S. Department of the Interior, Geological Survey; 2012 [Accessed July 7, 2016]). Available from: http://pubs.usgs.gov/gip/dynamic/historical.html.
8. C. D. Bacon, D. Silvestro, C. Jaramillo, et al. Biological Evidence Supports an Early and Complex Emergence of the Isthmus of Panama. *Proceedings of the National Academy of Sciences*. 2015;112(19):6110–15.
9. S. Eldredge, B. Biek. *Ice Ages – What Are They and What Causes Them?* (Utah Geological Survey, 2010 [Accessed March 22, 2016]). Available from: http://geology.utah.gov/map-pub/survey-notes/glad-you-asked/ice-ages-what-are-they-and-what-causes-them/.
10. B. Saltzman. *Dynamical Paleoclimatology: Generalized Theory of Global Climate Change*. (Academic Press, 2002).
11. M. J. Siegert. *Ice Sheets and Late Quaternary Environmental Change*. (John Wiley and Sons Inc., 2001).
12. D. Huddart, T. Stott. *Earth Environments: Past, Present, and Future*. (Wiley-Blackwell, 2012).
13. L. Augustin, C. Barbante, P. R. Barnes, et al. Eight Glacial Cycles from an Antarctic Ice Core. *Nature*. 2004;429(6992):623–8.

14. L. E. Lisiecki, M. E. Raymo. A Pliocene–Pleistocene Stack of 57 Globally Distributed Benthic Δ18o Records. *Paleoceanography.* 2005;20(1).

15. J.-R. Petit, J. Jouzel, D. Raynaud, et al. Climate and Atmospheric History of the Past 420,000 Years from the Vostok Ice Core, Antarctica. *Nature.* 1999;399(6735):429–36.

16. R. A. Rhode. *File:Ice Age Temperature.png* [Licensed under Creative Commons; Permission under GNU Free Documentation License]. (Wikimedia Commons, 2012 [Accessed April 9, 2017]). Available from: https://commons.wikimedia.org/wiki/File:Ice_Age_Temperature.png #filelinks.

17. B. M. Fagan. *The Little Ice Age: How Climate Made History, 1300–1850.* (Basic Books, 2000).

18. M. Maslin. The Climatic Rollercoaster, in Fagan B., editor, *The Complete Ice Age.* (Thames and Hudson, 2009): pp. 62–91.

19. P. U. Clark, A. C. Mix. Ice Sheets and Sea Level of the Last Glacial Maximum. *Quaternary Science Reviews.* 2002;21(1):1–7.

20. M. J. O'Leary, P. J. Hearty, W. G. Thompson, et al. Ice Sheet Collapse Following a Prolonged Period of Stable Sea Level During the Last Interglacial. *Nature Geoscience.* 2013;6 (9):796–800.

21. N. Shackleton. Oxygen Isotopes, Ice Volume and Sea Level. *Quaternary Science Reviews.* 1987;6(3):183–90.

22. J. M. Adams. *Global Land Environments since the Last Interglaciation.* (Biological Sciences, Seoul National University, 2002 [Accessed April 10, 2016]). Available from: http://geoecho.snu .ac.kr/nerc.html.

23. J. M. Adams, H. Faure. Preliminary Vegetation Maps of the World since the Last Glacial Maximum: An Aid to Archaeological Understanding. *Journal of Archaeological Science.* 1997;24(7):623–47.

24. N. Ray, J. Adams. A GIS-Based Vegetation Map of the World at the Last Glacial Maximum (25, 000–15,000 Bp). *Internet Archaeology.* 2001;11.

25. J. W. Williams. Quaternary Vegetation Distributions. *Encyclopedia of Paleoclimatology and Ancient Environments.* (Springer, 2009), pp. 856–62.

26. W. K. Hamblin, E. H. Christiansen. *The Earth's Dynamic Systems*, tenth edition. (Prentice Hall, 2001).

27. R. S. Thompson, K. H. Anderson. Biomes of Western North America at 18,000, 6000 and 0 14C Yr bp Reconstructed from Pollen and Packrat Midden Data. *Journal of Biogeography.* 2000;27 (3):555–84.

28. S. Yafeng, C. Zhijiu, S. Zhen. *The Quaternary Glaciations and Environmental Variations in China.* (Heibei Science and Technology Publishing House, 2005).

29. M. A. Huston. *Biological Diversity: The Coexistence of Species.* (Cambridge University Press, 1994).

30. M. A. Maslin, Y. Mahli, O. Phillips, S. Cowling. New Views on an Old Forest: Assessing the Longevity, Resilience, and Future of the Amazon Rainforest. *Transactions of the Institute of British Geographers.* 2005;30(4):390–401.

31. E. C. Pielou. *After the Ice Age.* (University of Chicago Press, 1991).

32. G. M. MacDonald, A. A. Velichko, C. V. Kremenetski, et al. Holocene Treeline History and Climate Change across Northern Eurasia. *Quaternary Research.* 2000;53(3):302–11.

33. T. Stocker, D. Qin, G. Plattner, et al. Climate Change 2013: The Physical Science Basis. In *Intergovernmental Panel on Climate Change.* (Cambridge University Press, 2013).

34. T. Flannery. *The Eternal Frontier: An Ecological History of North America and Its Peoples.* (Atlantic Monthly Press, 2001).

35. US National Oceanic and Atmospheric Administration. *The Younger Dryas.* (NOAA National Climatic Data Center, 2008 [Accessed March 22, 2016]). Available from: www.ncdc.noaa.gov /paleo/abrupt/data4.html.

36. D. E. Anderson. Younger Dryas Research and Its Implications for Understanding Abrupt Climatic Change. *Progress in Physical Geography.* 1997;21(2):230–49.

37. US National Oceanic and Atmospheric Administration. *Post-Glacial Cooling 8,200 Years Ago.* (NOAA National Climatic Data Center, 2008 [Accessed March 22, 2016]). Available from: www.ncdc.noaa.gov/paleo/abrupt/data5.html.

38. R. B. Alley. Ice-Core Evidence of Abrupt Climate Changes. *Proceedings of the National Academy of Sciences.* 2000;97(4):1331–4.

39. A. E. Carlson. What Caused the Younger Dryas Cold Event? *Geology.* 2010;38(4):383–4.

40. J. Mangerud, M. Jakobsson, H. Alexanderson, et al. Ice-Dammed Lakes and Rerouting of the Drainage of Northern Eurasia During the Last Glaciation. *Quaternary Science Reviews.* 2004;23 (11):1313–32.

41. A. N. Rudoy. Glacier-Dammed Lakes and Geological Work of Glacial Superfloods in the Late Pleistocene, Southern Siberia, Altai Mountains. *Quaternary International.* 2002;87(1): 119–40.

42. H. H. Lamb. *Climate, History and the Modern World.* (Routledge, 2002).

43. H. H. Lamb. *Climate: Present, Past and Future (Routledge Revivals): Volume 1: Fundamentals and Climate Now.* (Routledge, 2013).

44. H. H. Lamb. *Climate: Present, Past and Future (Routledge Revivals): Volume 2: Climatic History and the Future.* (Routledge, 2013).

45. R. Kumar, K. K. Kumar, V. Prasanna, K. Kamala, N. Despand. Future Climate Scenarios. *Climate Change and India: Vulnerability Assessment and Adaptation.* (Universities Press, Hyderabad, 2003), pp. 69–127.

46. S. L. Shafer, P. J. Bartlein, R. S. Thompson. Potential Changes in the Distributions of Western North America Tree and Shrub Taxa under Future Climate Scenarios. *Ecosystems.* 2001;4(3): 200–15.

47. W. Behringer. *A Cultural History of Climate.* (Polity Press, 2010).

48. P. H. Gleick. Water and Conflict: Fresh Water Resources and International Security. *International Security.* 1993;18(1):79–112.

49. H. H. Lamb. The Early Medieval Warm Epoch and Its Sequel. *Palaeogeography, Palaeoclimatology, Palaeoecology.* 1965;1:13–37.

50. IPCC. *Climate Change 2014: Synthesis Report.* (Intergovernmental Panel on Climate Change, 2014).

51. IPCC. *1.2.1.3.1. Inter-Annual and Decadal Variability of Atmospheric CO_2 Concentrations.* (Intergovernmental Panel on Climate Change, 2015 [Accessed July 8, 2016]). Available from: www.ipcc.ch/ipccreports/sres/land_use/index.php?idp=20#table1-2.

52. UN FAO. *FLUDE: The Forest Land Use Data Explorer.* (United Nations Food and Agriculture Organization, 2015 [Accessed May 25, 2016]). Available from: www.fao.org/forest-resources-assessment/explore-data/en/.

53. S. Luyssaert, I. Inglima, M. Jung, et al. CO_2 Balance of Boreal, Temperate, and Tropical Forests Derived from a Global Database. *Global Change Biology.* 2007;13(12):2509–37.

54. C. D. Oliver, N. T. Nassar, B. R. Lippke, J. B. McCarter. Carbon, Fossil Fuel, and Biodiversity Mitigation with Wood and Forests. *Journal of Sustainable Forestry.* 2014;33 (3):248–75.

55. P. Fairley. Introduction: Next Generation Biofuels. *Nature.* 2011;474(7352):S2–S5.

56. B. Lippke, R. Gustafson, R. Venditti, et al. Sustainable Biofuel Contributions to Carbon Mitigation and Energy Independence. *Forests.* 2011;2(4):861–74.

57. J. R. Regalbuto. Cellulosic Biofuels – Got Gasoline? *Science.* 2009;325(5942):822–4.

58. IPCC. *Terrestrial Carbon Processes: Background.* (Intergovernmental Panel on Climate Change, 2013 [Accessed September 6, 2017]). Available from: www.ipcc.ch/ipccreports/tar/wg1/099.htm.

59. UN FAO. *Global Forest Resources Assessment 2015.* (United Nations, 2015).

60. L. Poorter, F. Bongers, T. M. Aide, et al. Biomass Resilience of Neotropical Secondary Forests. *Nature.* 2016;530(7589):211–14.

61. R. L. Chazdon, E. N. Broadbent, D. M. Rozendaal, et al. Carbon Sequestration Potential of Second-Growth Forest Regeneration in the Latin American Tropics. *Science Advances.* 2016;2 (5):e1501639.

62. R. Slade, R. G. Saunders, A. Bauen. *Energy from Biomass: The Size of the Global Resource.* (Imperial College Centre for Energy Policy and Technology UK Energy Research Centre, 2011).
63. C. Wiedinmyer, J. Neff. Estimates of CO_2 from Fires in the United States: Implications for Carbon Management. *Carbon Balance and Management.* 2007;2(10).
64. R. Mileham. Science Geoengineering – Biting the Bullet. *Engineering & Technology.* 2007;2 (8):28–31.

Part IV
Landforms

10

Landforms and Soils

10.1 Distribution, Soils, and Management

A landform is a large, contiguous area of similar geomorphologic history and chemical and physical properties that distinguish it from other landforms [1]. About fifteen kinds of landforms exist, distributed irregularly in various sizes from quite small to larger than the country of Spain. They adjoin each other and continuously cover the Earth, including beneath oceans and seas. Each landform has similar properties wherever it occurs that make it useful for some things and difficult for others (Table 10.1). Along with climate, landforms strongly determine the biodiversity, lifestyles, and resources of people living on them.

The study of landforms emerged from both soil science and geomorphology. Soil science first focused on the productivity of each field for crop growth. Much was learned; but the perspective was narrow for integrating processes across larger areas for water, agriculture, buildings, land use allocation, mineral extraction, and other purposes. Consequently, conservationists integrated soil science with geomorphology to manage broader areas of similar properties for many uses.

Soils develop from "parent material" – either bedrock or usually inorganic material that is transported from elsewhere by wind, water, or ice and covers the bedrock deeply. Parent materials have a variety of characteristics that influence the developing soil properties. Landforms and their soils support agriculture and biodiversity, moderate water movement, provide building materials, and serve as stable or unstable places for living or traveling. This chapter will first discuss soils and then general characteristics of landforms. Subsequent chapters will discuss different landforms in detail.

10.2 Soil Properties and Variations

Igneous, metamorphic, and/or sedimentary rocks in various stages of weathering underlie all landforms (Chapter 26). In places, overlying deposits from elsewhere characterize the landform. Where these overlying deposits are absent or thin, the landform is characterized by the "bedrock," which is the hard layer of solid rock of different types covering the Earth. Bedrock is usually covered by loose rocks and soil (Figure 11.3a), but is sometimes exposed

Table 10.1 *Some characteristics of landforms. L = Low, M = Medium, H = High, V = Varied. L–H = range, with first letter predominant*

Bedrock-type landform	Productivity	Ease of erosion	Ease of compaction	Gravel/rock availability
Weathering in place				
Shield	L–M	L–M	M	L
Igneous	L–H	M–H	M–H	H
Basalt	L–H	L–M	L–M	H
Metamorphic/ sedimentary				
Metamorphic	L–H	L–M	M–H	H
Coastal plains	L–H	L	L–M	L
Peatlands	L	L–M	H	L
Karst	L–H	L–M	M–H	H
Transported				
Loess	H	H	H	L
Sand dunes	L	L	L	L
Volcanic ash	L–H	L–M	M–H	L
Alluvial floodplains	H–L	H	H	L
Peatlands	L	L	M–H	L
Permafrost	L	L	H–L	M
Glaciated				
Glacial till	L–M	L–M	M	H
Glacial outwash	L	L	L	H
Lacustrine	L–M	M	H	M
Mountains	V	M–H	M–H	H

(Figures 11.2a–b). The covering of loose rocks, sand, silt, and clay is the soil's "parent material." It can be either decomposed ("weathered") from bedrock or transported from elsewhere.

Rocks break apart physically and chemically and form soils, a property common to all landforms known as "weathering" [2–5]. Weathering becomes deeper with time, with soils many hundreds of millions of years old being 6 meters or more deep (Figure 11.3a). Soils differ in moisture, particle size distribution, structure, chemical composition, tendency to erode, angle at which they maintain slope stability ("stable angle of repose"), and other factors. The differences are caused by the parent rock material and how they were physically and chemically converted to soils. These differences cause landforms to be useful and hazardous in specific ways that can be anticipated and managed.

Soil is "the unconsolidated mineral" and/or "organic material on the immediate surface of the Earth that serves as a natural medium for the growth of land plants" [6]. It commonly consists of decomposing organic matter, living plant roots, microorganisms, animals,

Table 10.2 *Elements required by plants, the form required, and their relative proportions [9, 10]*

Element	Chemical symbol	Form required by plants	Adequate concentration in plants (%)
Oxygen	O	O_2, H_2O	45%
Carbon	C	CO_2	45%
Hydrogen	H	H_2O	6%
Nitrogen	N	NO_3-, NH_4+	1.5%
Potassium	K	$K+$	1.0%
Calcium	Ca	$Ca++$	0.5%
Phosphorus	P	H_2PO_4-, HPO_4-	0.2%
Magnesium	Mg	$Mg++$	0.2%
Sulfur	S	SO_4-	0.1%
Iron	Fe	$Fe+++$, $Fe++$	0.01%
Chlorine	Cl	$Cl-$	0.01%
Manganese	Mn	$Mn++$	< 0.005%
Zinc	Zn	$Zn++$	< 0.005%
Boron	Bo	Bo_3-, Bo_{47}	< 0.005%
Copper	Cu	$-Cu+$, $Cu++$	< 0.005%
Molybdenum	Mo	MoO_4-	< 0.005%
Nickel	Ni	$Ni++$	< 0.005%

elements (Table 10.2) and other cations and anions, air, water, and mineral particles. It exists in the surficial one to several meters of land. Soils vary dramatically in productivity for plant growth because of differences in moisture holding capacity, aeration, elements, and other factors.

10.2.1 Soil Moisture-Holding Capacity, Texture, and Structure

A soil's moisture holding capacity is based on its structure and texture. Texture is the distribution of mineral particle sizes [7]. Clays have the smallest diameter (<0.002 mm), silt next (0.002–0.02 mm), sand (0.02–2 mm), then various rocks (pebbles, cobbles, and boulders). Larger soil particles leave larger spaces (pores) between them when packed. This pore space is necessary for plant roots to get both air and water, allowing rain and irrigation water to infiltrate the soil instead of rapidly washing off its surface, and allowing soils to drain and restore air.

Moisture is held in pores between soil particles and is available for drainage and plant growth based on the pore size [2]. Water adheres to soil particles and crevices between particles in varying magnitudes of adhesive force. Water molecules adjacent to the particle (hygroscopic water) are held so tightly that very high pressures are needed to remove them. At increasing distances from the particle, the water molecules are held less tightly but not so

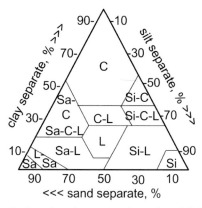

Figure 10.1 Soil texture triangle chart, based on proportions of sand (Sa), silt (Si), and clay (C) in the soil; L=loam [7].

loose that they drain with gravity. Finally, water at greater distances drains with the low force of gravity ("gravity water," about 0.5 atmospheres of pressure). Water retained in the soil after this gravity water has drained out is referred to as "field capacity."

Plants grow by removing water from soil between about 0.5 and 6 atmospheres of pressure ("surface tension") although some plants survive but grow little at over 15 atmospheres. Pore size determines how much water is available under different surface tensions. Soils with large pores, such as sandy soils, can take a lot of water into the pore spaces following a rain; but this water is held so loosely that it drains rapidly with gravity, leaving little left at field capacity for plant growth. Plants growing on such sandy soils commonly experience water shortages within one or several days following rain or irrigation.

Conversely, pores between clay particles are small and numerous. They can take in moderate amounts of water, but the water adheres tightly to the particles and does not drain away, nor is it available for plant growth. Soils with small pore spaces can remain saturated for a long time; no oxygen-containing air can get into the soil because the pores are full of water instead of air; and plant roots of most species suffer. Silt soils generally contain ideal pore sizes (not too large or small) so there is some air space and much water available for plant growth at field capacity and many days afterward. Soils occur in mixtures of textures (Figure 10.1).

Soils of pure clay or with enough clay to fill all pore spaces between any sand or silt that may be present are not good for plant growth unless they have a "structure" with suitable pore sizes. Roots, small animals, fungi, and frost penetrate clay and other soils and create holes; insert organic matter; bind soil particles together with roots, fungal hyphae, or sticky organic material; and so form larger pores than would otherwise be found in tightly packed clay soils. This arrangement is referred to as the soil's "structure" and can provide pores of appropriate size for good growth and moisture drainage. A soil's structure can be ruined by compacting it – grazing, driving heavy machinery, or otherwise manipulating it when wet. A ruined soil structure can be redeveloped by consciously plowing in organic matter or allowing regrowth of a forest, whose tree roots and soil flora and fauna can redevelop a structure over decades.

Sandy and gravelly soils are less prone to compaction and loss of structure. All soils can become "fluffed" – full of air – so that the pore spaces are too large to retain water. Natural fluffing can occur in windblown sand and silt, and the structure can later collapse. Fluffing is sometimes induced by plowing to reduce water evaporation and prevent weed growth.

Soils can be deficient in air in extremely rainy or poorly drained areas where the pore space is constantly refilled or elsewhere if clays and a poor structure keep the soil saturated for extended periods. This airless condition kills roots and prevents the plants from obtaining water, thus leading to the physiological equivalent of a drought. Plant species vary in their ability to endure different times of both water-saturated and dry soils (Chapter 13).

Depending on the rain intensity, available soil pore space, and surface slope, water can flow across the top of a soil; infiltrate the soil and be taken up by plants and evapotranspired into the atmosphere; or be allowed to percolate deeper to subsurface water where it remains, moves as subsurface flow into streams and rivers, or moves as groundwater to aquifers (Figure 17.1, Chapter 17).

Soils of different structures and textures will remain stable on slopes of different steepness – different "angles of repose." Loess soils generally remain stable at nearly vertical angles of repose (Figure 12.4a), while sands have quite flat stable angles of repose (Figures 12.3a, 14.7a). A slope is highly likely to fail (slump) if it exceeds its stable angle of repose.

Water and air move soil particles through suspension and "pushing, rolling and scraping" along the ground surface or river bed – a "bed load." In both methods, the size and volume of soil material moved is directly proportional to the velocity and volume of water or air (Lane's relationship) [8]. Clay particles on steep slopes are most easily eroded, and flat gravelly areas are difficult to erode. Soil held together by plant roots increases the effective size of the "particle" and makes erosion more difficult.

10.2.2 Variations in Soil Uses

Each soil texture and structure has advantages and disadvantages for different uses.

- Agriculture and forest crops such as peanuts and some pines are effectively grown on sandy soils, while soybeans and cottonwoods only grow well on moist, well-drained soils of finer texture.
- Moist, well-drained sandy clay loam soils are generally highly productive for agriculture. These soils need to be carefully tended to ensure they do not lose their structure, organic matter, and fertility.
- Most forest tree species grow fastest on the most productive soils; however, these soils are generally so valuable for agriculture that forests are cleared. Forests are generally grown commercially on slightly less productive soils (Chapter 28).

- Clay soils can have difficulty retaining their structure if plowed or grazed when wet and so are often poor for agriculture.
- Clay soils can be useful for dams, levees, and fish ponds if their structure is destroyed so they become compact and nonporous.
- Some clay soils can be used for making bricks, china, and other ceramics when the upper, organic layers are removed (Chapter 27).
- Forests are commonly harvested throughout the year. Harvesting is done on sandy and gravelly soils during wet seasons, and the more productive but compactible soils are operated on when dry and less prone to compaction.
- Sandy soils are preferred for golfing, horse riding, and similar recreation because the soils drain well and can be accessed soon after rains.
- Road construction is generally easier on sandy and gravelly soils; and gravel is often extracted to make roads and concrete. Landforms absent of sand and gravel have difficulty constructing roads or buildings unless the gravel and sand are transported from elsewhere.
- Cement is made largely from limestone, which is common in karst landforms but found in smaller deposits elsewhere as well.

10.3 Soil Chemical Composition and Development

Seventeen elements are needed for plant growth (Table 10.2). All but four – carbon, hydrogen, oxygen, and nitrogen – originate in the soil. Nitrogen originates in the air; but some can be found in developed soils, certain kinds of rocks, and organic matter (Table 8.1). Before other plants can use it, nitrogen often needs to be incorporated into soil in appropriate forms (Table 10.2) by a few plant groups (Legume family ["pulses" of Chapter 19] and *Alder* genus) or artificial fertilizers. Oxygen comes from the atmosphere and is used by plants in carbon dioxide and diatomic oxygen molecule forms.

Other elements come from the soil's parent material. Parent material is composed of rocks and minerals of various ionic compounds, covalent molecules, crystalloids, crystals, and other solids (Chapter 26). It consists primarily of oxygen, silicon, aluminum, with much smaller amounts of iron, calcium, magnesium, sodium, potassium, and other elements (Figure 26.2). Silicate (for example, quartz) and aluminum oxide (bauxite) rocks form the bulk, with carbonate rocks (limestone, gypsum, and marble) together forming lesser amounts (Chapter 26). Other elements, including the ones needed by plants (Table 10.2), are found in smaller amounts bonded within the parent material.

Parent material decomposes through both chemical weathering with rain and organic acids from roots and physical weathering such as frost cracking. Mineral aggregates weather as their weaker molecules break apart, fragmenting the rocks, and leaving intact harder molecules such as quartz (Figure 11.3b). Plants, microbes, arthropods, and small mammals grow within the fragments and further decompose (weather) them, creating soils (Figure 10.2) [3, 11, 12]. The weathering frees cations and anions that are then held in loose ionic bonds within the soil, taken up by plants, or leached into the groundwater. Elements in organic matter within the soil can become available to plants as the organic matter decomposes [2–5, 11]. Potassium, magnesium, calcium, phosphorus and other elements needed for plant growth leach through the soil with rainwater. In rainy climates, they may leach into the water table and beyond, making soils unfertile.

Soils commonly turn red as they age because of the dominating color of even small amounts of iron in the oxygenated condition (ferric ion); they can be blue when unoxygenated conditions turn the iron to its ferrous condition. Despite their similar red appearances (Figures 11.1, 11.3), temperate and tropical soils behave quite differently.

Silicate minerals decompose into clays that help arrest the leaching of mineral elements by loosely adhering them. Soil organic matter similarly arrests leaching minerals. The adhering ability of a soil is known as its "cation exchange capacity." The amount of clay increases as temperate soils age. Different parent materials and weathering patterns generate soils with different cation exchange capacities.

Different clays can expand when wet, reduce the pore space, and make some temperate soils saturated, prone to slumping, prone to losing their structure if plowed or grazed when wet, and prone to erosion as water washes away the clays. Some clays shrink dramatically when drying, creating problems with overlying roads and buildings.

Typical tropical conditions of high temperatures and rains dissolve the silicate clays and leave porous soils of very few elements but generally with less tendency to erode or become saturated. These soils are known as "laterites" and can form porous, hardened rocks or can remain soft (Figure 11.1). Their pore spaces can be eliminated with compaction, especially in surface soils. They are poor in elements, but can become more productive by using fertilizers and organic matter to increase their fertility and cation exchange capacity.

Elements can leach somewhat with rain or irrigation and accumulate at lower soil depths or in aquifers, creating such salty soils and aquifers that plant growth and human use are inhibited (Figure 17.4). (See "Goulburn–Broken Catchment" Case Study, Chapter 3.) Very old soils (Figure 11.1) tend to be poor in elements and sometimes have salty aquifers both because of the long period of leaching and because old landforms probably passed through tropical, lateritic-forming processes at some past time (Figures 10.4–5).

A soil's pH (relative acidity) is the result of the parent material's chemical composition, the quantity of cations within the soil, and the organic matter. Forest soils generally are acidic. Agriculture soils can be less acidic and often basic. Generally, broadleaf trees live in less acidic soils than do conifers.

Different parent materials contain more of some elements than others. Each element is taken into plants in only certain forms (Table 10.2), none of which are organic. The molecules can dissolve into the soil solution and be "sucked" (passively or actively) into the plant root with the soil water (Figure 10.2). Roots also pluck elements directly from the parent material [13].

Each year elements (Table 10.2) and organic matter containing carbohydrates (celluloses, sugars, starches) fall to the ground in plant litter or rainwater. They are taken up by soil microorganisms, which sometimes transport elements to the plants. Most

Figure 10.2 Pine tree showing roots penetrating and utilizing a gradient from well-developed soil downward to parent material and bedrock (Armenia).

microorganisms keep the elements for themselves, grow, and consume and respire the carbohydrates until none are left. Then, the microorganisms die and release the elements to the soil solution where they are taken up by plants, adhere to clay or organic particles, or wash into the subsurface water (Figure 17.1).

Elements can be unavailable to plants because they are bound in organic molecules in cold or wet soils where organic matter decomposition is slow (Chapter 12).

Microbes decompose organic matter extremely rapidly in hot climates with sufficient soil oxygen, making the organic cation exchange capacity low; there, elements not rapidly taken up by trees can leach out of the soil. Disturbances such as fires, forest harvesting, or agricultural production in hot climates or elsewhere in soils with low cation exchange capacities can leach elements from the soil. The disturbance usually increases direct sunlight and heats the soil, further increasing microbial decomposition of organic matter. It also removes living plants that take up elements before they can be leached and that add organic matter. Agriculture, grazing, and other activities that do not allow vegetation to regrow cause element losses. With soil element loss, plant growth can be poor unless fertilized or until further decomposition of the parent rock material replaces the elements, often many decades later.

Soil nitrogen is lost by volatilization in fires, leaching following disturbances, and harvesting of the leaves, fruits, and bark – the plant parts with highest nitrogen concentrations. Nitrogen is part of an element flush to the groundwater and streams following a disturbance; however, it is uncertain how much "flushed" nitrogen was in the soil or plants from before the disturbance and how much was newly added by nitrogen-fixing plants that invaded after the disturbance.

Bedrocks characteristic of different landforms can contain unusual concentrations of precious metals, crystals, and other minerals (Chapter 27). The concentrations are often created as the rocks form. Cracks of saturated water can lead to quartz and other crystal and crystalloid formations (Figure 11.3b; Chapter 27). Differential temperatures of melting and solidifying of bedrock concentrate minerals such as gold. Some landforms have concentrations of radioactive materials, with local areas being quite radioactive.

Despite variations, nearly all soils develop "horizons," or layers, parallel to the soil surface where various root and animal penetrations, leachings, and element accumulations occur. The horizons have specific characteristics and appearances [14]. Soils have been organized into hierarchical classification systems that are internationally standardized [15, 16].

10.4 Distribution of Landforms

There are patterns to landform locations based on the world's geomorphology. Bedrocks form, move, and disappear with geologic processes as parts of "geologic plates" (Figures 9.1, 10.3). Geologic plates consist of above- and below-water areas of very old rocks ("shields"), younger rocks formed from molten materials beneath ("igneous"), and rocks reformed from surface deposits hardened by pressure and heat ("metamorphic") [4]. The moving plates run into, over, and beneath each other; scrape past each other; and split apart [4, 5]. Instead of colliding, an ocean crust will commonly dip beneath the continent when they meet, creating a "subduction zone" along the continent and a volcanic mountain chain nearby within the continent. Active non-collision subduction zones currently exist near the Asian and American coasts of the Pacific Ocean, creating many active volcanoes (the "ring of fire," Figure 10.3).

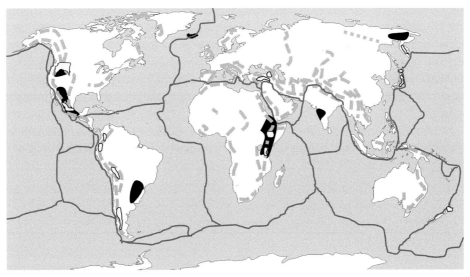

Figure 10.3 Predominant faults (thin dark lines), mountain areas (thick gray dashes), ash soils (gray with black borders), basalt flows (black) [3, 5, 20–22]. (A black and white version of this figure will appear in some formats. For the color version, please refer to the plate section. Color plate 2.)

Two continents can collide when the ocean between them is fully subducted; these can buckle and form mountains, with earthquakes but much less volcanic activity than during subductions because the rocks fold or gradually uplift and sink. Such activities are occurring in the Pyrenees, Alps, Carpathians, Pontus and Taurus, Caucasus, Zagros, Himalayas, Serra do Mar Brazil, and Great Dividing Range of eastern Australia.

Plates scraping past each other can also cause earthquakes, such as along the San Andreas fault of California.

The plates can also split apart, allowing hot rock from within the Earth to rise, cool, and form mountains, such as is in the middle of the Atlantic Ocean. The splitting can also create "rift" valleys such as the Rhine Valley in Germany, the Rio Grande Rift in south-western USA, the Baikal Rift of Russia, the submerged rift within the mid-Atlantic ridge, and the East African Rift – not to be confused with the Rif Mountains in northwestern Africa [3].

"Holes" can also occur in the plates, where the molten rock comes near the surface and forms hot springs and/or volcanoes. The basalt flows in eastern Oregon and Washington, USA about 16.5 million years ago were such a "geologic hotspot." The North American crust has now moved westward and that hot spot formerly in eastern Oregon and Washington erupted as a volcano many times before arriving at what is now Yellowstone National Park 2.1 million years ago. Since then, it has had three extremely large, explosive eruptions and is due for another (Chapter 12) [17].

Gravity, geologic uplift, wind, water, volcanos, or glaciers can move broken bedrock material. Eroded soil can slowly create new landforms where it collects as alluvial floodplains, coastal plains, sand dunes, or sea beds from water erosion, or as loess hills, ash-cap soils, or sand dunes from wind erosion. The sea deposits can slowly form sedimentary rocks and then metamorphic rocks as they compact, harden, and chemically alter. In aggregate, these changes maintain the large array of landforms [2–5]. The circulation of rock particles, other landform materials, and dead plants and animal in some cases, keeps landforms more capable of providing resources than if landforms were static.

Landforms have been classified in different ways [18, 19]. For this book, they will be divided into fifteen types (Figures 10.3–9). Five occur from direct weathering of the underlying bedrock:

1) weathered shield bedrock,
2) other recent igneous and very old metamorphic bedrock,
3) other old sedimentary and recent metamorphic bedrock (excluding karst),
4) large basalt flow bedrock,
5) karst landforms.

Seven landforms are caused by surface mineral layers transported and deposited from elsewhere, and so are different from the underlying bedrock:

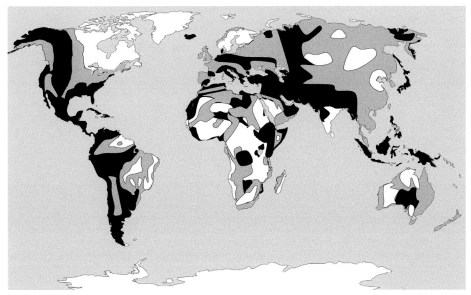

Figure 10.4 Predominant bedrock of different ages [2, 3, 20]. Shields (older than 500 million years) = white, Paleozoic and Mesozoic (500 million to 65 million years) = gray, Cenozoic (younger than 65 million years) = black.

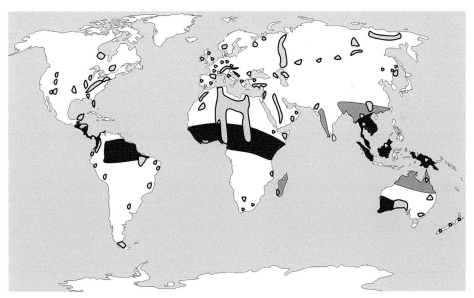

Figure 10.5 Predominant karst (light gray, heavy borders), relict lateritic soils (dark gray) and current laterites (black) [3, 11].

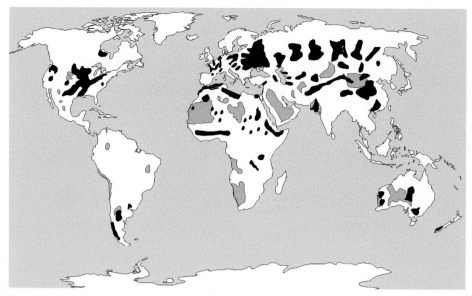

Figure 10.6 Loess deposits (black) and sand dunes (gray) [3, 5, 23–25].

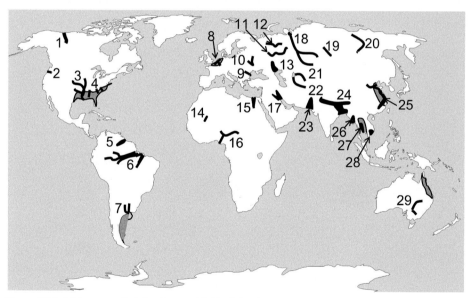

Figure 10.7 Major coastal plains (gray with black borders) [26, 27] and alluvial floodplains (black) [5, 12, 28] Table 10.3 is river legend.

6) volcanic ash and tuff,
7) sand dunes,
8) loess,
9) alluvial floodplains,

Table 10.3 *Legend for major alluvial floodplains of Figure 10.7*

1	Mackenzie River, Canada	16	Niger River, Nigeria
2	Columbia River, USA	17	Mesopotamia (Tigris and Euphrates Rivers), Iraq and Kuwait
3	Mississippi–Missouri–Arkansas Red Rivers, USA	18	Ob River, Russia
4	Southeastern rivers, USA	19	Jenesej (Yenisey) River, Russia
5	Orinoco River, Venezuela	20	Lena River, Russia
6	Amazon & Tocantins Rivers, Brazil	21	Syr Darya, Uzbekistan
7	Parana and Uruguay Rivers, Argentina and Uruguay	22	Amu Darya, Uzbekistan
8	Rhine, Weser, Ems Rivers, European Union	23	Indus River, Pakistan
9	Danube River, Romania	24	Ganga, Bhramaputra, Jamuna, Padma, and other rivers, India and Bangladesh
10	Dnieper River, Ukraine	25	Yellow (Huang) and Yangtze (Huang) Rivers, China
11	Severnaja River, Russia	26	Ayeyarwady River, Myanmar
12	Pecora River, Russia	27	Chao Phrayn and other Rivers, Thailand
13	Volga River, Russia	28	Mekong River, Cambodia and Viet Nam
14	Niger River, Sudan	29	Murray–Darling River Basin, Australia
15	Nile River, Egypt		

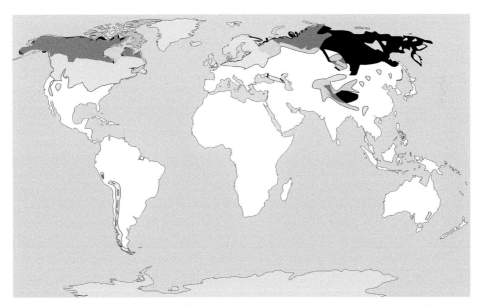

Figure 10.8 Recently glaciated areas (light gray) [3, 5, 24, 29], current permafrost (black) [5, 20–22, 30], and overlap (dark gray).

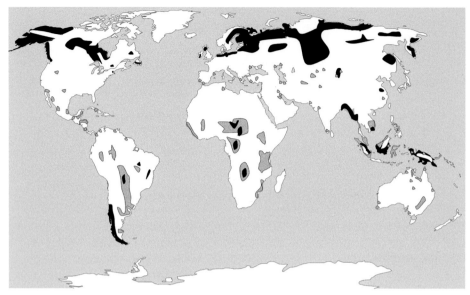

Figure 10.9 Major areas of bogs (black) and wetlands (gray) [31–33].

10) coastal plains,
11) glaciated areas,
12) bogs.

Three other landforms have unique features:

13) permafrost areas,
14) mountains,
15) wetlands.

   Globally, landforms are not consistently mapped, probably because various mapmakers are more knowledgeable about different places. The maps above are coarse amalgamations of different sources into an approximation of world landform locations.

   Asia and South America seem to have the most balanced landform distribution (Figure 10.10). Europe, Asia, and North America have the greatest proportion of mountains, with Africa and Australia having the least. Recently glaciated areas are primarily in Europe, Asia, and North America. Large shield (very old – Precambrian) areas can be found in North America, Africa, and Oceania – especially Australia. These are generally highly weathered, unproductive soils. Precambrian areas of South America, Africa, Australia, and Asia have a strong impact on resources because they lie in potentially productive climates and, unlike North America and Europe, have not been modified by glaciers. Europe contains the greatest proportion of loess and productive Mediterranean and temperate oceanic forest climates (Chapter 8); consequently, it is agriculturally productive except in the far northern and mountainous areas. Ash soils are found downwind from present or

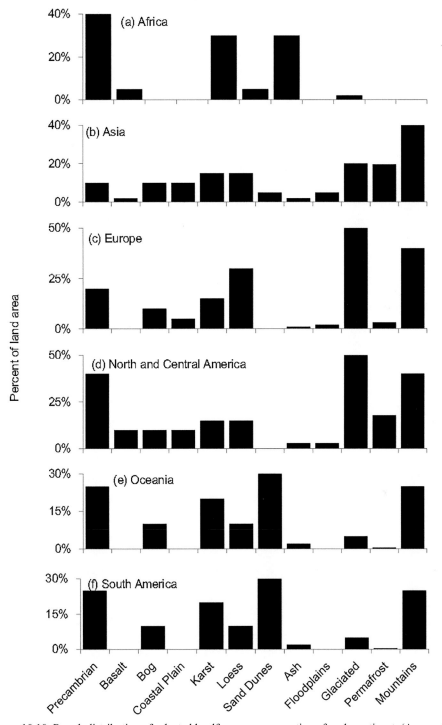

Figure 10.10 Rough distribution of selected landforms as a proportion of each continent. (Areas add up to more than 100% because transported soils cover bedrock in many places.) Estimated from Figures 10.3–9.

former volcanic areas, primarily along the "ring of fire" (Figure 10.3) in Asia and the Americas; ash soils are also found in Iceland and Asia Minor. Africa and Oceania (Australia) contain the greatest proportions of sand dunes. Karst landforms are most dominant in Africa and Oceania and least dominant in South America. Coastal Plains are present in all continents, but only in small amounts in Africa. Africa and Oceania also have the least productive soils for agriculture, with large amounts of low productivity Precambrian and sand dune formations and relatively little productive alluvial floodplains. Unlike Africa, Oceania (mainly Australia) has considerable loess; however, it is in dry areas and so is not too productive.

Bog formations usually contain peat and are common on cold or water-saturated soils – in formerly glaciated areas, coastal plains, and river floodplains and the islands of southeast Asia. Permafrost is found primarily in Asia and North America, with some in Europe and very small areas in South American and Australian mountains.

Conditions within a landform are not always uniform; but the variations in soil, topography, drainage, and other factors can be anticipated. The variations can often be traced to events that occurred during bedrock formation; redeposition; or subsequent disturbances, weathering patterns, or land use history [34, 35].

Soils generally accumulate in valleys by eroding, washing, creeping, or slumping from surrounding hills. Consequently, ridges generally contain shallow soils and exposed bedrock while valleys are more productive because of their deeper soils and water drained from the slopes. Exceptions are glaciated and alluvial areas. Glaciated valley bottoms tend to be infertile because they contain coarse sand and gravel. Alluvial floodplains are generally flat, but their lower areas (often less than a meter lower) commonly contain clay soils that are less productive because of little aeration.

Elevation differences within a landform create different climates; hills and valleys; and rockfalls, soil slumping, and snow avalanches.

Landforms can have small amounts of other landforms interspersed within them. Alluvial floodplains and coastal plains can contain peat bogs; and metamorphic soils can contain karst, volcanic, and loess areas. The variations can be locally mapped.

10.5 Managing Landforms

All landforms change, each in characteristic ways [3]. They often change slowly or infrequently and so appear static – and are commonly treated as such (Figure 1.3). River channels [36], shorelines [37, 38], city locations [39], and harbors [40, 41] occasionally change or disappear. These transitions are generally considered harmful because people are not prepared for them (Chapter 3); however, they sometimes provide future opportunities.

For example, abandonment of historic Ephesus (Figure 10.11) is sometimes considered an environmental tragedy. Its harbor was silted in by erosion caused by heavy farming and grazing upstream about 2,900 to 800 years ago [40]. Despite mitigating efforts, the city was eventually abandoned because of lack of harbor access and mosquitos in the silted harbor.

Figure 10.11 (a) Ancient Ephesus (circled) and vicinity (Turkey). Rich, alluvial soil where bay and harbor were silted is shaded. (b) Present, partially restored ruins of Ephesus center. Figure 10.11a courtesy of Google Earth.

The former harbor has now become highly productive alluvial farmland that probably provides more food at greater efficiency than the former uplands. Also, Ephesus' harbor would have been far too small to accommodate modern ships. Non-sustained management of Ephesus' harbor may have been appropriate and efforts to sustain it may have been wasted in retrospect.

Each landform has unique patterns of stability and change that can be anticipated and that offers opportunities to adjust the geologic system to provide more ecosystem services through infrastructures, including dams, levees, bridges, plowed soils, terraces, irrigation systems, tunnels, highways, and others [42]. Infrastructures are generally used to make landforms safer or more useful; and people have been adjusting landforms for millennia (Figure 18.7a). Unlike climate and weather systems, many benefits have come from modifying landform systems.

Landform features often appear unchanging until they reach a threshold and undergo a rapid and sometimes violent change (Figure 3.2b). Infrastructures such as grouting for sinkholes (karst), terraces for erosion (loess), levees and dams for floods (alluvial), and dune stabilizers can avoid or mitigate violent changes. Other infrastructures can mitigate landslides, snow avalanches, and GLOFs (Chapters 12, 19). Infrastructures that are poorly installed, maintained, replaced, or dismantled can create dangers – such as collapsing dams [43–45], collapsing bridges [46], collapsing agricultural terraces [47], sediment entrapment and eroding floodplains [48], failing levees [49], and eroding agriculture fields [50].

Infrastructures can increase people's wellbeing, such as irrigation reservoirs and ditches, dams, levees, and terraces providing more food per hectare – and thus saving energy as well as biodiversity by concentrating agriculture and not clearing more forests. Hydroelectric dams presently provide as much renewable energy to the world as all other renewable energy combined except nuclear power (Figure 25.6). Ditches that drained swamps or straightened river channels in alluvial floodplains reduced the danger of malaria [51]. Remote cattle drinking troughs can provide water for wildlife species and so extend their ranges. Reservoirs can create new wetlands and aquatic habitats even as they are destroying other areas. Highways, bridges, and tunnels enable people, food, and other things to move rapidly.

On the other hand, infrastructures can have negative effects (Chapter 19). Dams harm or destroy some habitats and species such as migrating fish. Dams and reservoirs can also displace villages in areas to be dammed. Bridges, tunnels, and highways can fragment habitats and enable human encroachment on biodiverse areas. Agriculture ditches and straightened river channels that allow rapid drainage can reduce the amount of water that percolates to an aquifer (Figure 17.1). New infrastructures often appear unattractive; over time, they sometimes become locally appreciated as historical relics (Figures 10.11, 18.7a).

People are now faced with several decisions relative to infrastructures:

• How are dangerous or harmful infrastructures made by past, poor decisions rectified?
• What is to be done with an old, declining infrastructure that will not last much longer?
• What new infrastructures should be added to give more people safety and well-being?

References

1. D. J. Easterbrook. *Surface Processes and Landforms*. (Pearson College Division, 1999).
2. D. Huddart, T. Stott. *Earth Environments: Present, Past, and Future*. (Wiley-Blackwell, 2012).
3. J. S. Monroe, R. Wicander. *Physical Geology*. (Brooks/Cole, 2001).
4. B. J. Skinner, S. C. Porter, J. Park. *Dynamic Earth*. (Wiley, 2004).
5. C. C. Plummer, D. McGeary, D. H. Carlson. *Physical Geology*, ninth edition (McGraw-Hill, 2003).
6. Soil Science Society of America. *Soil*, 2016 (Accessed March 24, 2016). Available from: www.soils.org/publications/soils-glossary.
7. US Natural Resource Conservation Service. *Guide to Texture by Feel* (US Department of Agriculture, 2015 [Accessed March 24, 2016]). Available from: www.nrcs.usda.gov/wps/portal/nrcs/detail/soils/edu/?cid=nrcs142p2_054311.
8. C. D. Oliver, B. C. Larson. *Forest Stand Dynamics*, update edition. (John Wiley, 1996).
9. F. Salisbury, C. Ross. *Plant Physiology*. (Wadsworth Pub. Co., 1992).
10. P. Barak. *Macronutrients and Micronutrients* (University of Wisconsin, 2003 [Accessed January 18, 2017]). Available from: http://soils.wisc.edu/facstaff/barak/soilscience326/listofel.htm.

11. J. R. Craig, D. J. Vaughan, B. J. Skinner. *Resources of the Earth: Origin, Use, and Environmental Impacts*, fourth edition. (Prentice Hall, 2011).
12. W. K. Hamblin, E. H. Christiansen. *The Earth's Dynamic Systems*, tenth edition. (Prentice Hall, 2001).
13. G. Voigt. Variation in Nutrient Uptake by Trees. In *Forest Fertilization Theory and Practice*. (Tennessee Valley Authority, 1968): pp. 20–7.
14. US Natural Resources Conservation Service. *Soil Education*. (U.S. Department of Agriculture, 2015 [Accessed March 24, 2016]). Available from: www.nrcs.usda.gov/wps/portal/nrcs/main/soils/edu/.
15. Soil Survey Staff. *Illustrated Guide to Soil Taxonomy, Version 2*. (USDA NRCS National Soil Survey Center, 2015).
16. UN FAO/UNESCO International Soil Reference and Information Center. *Soil Map of the World, Revised Legend*. (UN FAO, 1988).
17. Yellowstone Volcano Observatory. *Questions About Yellowstone Volcanic History*. (US Geological Survey, 2012 [Accessed March 21, 2016]). Available from: http://volcanoes.usgs.gov/volcanoes/yellowstone/yellowstone_sub_page_54.html.
18. J. A. Shimer. *Field Guide to Landforms in the United States*. (Macmillan, 1972).
19. World Landforms. *Types of Landforms and Definitions*. (WorldLandForms, 2015 [Accessed August 4, 2016]). Available from: http://worldlandforms.com/landforms/list-of-all-landforms/.
20. UN FAO. *Lecture Notes on the Major Soils of the World*. (UN FAO, 2015 [Accessed March 24, 2016]). Available from: www.fao.org/docrep/003/y1899e/y1899e04.htm#TopOfPage.
21. F. Niu, G. Cheng, Y. Niu, et al. A Naturally-Occurring "Cold Earth" Spot in Northern China. *Scientific Reports*. 2016;6.
22. Y. Shur, M. Jorgenson. Patterns of Permafrost Formation and Degradation in Relation to Climate and Ecosystems. *Permafrost and Periglacial Processes*. 2007;18(1):7–19.
23. G. T. Davidson. Where Were the Ice Age Glaciers? in *Climate History and Geology*. (University of Minnesota Digital Conservancy, 2010).
24. M. Maslin. The Climatic Rollercoaster, in Fagan B., editor, *The Complete Ice Age*. (Thames and Hudson, 2009): pp. 62–91.
25. US Natural Resources Conservation Service. *Global Soil Regions*. (US Department of Agriculture, Natural Resources Conservation Service, Soil Survey Division, 2005 [Accessed October 4, 2013]). Available from: soils.usda.gov/use/worldsoils.
26. M. H. Monroe. *Major Landform Regions of Australia*, 2011 (Accessed August 3, 2017). Available from: http://austhrutime.com/major_landform_regions.htm.
27. National Geographic Society. *Coastal Plains*, 2017 (Accessed March 25, 2017). Available from: www.nationalgeographic.org/encyclopedia/coastal-plain/.
28. World Landforms. *Floodplain Landforms*, 2015 (Accessed December 4, 2016). Available from: http://worldlandforms.com/landforms/floodplain/.
29. S. Yafeng, C. Zhijiu, S. Zhen. *The Quaternary Glaciations and Environmental Variations in China*. (Heibei Science and Technology Publishing House, 2005).
30. J. A. Heginbottom, J. Brown, O. Humlum, H. Svensson. Permafrost and Periglacial Environments, in Williams R. S. J., Gerrigno J. G., editors, *Satellite Image Atlas of Glaciers of the World*. USGS Professional Paper. 1386-A. (US Geological Survey, 2012).
31. C. Prigent, F. Papa, F. Aires, et al. Changes in Land Surface Water Dynamics since the 1990s and Relation to Population Pressure. *Geophysical Research Letters*. 2012;39(8).
32. World Energy Council. Peat. *World Energy Resources: 2013 Survey*. 2013.
33. Ramsar. Climate Change and Wetlands: Impacts, Adaptation, and Mitigation. 2002. (Ramsar COP* DOC.11 [AccessedApril 26, 2016]). Available from: http://archive.ramsar.org/cda/es/ramsar-documents-standing-ramsar-cop8-doc-11/main/ramsar/1-31-41%5E17764_4000_2__.
34. J. Hjort, J. Gordon, M. Gray, M. Hunter Jr. Valuing the Stage: Why Geodiversity Matters. *Conservation Biology DOI*. 2015;10.

35. J. J. Lawler, D. D. Ackerly, C. M. Albano, et al. The Theory Behind, and the Challenges of, Conserving Nature's Stage in a Time of Rapid Change. *Conservation Biology.* 2015;29 (3):618–29.

36. M. Guccione, W. Prior, E. Rutledge. *South-Central Section of the Geological Society of America (Centennial Field Guide).* (Geological Society of America, 1988).

37. R. A. Morton. *National Assessment of Shoreline Change: Part 1: Historical Shoreline Changes and Associated Coastal Land Loss Along the Us Gulf of Mexico.* (DIANE Publishing, 2008).

38. O. E. Frihy. Nile Delta Shoreline Changes: Aerial Photographic Study of a 28-Year Period. *Journal of Coastal Research.* 1988:597–606.

39. E. De Carolis, G. Patricelli. *Vesuvius, Ad 79: The Destruction of Pompeii and Herculaneum.* (Getty Publications, 2003).

40. H. Delile, J. Blichert-Toft, J.-P. Goiran, et al. Demise of a Harbor: A Geochemical Chronicle from Ephesus. *Journal of Archaeological Science.* 2015;53:202–13.

41. J. McPhee. *The Control of Nature.* (Farrar, Strauss, and Giroux, 1989).

42. H. Petroski. *The Road Taken: The History and Future of America's Infrastructure.* (Bloomsbury, 2016).

43. W. F. Johnson. *History of the Johnstown Flood: Including All the Fearful Record, the Breaking of the South Fork Dam, the Sweeping out of the Conemaugh Valley, the Overthrow of Johnstown. . .* (Edgewood Publishing Company, 1889).

44. N. Adamo, N. Al-Ansari. Mosul Dam Full Story: Safety Evaluations of Mosul Dam. *Journal of Earth Sciences and Geotechnical Engineering.* 2016;6(3):185–212.

45. N. Adamo, N. Al-Ansari. Mosul Dam Full Story: What If the Dam Fails? *Journal of Earth Sciences and Geotechnical Engineering.* 2016;6(3):245–69.

46. A. Larsen. Aerodynamics of the Tacoma Narrows Bridge – 60 Years Later. *Structural Engineering International.* 2000;10(4):243–8.

47. X. Xiang-zhou, Z. Hong-wu, Z. Ouyang. Development of Check-Dam Systems in Gullies on the Loess Plateau, China. *Environmental Science & Policy.* 2004;7(2):79–86.

48. D. J. Stanley. Nile Delta: Extreme Case of Sediment Entrapment on a Delta Plain and Consequent Coastal Land Loss. *Marine Geology.* 1996;129(3):189–95.

49. G. Sills, N. Vroman, R. Wahl, N. Schwanz. Overview of New Orleans Levee Failures: Lessons Learned and Their Impact on National Levee Design and Assessment. *Journal of Geotechnical and Geoenvironmental Engineering.* 2008;134(5):556–65.

50. D. Pimentel, C. Harvey, P. Resosudarmo, et al. Environmental and Economic Costs of Soil Erosion and Conservation Benefits. *Science-AAAS-Weekly Paper Edition.* 1995;267 (5201):1117–22.

51. J. Keiser, B. H. Singer, J. Utzinger. Reducing the Burden of Malaria in Different Eco-Epidemiological Settings with Environmental Management: A Systematic Review. *The Lancet infectious diseases.* 2005;5(11):695–708.

11

Bedrock Landforms

11.1 Weathered Shield Bedrock

The oldest above-water landforms of the Earth's crust are Precambrian landforms, or "shields," and consist of rocks formed more than 500 million years before present (MYBP), as well as their weathered derivatives (Figure 10.4) [1, 2]. Shields in Canada and Scandinavia were recently covered with glaciers and have features of this overlying landform. Some others are buried beneath volcanic ash, sand dunes, or other deposits and take on those characteristics.

Where not covered by other deposits or moved by glaciers, surface layers of shield bedrock have slowly decomposed to soil. Weathering patterns are characteristic of the past and present climates that these bedrocks have passed through (Figures 9.1, 10.5).

The long period of weathering has made these soils quite deep – 6 meters or more before bedrock is reached. Much of the material has decomposed and recomposed to red soils or rocks (Figure 11.1).

Shields are not prone to earthquakes because faults do not run through them. There are generally no underground cisterns or caves in shield formation and no aquifers except for water in disconnected cracks in the bedrock (Chapter 17).

Any mountains in shield landforms are very old and weathered, with little remaining relief. Precious metals and fossil fuel deposits are not generally found here. Gravel, rocks, and similar building materials are generally absent except far below the surface. Shield landforms generally are not valuable for agriculture because of their low element content, and so support relatively low-value, extensive uses such as forestry or grazing unless fertilizers and cation exchange additives are used. These landforms have little earthquake or volcanic activity, except perhaps in geologic hot spots.

11.2 Recent Igneous and Old Metamorphic Bedrock

Younger, but still quite old, stable bedrock platforms also have formed, often appended to the shields during plate tectonic processes (Figure 9.1). These bedrocks can be igneous such as granites or very old metamorphic rocks (Figures 10.4 and 11.2a) [2–5] – between

Figure 11.1 (a) Deeply weathered, red laterite soils in a shield landform (excavation for a house), Ghana, Africa. Courtesy of Ms. Dora Cudjoe. (b) Fence of weathered, red rocks taken from beneath deeply laterized soils, southwestern India. (A black and white version of this figure will appear in some formats. For the color version, please refer to the plate section. Color plate 4.)

Figure 11.2 (a) Old exposed metamorphic bedrock forming Cumberland Plateau, Tennessee, USA. about 350 MYBP. (b) Exposed metamorphic fold, Tibet, People's Republic of China, about 35 MYBP.

500 MYBP and 65 MYBP (Paleozoic and Mesozoic eras). Old metamorphic bedrock can have characteristics in common with sedimentary and recent metamorphic bedrock landforms.

These landforms have a bedrock base whose surface has been slowly decomposing and recomposing to clay soil (Figure 11.3) [5]. These soils can be old, red, and more than 3 meters deep in places. Long-ago cooling and shrinking of the igneous bedrock and

Figure 11.3 (a) Decomposing granite bedrocks creating several meters of red clay soil; granite bedrock in center. (b) Two large, parallel quartz seams running diagonally from lower left. Other rock has decomposed to soil. Both from western South Carolina, USA. (A black and white version of this figure will appear in some formats. For the color version, please refer to the plate section. Color plate 5.)

metamorphosing of other bedrock created cracks and seams that slowly fill with enduring quartz (silicate crystal-like compounds [Figures 11.3b, 26.3]) and valuable minerals. Erosion can expose bedrock to the surface.

Soils formed from igneous bedrock in temperate regions can be quite deep with a lot of clay (Figures 11.3a–b); like shields, water commonly flows on the surface or through the soil, but not through the bedrock except in cracks. Soils formed from metamorphic bedrock can also have characteristics of its parent material, with sandstone areas still retaining quartz. Water can flow along on the surface, through the soil, and through seams or porous rocks to underground caves, cisterns, and aquifers in metamorphic areas.

Some igneous and old metamorphic landforms are not as infertile as shields because there has been less time for leaching. Compaction and erosion can be controlled with appropriate practices, so agriculture, grazing, and forest activities can be sustained.

The soils are not necessarily level. Rugged mountains occur in these landforms (Figure 11.2a). The high clay content makes erosion and soil slumping common, although the stable angle of repose is sometimes quite steep (Figure 11.3b). Gravel is commonly available in streams and rivers running through these landforms; and bedrock can be quarried from exposures. Where the soils are not deep, skyscrapers and other large buildings can be anchored to the bedrock. Many clay soils can be used as part of the fill for making roads.

Earthquake and volcanic activity depend on the location of this landform.

11.3 Recent Metamorphic and Old Sedimentary Bedrock

Younger sedimentary/metamorphic landforms (less than 65 MYBP, Cenozoic era) are common in mountainous areas (Figures 10.4, 11.2, 11.4). Karst landforms will be discussed separately.

Figure 11.4 Sloping metamorphic bedrock showing different soils on top and water seeping within some layers. Photo taken in Northeastern People's Republic of China.

A continuum of solidification from sedimentary to metamorphic rocks created by gravity, pressure, and heat when submerged beneath other rocks [2–5] exists. Sedimentary/metamorphic formations commonly formed on coastal plains that varied between being submerged and above water, allowing different sediment types to accumulate in generally broad, parallel layers. Each layer can be between a few centimeters to many meters thick and can be sandstones, siltstones, claystones, limestone formed from coral and marl beds (predominant in karst areas), coal formed from peat bogs, and others. Each layer exists in varying degrees of purity or mixture [3].

The rock layers weather and decompose at different rates, causing irregular features on the landscape such as a cave or overhang where a resistant rock lies atop an easily eroded one. The stable angle of repose varies among layers and rock types.

With uplifting, the rocks sometimes became folded, split, turned upside down, overlapped, tilted in various directions, and otherwise misaligned (Figure 11.2b). The topography can be rugged or gentle. For example, the Himalaya, Taurus, Alps, Cumberland, and Allegheny Mountains are quite rugged. The Shenandoah Valley, northeastern China, southwestern United States, southeastern Australia, and parts of the North American Great Plains consist of this landform but are topographically quite gentle; even the gentle areas can have abrupt cliffs, valleys, and ridges.

Water drainage in sedimentary/metamorphic landforms occur in various ways. Overland flows create streams and rivers. These often follow the folds and layers of rocks, creating rectangular, parallel, and trellis drainage patterns [6]. Water can disappear into, and reappear from, caves created in some layers. When rock layers lie at a slope, water can seep into porous bedrocks on the uphill side and emerge on the downhill side, creating dry and wet sides to hills and different vegetation and land uses (Figure 11.4). Water accumulating in a porous layer above a less porous one can both flow laterally and saturate higher layers, causing slipping at the layers. The overtopped porous layer can also form a "confined aquifer" (Chapter 17). Agriculture productivity of these landforms depends on which types of bedrock are at the surface. Sandstone at the soil surface would be moderately poor for growing crops; a limestone surface would be quite productive. Coal or oil at the surface can create toxic conditions unfavorable to vegetation growth. This coal or oil – and gas moving upward – can catch fire from natural or human causes, creating very long-lasting fires. Rocks with unusual chemical composition such as serpentine can appear at the surface and create distinctive colors as well as habitats for plants specifically adapted to survive in these soils.

Sedimentary/metamorphic landforms provide a variety of uses, depending on both the distribution and accessibility of different types of rock layers and the ruggedness of the topography. Rugged topography often results in "talus slopes," which are slopes covered with boulders that break off and roll down when a less resistant layer beneath erodes. Mechanized agriculture and, to a lesser degree, forestry are difficult because of the slopes' steepness and the boulders' presence with propensities to roll. Road building is difficult and dangerous because of these boulders and the possibility of suddenly uncovering caves beneath the work area. Slipping at the layers described earlier can be exacerbated when the rock face is cut at a roadside. Shearing of large sections of rock can also occur as water and seasonal ice accumulate between layers.

Building construction can sometimes be problematic for the reasons above. The possibility that a cave or cliff will collapse beneath a building is also a risk. Where underground mining for coal or other minerals creates shafts, crosscuts and other subsurface openings in addition to caves, the risk increases.

The landform can be useful where its topography is gentle and appropriate rock and soil layers are on the surface. Mixtures of limestone and siltstone on the surface that have decomposed/recomposed into soils can be quite fertile. The many elements in the calcareous rocks are continuously released to the soil, and these soils can be farmed for many years with relatively little concern for element depletion. The lands can also be farmed, grazed, or less intensively managed where the surface contains less fertile soils derived from sandstone or siltstone. Coal layers within the landform can be mined by shaft or "strip" mining. Oil sometimes emanates from the landform. Some of these oils have been valued for centuries for medicines and oil lamp fuels and are now valued by the petroleum industry.

Minerals are sometimes found in certain layers, and the redistribution of minerals during the metamorphic process can lead to valuable stone, crystal, and metal concentrations and seams (Chapter 26).

This landform is especially dangerous near geologic faults, since rocks tumbling down talus slopes as well as ground surfaces collapsing into caverns can be triggered by earthquakes. Minor earthquakes can reroute underground water flows and leave wells dry. Even far away from major fault lines, earthquakes can occur in metamorphic and sedimentary rocks as caverns collapse, setting off a chain reaction of other caverns also collapsing.

Figure 11.5 (a) Recent basalt flows covering parts of Iceland. (Highway shows scale.) (b) Basalt flows and thin soils in the Western Ghats, India. (c). Basalt columns and thin soils in eastern Washington, USA.

Table 11.1 *Ages of some basalt flows shown in Figure 10.3 [9, 10]*

Location	Time of flow (MYBP)	Location	Time of flow (MYBP)
South America	100	Northwestern North America	12–17
India, Deccan Traps	60–68	Iceland	16–18
Africa	30	Siberia	250

11.4 Large Basalt Flow Bedrock

Basalt is formed when molten rock within the Earth flows to the surface and cools rapidly (Figure 11.5) [2–5]. Unlike granite, which separates into large crystals of intermixed minerals in its rocks because it cools slowly far beneath the Earth's surface, basalt cools rapidly and so has very small crystals in its rock [3] and appears uniformly black in color.

Repeated basalt flows from geologic hot spots form areas with layers of flowed basalt (Figures 11.5a–b) covering tens of thousands of square kilometers in some places (Figure 10.3). A single layer can contain 50 cubic kilometers of basalt.

Large basalt flows may be related to extraterrestrial impacts [7, 8]. The flows range from several decades old in Iceland to over 100 MYBP in South America (Table 11.1).

Some basalt flow landforms can be distorted to different angles while other areas remain parallel to the Earth's surface. Characteristics of the landform also depend on the amount of weathering of the basalt, which depends on its age, slope position, and climate. Even in old basalt flows, the thin soils on convex slopes support little vegetation and have few uses. One has difficulty even digging a hole, and fence posts and orchard trees are commonly supported by surrounding them with piles of rocks instead of digging holes (Figure 11.6).

Basalt that flows onto dry land creates distinct vertical columns as it cools and contracts beginning at its surface ("column basalt"; Figure 11.5c). Basalt that flows underwater creates "pillow shapes" as the molten, billowing basalt surface rapidly cools, breaks open, and molten basalt within billows farther. These appearances are maintained for many millions of years. Basalt decomposes and becomes quite a fine-textured (often clay) soil. Where this soil accumulates, such as in concave slopes, valleys, and alluvial floodplains, the area can be used for grazing and agriculture [11, 12].

Older basalt areas such as in India and South America have larger areas of productive soil because the rocks are more decomposed. Water commonly flows through the soil and on the surface. Unless affected by other forces such as uplifting (mountain building), the flow creates a dendritic pattern to streams and rivers, or a parallel pattern if several basalt flows tilt and create water channels. Groundwater is found in cracks in basalt that are generally not connected, so there are not large aquifers but some groundwater is available for local wells.

With suitable climate, these formations can support forests or grasslands. Younger basalt landforms and ridges in older landforms are relatively accessible in wet seasons, and so can be advantageous for timber harvest even though growth may be slow. Basalt landforms may be good for building construction where the foundation can penetrate to bedrock. Road

Figure 11.6 Rock crib used to hold up fence post in basalt landform, eastern Oregon, USA. Courtesy of Mr. Charles Bettigole.

construction is generally not difficult in level basalt. Talus slopes can be used for stones and gravels for road building.

Caves are occasionally found between layers of basalt, and intricate buildings have been carved in basalt in India and in smaller basalt formations in Armenia.

Few fossil fuels or precious metals are found within basalt; however, gas and oil from older geologic deposits can be trapped beneath.

Basalt formations are generally quite stable relative to earthquakes, but are susceptible to new basalt flows, volcanic activities, flooding, debris avalanches, and other disturbances depending on their climate and topographic positions.

11.5 Karst Landforms

Karst landforms are a special case of sedimentary and metamorphic landforms described earlier; however, they contain a high degree of carbonate rocks, such as limestones and marble (Figures 11.7–8) [2–5, 13]. These rocks originate from coral reefs, marl beds, and similar sources that are uplifted (Figure 11.7). This landform is found in many parts of the world where a submerged coastal plain has been uplifted (Figure 10.5).

These rocks consist of calcium carbonate and other cations bonded with carbonate as well as impurities. Weak acids such as those produced by tree roots or rain water dissolve the rocks, producing carbon dioxide, which moves into the atmosphere, and a basic soil solution containing cations of calcium and other elements commonly needed by plants.

Karst soils can be very fertile where they still contain dissolving carbonate rocks that release cations [14]. Most agriculture crops grow in the mildly acidic or slightly basic soils created by these rocks. In areas of heavy rainfall, warm temperatures, and long periods of farming, the carbonate rocks may have disappeared from agriculture fields and the cations leached out.

Figure 11.7 Karst landform with (a) limestone rocks and (b) Cedars of Lebanon, Taurus Mountains, Turkey.

Karst landscapes where slopes are gentle and a lot of soil has formed above the bedrock can be very useful for mechanized agriculture. The residual soil can erode and does not have a very steep, stable angle of repose.

Other karst areas contain many rocks and boulders and so are unsuitable for mechanized agricultural operations (Figure 11.7). The soils here are often deep and rich in crevices between the boulders, and orcharding, grazing, and forestry can be practiced.

Carbonate rocks do not dissolve uniformly within a formation. Consequently, dissolution creates caves, underground caverns, cisterns, interesting patterns in rocks, and areas of soil interspersed with rocks (Figure 11.8) [5].

Water from karst areas generally contains dissolved cations and has a distinctive taste. Drainage patterns appear irregular, with rivers appearing from or vanishing into underground caves [3]. On a larger scale, the drainage pattern is contorted or radial [6]. The intricate underground crevices where water flows can change with dissolution and deposition of carbonate compounds. Consequently, it is not unusual for water flow patterns to change for little perceptible reason. Such changes can create problems of wells suddenly drying.

Many karst landforms are on geologic plate boundaries and consequently are subject to earthquakes induced by plate collisions and fault slippages, which can exacerbate problems of caves collapsing common to most metaphoric areas, described earlier.

Figure 11.8 Features of karst landforms: (a) Natural caves, often with underground lakes (Anamur–Mersin–Mediterranean, Turkey). (b) Sinkholes (Asia Minor, Turkey). (c) Caves (Crimea). 11.8a courtesy of Professor Melih Boydak.

Karst landforms can be associated with other sedimentary deposits; and coal, sand, and precious metals can sometimes be found. Mining lime for fertilizer and cement is common, as is mining of marble in carbonate rocks of appropriate quality.

Building and road construction is difficult in karst topography, not only for reasons described earlier for metamorphic landforms, but also because the rock dissolves easily. Carbonate rock can be used for gravel, although it is not ideal. Large sinkholes can suddenly and unexpectedly appear as cave roofs collapse from dissolution of the rock – as occurs in Florida, USA, with fatal results. On the other hand, some caves have been used for thousands of years and appear quite stable.

11.6 Mountains

Mountains are described as a distinct landform because their relief and steepness give them common properties that are different from non-mountainous areas. They contain other landforms and so have these landforms' features as well as those unique to mountains [2–5].

Most mountains are grouped into "belts" whose lengths are long compared to their widths [3].

Mountains are generally found in places of current or past geologic uplift at plate boundaries (Figure 10.3) [5]. Volcanic mountains can form quite rapidly, rising over hundreds of meters within a few decades (Figure 11.9); however, mountains caused by buckling of parent material usually form more slowly. Recently formed – and still forming – mountains can be quite high and rugged compared to very old ones.

Several features distinguish mountainous areas. Most are constantly uplifting while their valleys are being eroded by streams and rivers. Consequently, their slopes are becoming continuously steeper and a slope failure occurs. Inevitably, most mountain slopes will exceed their stable angle of repose and collapse at some time, although the long times between these collapses make mountains seem stable. The collapses can be anticipated.

The ranges of elevations, different aspects and inclinations of hillslopes creating various amounts of direct sunlight and warmth, and the differences in rain between aspects create a great diversity of climates within short distances [15, 16] (Chapter 13). The same vegetation and fauna may be found on all sides of a mountain, but often at lower elevations on the cool sides. During climate changes, species commonly move up or down mountains to a suitable climate, thus retaining a diversity of species of many plant communities in a small area. During climate warming, individuals of a species can move upward to many separated mountain peaks and thus become separated, such as in the "sky islands" of Arizona [17]. These separations can lead to genetic drift between populations and possibly to speciation (species pump, Chapter 13).

The same variety of climates in a short vertical distance has allowed efficient grazing systems, where herders and beekeepers move their animals up and down mountains and to

Figure 11.9 Volcanic mountain in Hokkaido, Japan, that began in about 1940 and rose to most of its present height in the first decade. Photo taken in 1995.

warm and cool sides as seasons change. People in mountainous areas have also responded to climate changes such as the Little Ice Age by moving vineyards and other crops to warmer places at lower elevations.

Valleys isolated by steep ridges have allowed many plants, animals, and human cultures to become isolated and protected within a small area. These isolations further amplify the species pump for plants and animals (Chapter 13) and maintain a diversity of cultures in human populations in such areas as the Smokey (USA), Balkan, Caucasus, Himalaya, and Andes Mountains. These isolations can also lead to human genetic inbreeding.

Mountains have commonly been used as places of refuge by people, flora, and fauna. Even though large-scale agriculture is difficult, the ready accessibility to irrigation water, the productive soils in valleys (except in glaciated areas), and the ability to terrace can make agriculture feasible. At the same time, the proximity of diverse climates makes grazing feasible and allows crop failure at one elevation to be mitigated by sending food from another elevation. Mountains are increasingly being used for recreation – hiking, snow skiing, rock climbing, and sightseeing.

Mountainous relief can create moisture (Figure 7.2) and even glacial conditions at low elevations that would not occur without the nearby tall mountains (Chapters 7 and 9). Water from mountains flows to low elevations where the water would not be found were the mountain not there. Lower slopes of mountains in non-glaciated areas are generally very productive for plant growth if water is sufficient.

Mountains are convenient for forest management because forestry does not compete with intensive agriculture for the land. The seasonal nature of forest operations and the cable systems designed for mountainous terrain make forest management feasible.

Mountainous areas are also convenient for water catchment. They often receive much precipitation, topographic boundaries for catchment area protection are quite clear, dams are easily constructed, and the low value and few uses of mountain lands mean they are not under great pressure to convert to other uses. Both snowpack and rainwater can be easily caught and stored.

Road building in mountains is generally difficult because of the many side cuts, tunnels, and bridges needed. Building construction is usually difficult because of the steep terrain; however, bedrock is generally quite near the surface, and so stable structures can be constructed. Various landform types within mountains can exacerbate or ease road and building construction. Obtaining rocks and gravels for construction are generally not difficult.

Mountainous areas have many hazards. They are susceptible to debris or rock avalanches, especially during earthquakes. They are also susceptible snow avalanches, slope failures, flash floods, fires that become especially hot as they move uphill, volcanoes, GLOFs (Glacial Lake Outburst Floods, Chapter 12), lahars (mudflows caused by melting of frozen soil when a volcano warms up), and "killer lakes" (tropical lakes that emit lethal carbon dioxide concentrations that flow down valleys when disturbed by landslides; Chapter 17) [18, 19].

References

1. P. Cattermole. *Building Planet Earth*. (Cambridge University Press, 2000).
2. B. J. Skinner, S. C. Porter, J. Park. *Dynamic Earth*. (Wiley, 2004).
3. C. C. Plummer, D. McGeary, D. H. Carlson. *Physical Geology*, ninth edition. (McGraw-Hill, 2003).
4. W. K. Hamblin, E. H. Christiansen. *The Earth's Dynamic Systems*. tenth edition. (Prentice Hall, 2001).
5. J. S. Monroe, R. Wicander. *Physical Geology*. (Brooks/Cole, 2001).
6. J. A. Shimer. *Field Guide to Landforms in the United States*. (Macmillan, 1972).
7. L. W. Alvarez. Experimental Evidence That an Asteroid Impact Led to the Extinction of Many Species 65 Million Years Ago. *Proceedings of the National Academy of Sciences*. 1983;80(2): 627–42.
8. K. O. Pope, S. L. D'Hondt, C. R. Marshall. Meteorite Impact and the Mass Extinction of Species at the Cretaceous/Tertiary Boundary. *Proceedings of the National Academy of Sciences*. 1998;95 (19):11028–9.
9. V. E. Courtillot, P. R. Renne. On the Ages of Flood Basalt Events. *Comptes Rendus Geoscience*. 2003;335(1):113–40.
10. M. K. Reichow, M. Pringle, A. Al'Mukhamedov, et al. The Timing and Extent of the Eruption of the Siberian Traps Large Igneous Province: Implications for the End-Permian Environmental Crisis. *Earth and Planetary Science Letters*. 2009;277(1):9–20.
11. A. Webb, A. Dowling. Characterization of Basaltic Clay Soils (Vertisols) from the Oxford Land System in Central Queensland. *Soil Research*. 1990;28(6):841–56.
12. B. Jenkins, D. Morand. A Comparison of Basaltic Soils and Associated Vegetation Patterns in Contrasting Climatic Environments, in Roach I., editor, *Regolith and Landscapes in Eastern Australia*. (CRC-LEME: Cooperative Research Center for Landscape Environments and Mineral Exploration, 2002).
13. D. Huddart, T. Stott. *Earth Environments: Present, Past, and Future*. (Wiley-Blackwell, 2012).
14. C. Weisbach, H. Tiessen, J. J. Jimenez-Osornio. Soil Fertility During Shifting Cultivation in the Tropical Karst Soils of Yucatan. *Agronomie*. 2002;22(3):253–63.
15. J. Hjort, J. Gordon, M. Gray, M. Hunter Jr. Valuing the Stage: Why Geodiversity Matters. *Conservation Biology DOI*. 2015;10.
16. J. J. Lawler, D. D. Ackerly, C. M. Albano, M. G. Anderson, S. Z. Dobrowski, J. L. Gill, et al. The Theory Behind, and the Challenges of, Conserving Nature's Stage in a Time of Rapid Change. *Conservation Biology*. 2015;29(3):618–29.
17. J. E. McCormack, H. Huang, L. L. Knowles, et al. *Encyclopedia of Islands*. 2009;4:841–3.
18. J. McPhee. *The Control of Nature*. (Farrar, Strauss, and Giroux, 1989).
19. C. D. Oliver, B. C. Larson. *Forest Stand Dynamics*, updated edition. (John Wiley, 1996).

12

Landforms of Transported Materials

12.1 Volcanic Ash and Tuff

Explosive volcanic eruptions throw out "tephra" – ash, tuff, and similar particles – which generally falls in the prevailing downwind direction, usually from near the boundaries of tectonic plates (Figure 10.3). The Cascade Mountains, USA, contain large areas of ash and tuff to their east, as do parts of eastern Asia Minor. Both areas have prevailing Westerly Winds. Deposits can be found in other directions if the volcanic blast emitted them horizontally (Mt. St. Helens, 1980 eruption) or the winds were blowing elsewhere at the time of eruption.

Ash is both pulverized bedrock and decimated pumice (Figure 12.1) [1–5]. This material is usually hurled high into the air, and falls at different distances from its source [5]. Ash layers can form soils several meters thick, and ash layers in the soil three cm thick have been found 500 km from its volcanic source. Large particles commonly fall close to the volcano. If the eruption is very strong, some fine ash is pushed into the stratosphere and encircles the Earth for several years afterward, increasing the solar reflectance and thus creating cooler temperatures [4].

Some particles fall to earth before becoming completely solidified. On the ground they stick together and form a hard or soft rock known as "tuff" [5]. Other particles are individually solidified as ash and act more like individual soil particles of different textures.

Where tuff hardens into firm rock, the landform has characteristics similar to other igneous or very old metamorphic bedrocks. Hard rocks can be used for building, and caves have been dug in the soft rock. Soft rock formed from tuff can create a landform that is easily weathered and eroded. Soils that are dominated by ash or soft tuff are known as "ash-cap" soils.

Ash deposits vary in particle size, shape, and chemical composition. Particles landing from a single volcanic eruption tend to be sorted by the wind and so are similar in size. As the wind slows, first heavy particles drop, followed by increasingly lighter ones, creating areas of uniform particle size (Lane's relation, Chapter 10). Repeated volcanic eruptions can create layers of ash of different textures (Figure 12.1b).

The ash can be coarse like sand or gravel, medium, or fine like clay. Because the ash is often liquid rock that exploded and solidified in midair, it can be brittle and contain hollow

Figure 12.1 Ash deposits (a) close to the volcano shortly after the 1980 eruption of Mt. St. Helens, USA, and (b) forming deep layers and soil, as seen in a road cut in the Andes Mountains, Ecuador.

pores. These pores allow coarse ash to hold more water at field capacity than sand or gravel of similar sizes do. On the other hand, the brittleness means that heavy machinery can crush and compact the ash, destroying both the soil structure and turning the ash into finer particles that hold less water.

The chemical composition of ash depends largely on the parent material.

Volcanic ash can increase the depth of soil material, giving greater soil moisture-holding capacity and root space above bedrock. Although productive, problems with crushing ash soils mean they may not be well suited for agriculture, with its frequent plowing, weeding, and harvesting. Where there is sufficient rainfall, ash-cap soils are productive for forest growth; however, they are commonly carried to the dry rain shadows on leeward sides of volcanic mountains by prevailing winds.

Ash-cap soils are not particularly well suited for construction, and stable building foundations are usually set in the bedrock beneath. Ash soils generally have a high stable angle of repose (Figure 12.1b).

Tuff areas have been used to excavate elaborate caves and hideouts for millennia in eastern Asia Minor (Cappadocia). Slightly harder tuff is still softer than granite or marble and has also been used for construction in eastern Asia Minor (for example, Armenia).

12.2 Sand Dunes

Sand dunes can be very large areas in deserts termed "sand seas," where sand is not held together by plant roots (Figures 10.6, 12.2, 14.7a). They can also be smaller areas along ocean, lake, or river shores. Sand dunes form and move if there is just enough initial relief for the sand to accumulate but not become permanently stationary. They move in waves with the wind. The winds are generally just strong enough to push the particles and lift them about a meter, with occasional gusts lifting them higher [2, 4, 5].

Sand dunes can form where glaciers retreat and leave large areas of sandy outwash, discussed later. They can also form by overgrazing (Figure 14.7c), poor agriculture, or a drying climate. Sand dunes can originate on exposed ocean, lake (Aral Sea), or river beaches that have enough active water movement to sort the particles, enough variation in

Figure 12.2 Sand dunes (advancing toward viewer) beginning to cover road and river valley, Gobi Desert, People's Republic of China.

water level for the sand to be above water and subject to winds, and enough wind to move the particles (Figure 14.7b) [1–5].

Winds can move sand dunes great distances over many decades, centuries, or millennia. They can cover hundreds of square kilometers with a moving front many kilometers wide at a rate of 10 kilometers per year (Figure 12.2). Individual dunes can move more rapidly and have been known to bury and kill people, with the bodies exposed years later when the wind blows the dune onward. Predictably, dunes move with the prevailing winds to the east in temperate zones.

Moving dunes create problems both because they fill and cover whatever is in their way with sand – roads, railroad tracks, and buildings (Figure 12.2) – and because they create a thick layer of sand with very low water-holding capacity. They can travel into moist areas and destroy vegetation (Figure 14.7b). They can cover river beds and bury or alter river courses.

Very few streams or rivers are found in sand dunes, stabilized or otherwise. Rainwater rapidly percolates through the sand before much can evaporate, and then the sand keeps more from evaporating as the water moves downward and then horizontally along underlying, less permeable layers [6, 7]. The water emerges as moist areas and even lakes ("oases") where the sand is thin above the impermeable layer. An oasis can last for decades but can also disappear if water channels along the impermeable layer shift or blowing dunes cover the lake.

Efforts to keep dunes from moving and causing more destruction generally involve erecting short barriers that accumulate the sand and planting perennial plants to form a root network and litter layer to hold the sand (Figures 12.3a, 14.7b) [8]. Plants that can survive on sand dunes are used for stabilizing. They are deep-rooted trees and shrubs or shallow-rooted grasses that only evapotranspire immediately after rains. Stabilizing sand dunes with plants can increase evapotranspiration and reduce the water percolating to the impermeable

(a)

Figure 12.3 (a) Short barriers stabilizing sand dunes. (Sand blows from right to left.) Gobi Desert, People's Republic of China. (b) Dunes stabilized along a river long ago by planting a pine forest, Ukraine.

layer, causing less to flow to the oases. Some countries are now appreciating and protecting their few remaining unstabilized dunes because so many have been altered through stabilization, construction, and sand mining [9].

Sand dunes stabilized with forests, shrubs, or grasslands create droughty soils with vegetation prone to fires.

The chemical composition of sand dunes is generally a product of the material comprising sand particles, usually quartz; they are extremely poor in elements where composed of quartz sand.

Sand dunes have a low angle of repose and easily slump. Their loose structure makes travel in vehicles or on foot difficult without roads. Sand dunes that are stabilized by vegetation and compacted are extremely accessible even shortly after rains. Some sands are also valuable for industrial purposes, including mixing with cement to make concrete (Chapter 25).

12.3 Loess

Loess landforms are created by silt particles transported by the wind (Figure 10.6). These particles move through the air for longer distances than the heavier sand [1–5]. Where the silt falls to the ground, it can cover other geologic formations in layers up to many meters thick (Figure 12.4a).

Loess originates from areas of exposed soil material such as deserts, forests, and grasslands that have lost their vegetation covering through natural disturbances such as a glacier's retreat exposing sand/silt bars in braided rivers and drying lakebeds. Excessive farming and grazing can also expose the soil and generate wind-blown loess, especially in dry places. Most large loess deposits in Europe, North America, and South America are probably from the bare outwash areas created by retreating glaciers (Figures 9.5a, 12.10b). Asia's glacial history is less certain, and loess deposits there may be from recent or much older glaciers or the dry lands of central Asia.

Loess areas of thousands of square kilometers are common quite far to the leeward side of the loess-forming areas. "Brown winds" common in Korea and elsewhere result from the prevailing winds carrying silt that originated far to the east [10].

Where rainfall and temperatures are appropriate, loess can form soils that are very productive for agriculture and forest growth [2]. The soils can be deep, uniform in texture, free of rocks and boulders that impede machinery, and of good water holding capacity. Some of the best agriculture areas in the world are on loess landforms, such as the Palouse Hills and Midwestern Corn Belt (USA), central Ukraine, and the Loess Hills of China. Loess soils have the chemical properties of its parent silt material.

Its windblown nature makes loess soil particles adhere to each other peculiarly. They can have a very high stable angle of repose (Figure 12.4a). Caves have commonly been dug in them in Vicksburg, Mississippi, USA, for protection; in Ukraine for monasteries; and in China for residences.

Where vegetation does not hold the loess because of a dry climate, deforestation, over-grazing, or farming, loess can be very easily eroded [11, 12]. Eroded loess silts streams and rivers and creates dust clouds. The erosion and high stable angle of repose can result in deep gullies with vertical sides (Figure 12.4a). Loess soils can have local areas of loosely structured (not compacted) particles that can collapse, especially during rains, and create sink holes. Appropriate management such as maintaining cover crops, growing appropriate crops [11], planting tree "shelterbelts" for windbreaks, creating terraces and berms (Figure 12.4), and installing concrete structures and trees to slow water movement down slopes can combat this water and wind erosion.

(a)

(b)

Figure 12.4 (a) Deep, eroded loess soils showing steep angle of repose and recent terraces and berms in People's Republic of China. (b) Deep loess soils in Ukraine stabilized by berms, perennial vegetation, and tree shelterbelts.

Loess areas are not especially good for road and building construction. Gravel and sand are generally absent and need to be brought from another landform; and sinkholes can erode or collapse building foundations and roads. Bricks can be made from loess.

Loess soils contain few minerals for mining and generally no fossil fuels, although they may be present in underlying formations.

Most other geologic hazards such as earthquakes that loess areas may be subject to are based on their underlying geologic formation.

12.4 Alluvial Floodplains

Alluvial floodplains are found along most rivers that flow through level topography, with larger rivers and flatter topography creating floodplains that can be tens of kilometers wide (Figure 10.7). Many smaller floodplains also exist with similar properties along streams and small rivers but may be only one or several meters wide.

Streams and rivers incise through soils of bedrock parent material, ash, loess, and other deposits. They wash this material into the river channel confined by higher ground and move it downstream through a combination of suspension and pushing the material as bed load. Where the river channel becomes more level and/or wide and unconfined down-stream, the water flows less rapidly; and the material is dropped (Figure 12.5) [1–5].

As materials drop and block the channel, the water behind it creates another channel in this or nearby sediments. The action eventually forms a quite flat area incised by relatively small present and previous water channels (Figure 12.6) [2]. During times when much water is flowing, the water can spread out and cover the large flat area, slowing down and dropping its suspended sediment. The large, flat areas – alluvial floodplains [4] – have similar characteristics throughout the world. The water especially deposits sediment as it reaches an ocean or sea because it slows and the salt causes the clay to flocculate – coalesce and sink. The shifting stream or river channel gives a characteristic "delta" shape and name to many floodplains at the mouths of oceans or seas (Figure 12.5b).

Alluvial soil patterns can be understood from deposition and redeposition processes. When a stream or river rises above its banks (floods), the flooding water spreads over a larger area and slows its velocity. The slower water can hold less sediment, and so the larger particles and the greatest particle volume – the sand and gravel – are deposited along the banks (Lane's relation, Chapter 10). Water flowing beyond the banks away from the river channel continuously slows and deposits particles of decreasing sizes. The process creates relatively high areas ("natural levees") of coarse sand along both sides of the channel. From these natural levees, the ground slopes gradually downward away from the river and contains continuously finer textures. At the lowest parts, away from the river or stream, water that did not drain back into the river remains in shallow pools. Clay particles suspended in the water either fall out of suspension or remain after the water evaporates. Fish that flowed into these temporary ponds can die, adding local fertility.

A distinctive soil texture gradient from sand, to silt, and then to clay is created. This gradient can extend over a short distance or many kilometers. Different species of trees, shrubs, and herbs grow along this gradient. On larger floodplains, different agricultural crops are grown on appropriate soil textures at different distances from the river.

The pattern above is complicated by the river meandering within the floodplain and reworking the soil (Figure 12.6) [4]. The fastest moving water in the stream or river channel shifts from one side to the other in a sinuous pattern as it flows. The fast-moving water on one side picks up sediment, causing the channel to undercut the bank. Concurrently, the slow-moving water on the other side of the channel deposits sediment where it flows, increasingly creating a meandering channel even if it were initially straight [2].

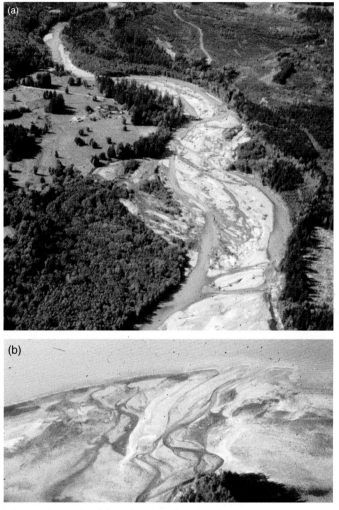

Figure 12.5 (a) Small alluvial floodplain showing changing river courses among deposited materials in Washington State, USA. (b) Small delta forming where river channel shifts around deposited materials in Alaska, USA.

The meandering river continuously removes earlier alluvial deposits with the fast current and redeposits them downstream in the slower currents. As river volume and velocity change, the sizes of particles deposited change; and a river bank on the slow (deposition) side can contain long, thin strips of soils of different textures that support different plant species. Consequently, floodplains consist of irregularly shaped areas of similar soil textures interspersed with areas of other textures.

Meandering rivers create increasingly larger arcs [4] that lengthen the river's or stream's course through the floodplain, further slowing the water and causing more sediment to be

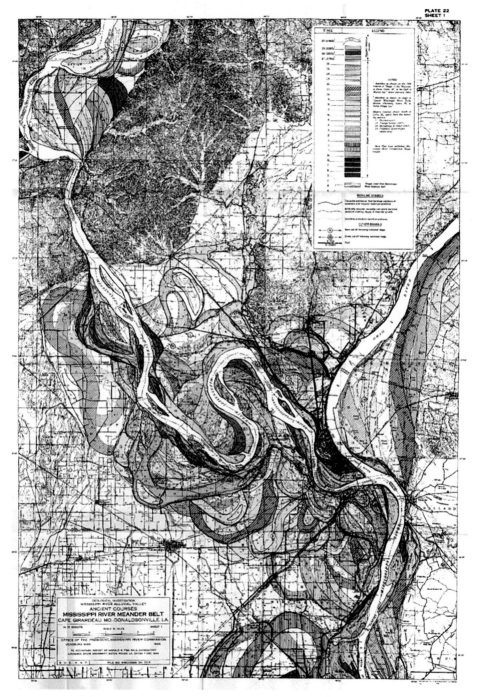

Figure 12.6 Mississippi River near Cairo, Illinois, USA, running through constrained channel (upper left), then meandering within broad floodplain, creating abandoned channels, "cutoffs," and "oxbow lakes." [13]

Figure 12.7 (a) Mississippi River floodplain forest. (b) Nearby area cleared for intensive, highly productive agriculture. Courtesy of Ms. Mariana Sarmiento.

deposited. These arcs become increasingly large until the river erodes the land between arcs, creating a passage ("cutoff") and isolated lakes known as "oxbow lakes." Oxbow lakes can become seasonally dry but fill during floods, commonly creating a lake floor rich in clay.

A lot of sediment moving down a floodplain, such as when glaciers melt, creates multiple channels. It exposes sand, silt, and clay elevated deposits ("bars") among the channels. A river with many channels and bars is known as a "braided" river (Figures 12.5a–b, 12.10b).

Lakes, cutoffs, natural levees, and channels are created and destroyed periodically in alluvial floodplains. Remnant features are sometimes found such as high terraces that are no longer flooded and loessial soils on the leeward sides of floodplains.

If a dynamic equilibrium exists in the amount of sediment flowing into and out of a floodplain, the floodplain will remain approximately stable in elevation and very flat (Figure 12.7). The excess sediment will be pushed out, eventually into a lake, sea, or ocean. There, it can slowly settle but also extend the land if there is a continental shelf. If water with reduced sediment flows into a floodplain, the sediment-poor, flowing water will pick up sediment from within the floodplain and carry it downstream, down-cutting the floodplain. This down-cutting creates high, remnant terraces.

Floodplain soils vary in productivity for agriculture depending on their soil texture and structure. Productive floodplains can grow about four to six times as much cotton per hectare as nearby soils from igneous formations. Even small floodplains of minor streams can be useful for agriculture. Very sandy or very clayey soils in floodplains are less useful for agriculture.

Alluvial floodplains also provide habitats for tree, shrub, and herb species rarely found elsewhere, leading to an interesting trade-off among values (Figure 12.7). On the one hand, concentrating agriculture fields in alluvial floodplains means that much larger areas than these fields can be converted back to forests or other vegetation on other landforms with an increase in overall biodiversity but no loss in food production. On the other hand, some

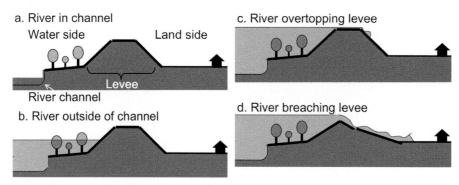

Figure 12.8 Cross section of a levee and its flood stages.

floodplain areas need to be kept in native vegetation if species dependent on floodplain habitats are to be maintained.

The annual flooding of alluvial landforms has created several problems and various solutions. When floodwaters overran river banks onto floodplains, they usually flooded large, flat areas with shallow water (Figure 12.7a). Deposited sediment often obscured property and field boundaries and deposited poor, sandy soils on top of previously productive fields of silt-loam. The flooding waters also commonly rerouted the river and washed away homes and roads. One solution has been to build high, artificial barriers ("artificial levees") parallel to the river's channel to prevent flooding water from washing onto the floodplain (Figure 12.8). Another solution has been to create upstream dams to prevent floods.

Artificial levees prevent the annual flooding; however, the confined floodwaters rise much higher than the previously unconfined floodwater and require extremely large, strong levees. They can also create disastrous floods if the levee breaks or is crested and erodes. In addition, sediment is no longer deposited on the floodplain and the river no longer meanders across it to rework the soil. Instead, the sediment is pushed farther downstream and the river is confined to a single channel. It is unclear if the absence of flooding and new soil in a protected former floodplain will eventually reduce the soil's productivity. Currently, fertilizers often help counteract negative effects.

Upstream dams and water diversions can minimize floods by regulating water flow. Until recently, dams were designed to stop all water, causing the calmed, confined water to drop its sediment within the reservoir (Figure 19.2). Water later released contains much less suspended sediment, so this clear water no longer deposits sediment in a roughly steady state as it picks up other sediment. Instead, the clear water from the reservoir only picks up sediment, undercutting the river channel and floodplain. The lack of new sediment may also cause coastal salt marshes to become submerged, especially as sea levels rise [14].

Another common problem is that people assume it is safe to build on floodplains. Consequently, they remove these very productive soils from the agriculture or biodiversity land base and replace them with buildings in jeopardy of being destroyed.

Floodplains are difficult for road and building construction. The silt and clay soils often expand and contract with wetting and drying, causing buildings and roads to tilt and buckle. Wide floodplains generally require rock and gravel to be transported from beyond it – often tens or hundreds of kilometers away. In earthquake-prone areas, construction on alluvial soils is particularly dangerous. Alluvial soils "shake like jello" during earthquakes and cause buildings to collapse much more readily than buildings anchored to bedrock.

12.5 Coastal Plains

Coastal Plains are relatively flat areas on the edges of continents where ocean or a combination of ocean and land-deposited sediments have built up (Figure 10.7), generally caused by a receding ocean or an uplifting land [1–5]. They can be quite narrow or stretch for 180 kilometers inland and 800 kilometers along the ocean shore, as in the case of the southeastern United States. Other large coastal plains are in southeastern South America, northern Europe (the "Low Countries"), eastern China, southeastern Asia, and northeastern Australia. Being largely ocean-deposited sediments, they generally display little topographic relief, although considerably more than alluvial floodplains. (Compare Figures 12.7 and 12.9.) For example, the coastal plain in the southeastern United States rises an average of less than one half-meter per kilometer for 180 kilometers inland.

Variations in coastal plain soils and slight relief reflect the area's local geomorphologic origin [2, 3, 5]. There are flat, calcareous areas generated by marl beds, flat clay areas that were lagoons, very sandy areas that were old beaches, and limestone deposits that were coral reefs. These coastal plains generally rise and submerge with plate tectonic and glacier-related changes in sea levels. Coastal plains can form the origin of sedimentary and metamorphic rocks.

Many rivers originate in coastal plains but are small. Because of the flat terrain and little erosion, they generally do not carry much sediment. Sometimes they can be a very dark color

Figure 12.9 Coastal plains showing variations in soil texture and drainage and slightly more relief than alluvial floodplains. (a) Rice paddies interspersed with drier lands in Thailand. (b) Forest operations on sandy soils in southeastern USA.

("black waters") caused by dissolved tannins and other organic material. Other rivers originate in mountains, hills, and other actively eroding areas farther inland and flow through coastal plains to the oceans. These rivers' sediments can build levees and floodplains that isolate them from the drainage of the surrounding coastal plain. Such rivers are sometimes referred to as "red rivers" or "yellow rivers" depending on the color of sediment they carry from uplands.

Coastal plain soils vary dramatically in texture. Texture differences combined with slight elevation differences can create poorly and well-drained areas (Figure 12.9). The lack of relief and lack of rocks, except for some weak calcareous rocks, make coastal plain areas of suitable drainage and soil texture used intensively for agriculture; and the diversity of soil textures allows growth of a diversity of crops. Poorly drained areas and excessively drained, sandy areas are generally left as forests and grassland or tree swamps. Some areas are so flat, low in elevation, and near the soil water table that water simply does not drain away. These areas can form bogs. Other areas remain saturated for much of the year and only become assessable during droughts.

The flat terrain allows rivers to be used for navigation, free of cataracts. Near the coast, the many rivers are often modified to form continuous canals for water transportation safe from ocean storms. Rivers near the ocean rise with fresh water when the ocean's high tide prevents the water from flowing out. The predictable fluctuations of these fresh "tidewater" rivers allow dependable irrigation of such crops as rice.

Road construction in coastal plains is made easy by the flat terrain, but difficult by the lack of drainage and weak calcareous rocks for gravel. Often extensive bridges are built over wet areas. The great distance below the surface to bedrock means buildings in coastal plains are not easily anchored, so are susceptible to winds and earthquakes. Calcareous layers in coastal plains can create sinkholes (karst, Chapter 11).

Coastal plains are highly susceptible to flooding caused by hurricanes and heavy rains as well as to tsunamis. They are also highly susceptible to droughts because the lower water table in the flat coastal plain makes large, contiguous areas uniformly dry. Then, they are also susceptible to fires.

Coastal plains allow much water to infiltrate into the different sedimentary layers, thus creating large aquifers.

12.6 Glaciated Areas

Continental and mountain glaciers occur periodically (Chapter 9), with the last major retreat occurring about 18 thousand to 12 thousand years ago [15]. They have smaller advances and retreats within the larger glacial periods. Glaciers leave distinctive landforms that can last for millions of years. The extent of the last glaciation is shown in Figures 9.5a and 10.8. Mountain glaciers also occur during glacial periods and some remnants of these still exist (Chapter 9).

An advancing glacier behaves somewhat like a giant bulldozer [1–5], but can be over one kilometer thick. It rides over hills and rivers pushing trees, other vegetation, rocks, and soil

Figure 12.10 (a) Glacial till, with rocks of various sizes visible, in Sweden following a fire that removed the forest litter. (b) Glacial outwash plain from mountain glacier in the Chugach Mountains, Alaska, USA.

in front of it; sometimes it rides over and compacts them [5]. Material is pushed up and to the sides and front of an advancing glacier. It also grinds some material into its ice.

As a glacier melts hundreds or thousands of years later [3, 4], it releases large quantities of water that wash much of the clay away; and the rock material that was ground into the ice then falls on top of the compacted material that was below the ice. The result on hills is material ("till") that is a random mixture of soil textures and rocks from parent materials from local sources as well as sources closer to the glacier's origin (Figure 12.10a) – but little clay. The material that was compacted beneath the glacier is known as "compact till," and the material that fell on top from within the glacier is "loose till" [2].

The large volume of water from a melting glacier carries suspended sediment with it. The water converges into streams and rivers in low-lying areas similar to floodplains, except the high sediment load makes the rivers braided. The high volume and velocity of water washes away the silt and clay, leaving only sand and gravel (Figure 12.10b). The tumbling of rocks and gravel against each other in the water makes them rounder and smoother. Variations in the velocity of water in the stream and river floodplains cause sands and gravels of the same sizes to fall out of suspension at the same place and time, creating small mosaics of similarly sized rocks, gravels, and sands. These water-washed, sorted glacial sediments are known as "outwash."

Continental glaciers spread over hundreds of kilometers away from mountains. Mountain glaciers sometimes follow valleys, leaving upper slopes and ridges unglaciated. Mountain glaciers push sand and rock material against the valley sides to form "lateral moraines." Both mountain and continental glaciers have a "terminal moraine" that marks where the glacier's advance stopped with a wall of rock and soil material it had pushed in front of it. Receding and re-advancing glaciers can move down valleys and form parallel lateral moraines and outwash along their sides, which become sand and gravel "kame terraces" on hillsides after the glacier melts [5].

A melting glacier can leave many features behind. Ice pieces many meters in diameter can be left in glacial till or outwash. When the ice melts, "potholes" are formed that can become lakes. Glaciers can change river courses or create lakes by damming the channel with ice or sediment. They can scour concentrations ("lodes") of gold, sending some gold into the glacial outwash. "Placer" gold mining that sorts the gold from the other outwash is common in previously glaciated areas (Chapter 27).

A glacier can partially melt and create large lakes that are dammed by hills, moraines, or the ice itself. Some suspended silt and clay falls, accumulating on the lake bottom. These lakes eventually drain, often violently enough to create floods, known as "jokuloups" or "GLOF"s (Glacial Lake Outburst Flood; Figure 12.11) [4, 16]. Six thousand years ago, an ice dam had created glacial Lake Missoula that extended through presently Idaho and Montana, USA. On several occasions, the ice dam burst, spreading water violently across Washington State and down the Columbia River to the Pacific Ocean [17]. Many wave and water erosion patterns in dry earth still bear witness to these large GLOFs. A similar flood occurred when former Lake Bonneville drained down the Snake River and out of the same Columbia River near the last glacial period's end. The Bonneville Salt Flats and Great Salt Lake, Utah, are remnants of this giant lake.

Smaller GLOFs are still occurring in periglacial areas of the Himalayas and Andes Mountains, Alaska, and Scandinavian countries, to the peril of local people. Alarm systems are being installed in mountain valleys that warn of an approaching GLOF, and proactive draining of dangerous glacial lakes is being undertaken in the Himalaya and Andes Mountains. A hydroelectric facility has been established on at least one proactively draining lake in the Andes Mountains.

Three soil types are common in formerly glaciated areas: glacial till, glacial outwash, and old lake ("lacustrine") deposits. Glacial till and outwash soils in areas of hard, element-poor

Figure 12.11 Results of a relatively small (but dangerous) GLOF in the Himalaya Mountains, Namche, Nepal. Footbridge in center shows scale.

bedrock are not very productive because of the innate poorness combined with the little time for the rocks to decompose, release elements, and form clays. Lacustrine deposits contain clays, although sometimes the fine clay texture can make the soils difficult to farm (Chapter 10). Glacial till soils can be quite productive if formed from element-rich, soft bedrock; and lacustrine soils can be productive if enough coarse particles or structure exists.

Tills are commonly found on uplands. Commonly farmed tills are the tops of north–south hills sculpted by the glaciers, known as "drumlins." Where numerous boulders do not inhibit plowing, they sometimes can be productively farmed. Post-glaciated soils bring up new rocks every spring through "frost heaving" – the lifting of large objects such as rocks and fence posts through the winter freezing/thawing soil actions. Farm fields in glaciated areas are characterized by rock walls because the annually emerging rocks are first tossed beneath the wood fences to avoid the plow and later stacked to replace the wood fences.

Except for old lakebeds, valleys in formerly glaciated landforms almost always contain sandy and gravelly outwash. They are commonly less productive than surrounding till slopes, in contrast to the most productive soils being in valleys in other landforms. The outwash is generally poor in elements and so poor in water holding capacity that streams flowing from surrounding slopes sometimes disappear beneath the surface when reaching the outwash.

Relatively few plant and animal species are found in glaciated areas because all living species had to migrate into this landform relatively recently.

Building roads in glacial till areas can be difficult where there are many large boulders. On the other hand, sand and gravel is abundant in nearby outwash; and roads are quite easy to build in outwash. Areas where glaciers have retreated during the past several hundred years can be extremely dangerous. Lateral and terminal moraines that were formed by the ice can be lying at unstable angles of repose once the ice has melted. A slight stimulation can create large rockslides.

12.7 Bogs

In cold climates or areas with little soil oxygen such as water-saturated soils, microorganisms consume the litter so slowly that the organic matter accumulates. Elements stay in the litter and so are unavailable to plants, often making trees and other plants die. Specialized plants that need very few elements cover and insulate the area, further inhibit organic matter decomposition, and form bogs (Figures 10.9 and 12.12, right). The most common specialized plant is sphagnum moss (genus *Sphagnum*) of which about 300 species exist throughout the world. Sphagnum gets its element needs from rainwater and grows from its top (Figure 12.12b). The living and dead moss holds stagnant water, excluding oxygen and building up tannins and other chemicals that inhibit further organic decomposition [3]. Peat bogs can become several meters thick and many hectares in size.

Peat bogs can form on well-drained areas in cold climates because microbes decompose the organic matter so slowly that there is a net organic matter accumulation. Peat bogs also form in lakes and other areas saturated by stagnant water in warm climates; here, the oxygen-deficient water prevents decomposition of the organic matter. Bogs on lakes are often floating and can contain trees. They first appear as islands and can eventually cover the lake. Walking on these bogs can feel strange because the surrounding vegetation quivers and one can fall through to the water beneath.

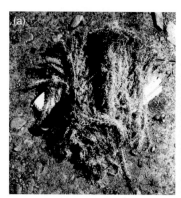 

Figure 12.12 (a) Sphagnum moss (peat moss), Fairbanks, Alaska, USA. (Underlying notepad shows scale.) (b) Extensive cold area changing from forest to peat bog as trees die with element tie-up in litter and growth of peat moss, Alaska, USA.

Peat bogs provide habitats for few plants and provide few elements (Figure 12.11b). Specialized plants such as the pitcher plant (*Sarraceniaceae, Nepenthaceae*, and other families) and the Venus fly trap (*Dionaea muscipula*) are found on them. Such plants get their elements by trapping and decomposing insects. Peat bogs are so inert that well-preserved human corpses buried thousands of years ago have been found.

Peat bogs do provide sphagnum moss, which is useful in nursery practices. Peat is sometimes cut for fuel. The world contains about 2 quintillion (10^{18}) BTUs of peat in estimated or proven recoverable reserves – enough to provide the world's energy needs for about 5 years.

Peat bogs have little use for farming unless they are burned when dry. Burning eliminates the organic matter and releases elements from the peat. The elements then accumulate in the underlying mineral soil, and the soil becomes quite productive.

Fires, windthrows, avalanches or other disturbances can mix soils tending to become bogs and stimulate organic matter decomposition, making more elements available and so reducing the soil's tendency to become a bog. Spruce hemlock forests in coastal Alaska owe their continued existence to periodic windstorms that uproot trees, stir the soil, let sunlight reach and warm the soil, and so prevent them from turning into bogs. Deep plowing has been done in Russia to transform peat soils and then increase forest growth. Road and building construction are generally difficult on peat.

During drought years, the water table can drop in peat bogs; and the dry surfaces can burn. The El Niño dry cycle created such drought conditions in bogs in Indonesia in 1997. Local people took advantage of these droughts and burned the bogs, leading to worldwide concerns of these fires and their smoke. Similar fires, but of nonintentional origin, occurred during the drought of 2002–04 in Florida.

12.8 Permafrost

Permafrost areas contain frozen soil as much as to 600 to 1,500 meters thick in very cold terrestrial areas and extending beneath oceans near both the North and South Poles (Figure 10.8) [3, 18]. Permafrost also exists on cold, high mountains even in the tropics [19]. In progressively warmer areas the permafrost becomes thinner, less continuous, and then absent. The global warming trend (Figure 7.5) is generally melting permafrost consistently in its southern fringes throughout the northern hemisphere.

Permafrost is believed to have begun during past glacial periods, largely in cold areas without glaciers to insulate the soil from extremely frigid air while geothermal energy warmed the ground from beneath. Permafrost is generally thinner beneath glaciers and formerly glaciated areas.

Permafrost can exist in several forms) [19, 20]:

- Buried ice ("ground ice") associated with wet areas;
- Ice-cemented permafrost, where water within a formation freezes among the minerals;
- Dry-frozen permafrost, where the water has sublimated from ice-cemented permafrost.

Overlying vegetation can play a large role in permafrost melting and reforming [20], as can winter snow depth [21]. Wet summer vegetation can conduct heat downward and help melt the permafrost; however, dry vegetation can insulate the ground and prevent the permafrost's melting in the summer heat [22]. Wet winter vegetation can conduct cold downward and deepen the permafrost. The permafrost also influences the vegetation. Summer warming can melt the top 20 to 70 cm of soil, creating saturated soils, slumping, soil creep – and habitat for very many mosquitos. Where the permafrost has melted in parts of the dry Tibetan plateau, the vegetation has shifted from a temperature-controlled grassland with sufficient melted soil water perched on the ice to an arid area because the moisture is no longer perched and has percolated beyond the rooting zone [23].

Permafrost creates unusual challenges for people. A heated house can create a warm spot that melts, and the house slumps or sinks into the saturated soil – or often a hole filled with water. A heavy vehicle can similarly sink into the permafrost. Such problems are addressed through a combination of mapping, seasonal uses of sensitive areas, and insulating construction to prevent foundations from melting the permafrost and sinking.

Melting permafrost may increase both photosynthesis and respiration, with a net release or uptake of carbon dioxide [24].

12.9 Wetlands

Wetlands are relatively small areas within many other landforms [25] but can be quite numerous (Figure 10.9). Included as "wetlands" are tidal lands, wet areas caused by depression and high water, and the interface of terrestrial and aquatic landforms. Wetlands can contain fresh, brackish, or salt water. Non-tidal wetlands are especially common in alluvial floodplains, coastal plains, peatlands, recently glaciated areas, and other places with high precipitation [26]; however, their presence in deserts and other arid places greatly extends the biodiversity (Figure 12.13, Chapter 15). Wetlands are also known as swamps, marshes, fens, shallow lakes, and bogs [27].

The temporary or permanent water saturation of their soils makes wetland vegetation, soils, and hydrology unique (Chapters 15, 17). Because the topography determines their water-saturated nature, they are difficult to drain and use for other purposes. Increasingly, people are beginning to appreciate their value for biodiversity, mediating floods, and recharging aquifers.

Many wetlands contain abundant sediment, elements, moisture, and oxygen from the common inflow of waters from uplands and their fluctuating water levels. The elements and sediment commonly fall to the bottom in the wetlands' calm waters. The rich waters enable growth of many plants and animals specifically adapted to the commonly water-saturated soils interspersed with dry conditions. These plants and animals establish a food web that concentrates harmful elements and chemicals in the higher trophic levels. The sediment precipitation also concentrates these elements and chemicals in the wetland floor.

Figure 12.13 Wetland within the Atacama Desert, Chile.

Many wetlands provide animals and plants that are eaten by animals living beyond the wetland, especially in tidal wetlands where the larger fish and other animals ultimately depend largely on these wetland creatures. Wetlands also provide places for animals to drink, thus enabling them to extend their territories farther from rivers and streams [27].

References

1. W. K. Hamblin, E. H. Christiansen. *The Earth's Dynamic Systems*, tenth edition. (Prentice Hall, 2001).
2. J. S. Monroe, R. Wicander. *Physical Geology*. (Brooks/Cole, 2001).
3. B. J. Skinner, S. C. Porter, J. Park. *Dynamic Earth*. (Wiley, 2004).
4. C. C. Plummer, D. McGeary, D. H. Carlson. *Physical Geology*, ninth edition. (McGraw-Hill, 2003).
5. D. Huddart, T. Stott. *Earth Environments: Present, Past, and Future*. (Wiley-Blackwell, 2012).
6. C. D. Oliver. Subsurface Geologic Formations and Site Variation in Upper Sand Hills of South Carolina. *Journal of Forestry*. 1978;76:352–4.
7. J. Munoz-Reinoso. Vegetation Changes and Groundwater Abstraction in Sw Donana, Spain. *Journal of Hydrology*. 2001;242(3):197–209.
8. F. Van der Meulen, A. Salman. Management of Mediterranean Coastal Dunes. *Ocean & Coastal Management*. 1996;30(2):177–95.
9. G. Gómez-Pina, J. J. Muñoz-Pérez, J. L. Ramírez, C. Ley. Sand Dune Management Problems and Techniques, Spain. *Journal of Coastal Research*. 2002;36(36):325–32.
10. Y. Chung, H. Kim, D. Jugder, L. Natsagdorj, S. Chen. On Sand and Duststorms and Associated Significant Dustfall Observed in Chongju-Chongwon, Korea During 1997–2000. *Water, Air and Soil Pollution: Focus*. 2003;3(2):5–19.

11. T. Wang, J. Wu, X. Kou, et al. Ecologically Asynchronous Agricultural Practice Erodes Sustainability of the Loess Plateau of China. *Ecological Applications*. 2010;20(4):1126–35.
12. J.-Q. Fang, Z. Xie. Deforestation in Preindustrial China: The Loess Plateau Region as an Example. *Chemosphere*. 1994;29(5):983–99.
13. H. N. Fisk. *Geological Investigation of the Alluvial Valley of the Lower Mississippi River.* (US Dept. of the Army, 1944).
14. S. C. Anisfeld, T. D. Hill, D. R. Cahoon. Elevation Dynamics in a Restored Versus a Submerging Salt Marsh in Long Island Sound. *Estuarine, Coastal and Shelf Science*. 2016;170:145–54.
15. R. F. Flint. *Glacial and Quaternary Geology*. (Wiley, 1971).
16. M. F. Meier. Ice and Glaciers, in Chow V. T., editor, *Handbook of Applied Hydrology*. (McGraw-Hill, 1964): pp. 16.1–.32.
17. V. R. Baker. The Spokane Flood Controversy and the Martian Outflow Channels. *Science*. 1978;202:1249–56.
18. J. A. Heginbottom, J. Brown, O. Humlum, H. Svensson. Permafrost and Periglacial Environments, in Williams R. S. J., Gerrigno J. G., editors, *Satellite Image Atlas of Glaciers of the World*. USGS Professional Paper. 1386-A. (US Geological Survey, 2012).
19. J. G. Bockheim, I. B. Campbell, M. McLeod. Permafrost Distribution and Active-Layer Depths in the McMurdo Dry Valleys, Antarctica. *Permafrost and Periglacial Processes*. 2007;18(3): 217–27.
20. Y. Shur, M. Jorgenson. Patterns of Permafrost Formation and Degradation in Relation to Climate and Ecosystems. *Permafrost and Periglacial Processes*. 2007;18(1):7–19.
21. V. Romanovsky, T. Sazonova, V. Balobaev, N. Shender, D. Sergueev. Past and Recent Changes in Air and Permafrost Temperatures in Eastern Siberia. *Global and Planetary Change*. 2007;56 (3):399–413.
22. F. Niu, G. Cheng, Y. Niu, et al. A Naturally-Occurring "Cold Earth" Spot in Northern China. *Scientific Reports*. 2016;6.
23. S. Yi, Z. Zhou, S. Ren, et al. Effects of Permafrost Degradation on Alpine Grassland in a Semi-Arid Basin on the Qinghai–Tibetan Plateau. *Environmental Research Letters*. 2011;6(4):045403.
24. E. A. Schuur, J. Bockheim, J. G. Canadell, et al. Vulnerability of Permafrost Carbon to Climate Change: Implications for the Global Carbon Cycle. *BioScience*. 2008;58(8):701–14.
25. I. Aselmann, P. Crutzen. Global Distribution of Natural Freshwater Wetlands and Rice Paddies, Their Net Primary Productivity, Seasonality and Possible Methane Emissions. *Journal of Atmospheric Chemistry*. 1989;8(4):307–58.
26. C. Prigent, F. Papa, F. Aires, et al. Changes in Land Surface Water Dynamics since the 1990s and Relation to Population Pressure. *Geophysical Research Letters*. 2012;39(8).
27. US Environmental Protection Agency. What Is a Wetland? 2016 (Accessed November 22, 2016). Available from: www.epa.gov/wetlands/what-wetland.

Part V
Biodiversity

13

Biodiversity: Individual Species

13.1 Biodiversity Overview

"Biodiversity" – a contraction of "biological diversity" – has many definitions [1–4] but refers to the genetic diversity of life – the species, variations within species, and the complex systems within which species exist – known as "ecosystems" [5]. The numerous, interacting species can create ecosystems that enable people to live.

Biodiversity is addressed at both community and species levels. For this book, a "community" is a contiguous area of similar plants and animals, soils, climate, and history (Chapter 14). Species are integral parts of communities, and there are so many species that each one cannot be addressed individually. Alternatively, many properties of biodiversity can be studied and most species can be conserved by concentrating on communities. Management and research at the community level are known as "coarse filter" biodiversity [6, 7]. "Fine filter" biodiversity addresses individual species, usually those that cannot be studied or conserved using coarse filter biodiversity – species that "fall through the sieve" of the coarse filter.

At least four organizations have inventory systems for species, biodiversity, and/or endangered species:

- International Union for the Conservation of Nature (IUCN) [8],
- Convention on International Trade in Endangered Species (CITES) [9],
- United Nations Food and Agriculture Organization Forestry Division (UN FAO Forestry) [10], and
- Intergovernmental Science-Policy Platform on Biodiversity and Ecosystem Services (IPBES) [11].

These organizations are collecting data that are becoming increasingly useful and accurate. However, all data are subject to errors, and endangered species observations and analyses should be treated as preliminary. A premise of this book is that preliminary data at least give the reader trends and ways to analyze the data, trusting that the reader will account for their imperfections. This book will primarily rely on UN FAO Forestry [10] data and what will be referred to as "UNFRA groups." These data are for each country and for total species as well as endangered and endemic species for seven species groups: amphibians, birds, ferns, mammals, palms, reptiles, and trees – further identified as "forest" and "non-forest" species. Cursory analyses suggest the data are compatible with IUCN and CITES.

Species' vulnerabilities to extinction are noted in several ways. The IUCN "Red Book" [8] defines nine relative degrees of danger of extinction, from "Not Evaluated" and "Least Concern" to "Critically Endangered" and "Extinct." Its list contains approximately 44,500 threatened species. "Threatened" refers to species that are "Critically Endangered," "Endangered," or "Vulnerable" (categories 3 through 5). "Endemic" refers to species found in only one place in the world [8].

CITES maintains three lists for each country:

- List #1: (931species for all countries) threatened with extinction;
- List #2: (34,419 species for all countries) not threatened with immediate extinction, but trade would threaten their survival;
- List #3: (147 species for all countries) one or more countries have requested CITES to assist with controlling their trade [9].

Neither species nor communities are very distinct. Individuals of a species can be a single cell or millions of cells. Some groups of species interact so intimately that it is difficult to consider them as individual species. Some species interbreed and others do not. Community boundaries can also be vague.

This chapter will focus on the individual organism/species – the fine filter. Following chapters will discuss communities – the coarse filter – the global distribution of biodiversity, and considerations for managing biodiversity.

13.1.1 Origin of Life and Development of Biodiversity

Life began about 3.5 to 4 billion years ago when aggregations of molecules containing carbon extracted energy from other molecules, created membranes around themselves, divided, and multiplied as cells. Information coded into specialized molecules such as RNA (ribonucleic acid) and DNA (deoxyribonucleic acid) enabled new cells to replicate their parent cell. Dividing cells occasionally created a change in the information-bearing molecules – a "mutation." Successful mutations led to multiple cell organisms and then to cell specialization. The chlorophyll molecule became part of some of these cells and enabled them to photosynthesize – gain energy from sunlight, in the process breaking down atmospheric carbon dioxide and releasing oxygen to the atmosphere. Other cells preyed on these photosynthesizing plants for energy.

Some mutations led to "genetic reproduction" (mating) – the sharing of cell information as DNA or similar molecules and production of offspring. The offspring is slightly different from each parent, since it combines some characteristic information of each. These differences are "genetic variations." The DNA information of an individual is the individual's "genotype." The behaviors and appearances of an individual are its "phenotype." Phenotypic characteristics are often compromises of conflicting information from both parents or dominances of information from one parent.

Presumably, early individuals expanded their populations until organisms occupied all suitable environments. When the environment became crowded and limiting, the organisms competed to survive [12]. Organisms with more competitive phenotypes or good luck survived and reproduced. Consequently, the population shifted in phenotype and genotype through a process of "genetic selection."

Progeny from a small population breeding within itself does not inherit counterbalancing characteristics from diverse parents. Such "inbreeding" can produce extremely superior and inferior characteristics, but usually leads to the population's eventual decline and extinction.

Mutations and genetic selection also produced organisms different enough to survive and exploit new environments better than other ones, a process known as "genetic differentiation" of the population. Differentiation can lead to new species.

Several things can lead to rapid differentiation [13, 14]:

- Environmental changes;
- Genetic diversity within the population;
- Colonizing an unoccupied environment;
- Isolation and "genetic drift": random differences that occur after a population becomes isolated;
- "Founder effect": an inherited characteristic that is absent or abundant at the time a population becomes isolated, so the characteristic is present or absent in most progeny.

At times, organisms can assist each other in surviving, first without becoming dependent on the assistance ("proto-cooperation" or "facilitation" [15]) and eventually becoming mutually interdependent ("mutualism").

Some genetic groups have remained single-cell species. Others have formed organisms of many cells and specialization within the cells – part of the continuing process of "evolution," or continued genetic differentiation.

13.1.2 Importance of Biodiversity

The increasingly diverse species – biodiversity – act as building blocks that create increasingly complex, dynamic ecosystems (Figures 1.1, 2.2). People and societies intimately depend on complex ecosystems and their underlying biodiversity. People host other organisms living on and within them. People use and affect many others [16–18]. Ecosystems contains multiple interdependencies and feedbacks, such as the dependencies of living plants on microorganisms and arthropods to release essential elements from dead animals and plants. Similarly, many flowering plants – and so animals – depend on insects to pollinate them. Animals depend on plants for oxygen produced by photosynthesis.

The abundance of species enables ecosystems to have complex hierarchies and networks (Chapter 2). The abundance maintains the vibrancy that enables life through the continuous interactions of different species from the rich pool that is biodiversity. The currently great richness of species (Figure 2.6) is probably necessary for people to survive. After they extinguished the trees, the human population of Easter Island survived in the simplified ecosystem, but with a diminished population and lifestyle [19]. When wolves disappeared from Yellowstone, the food network changed, with some plants and animals becoming more abundant and others scarce [20, 21].

Ecosystems and their value to people are always changing; and people modify ecosystems. They have increased global food availability by genetically modifying, relocating, and favoring species; and they have reduced deaths from malaria by controlling certain organisms and their habitats. On the other hand, people have inadvertently caused species extinctions; put poisons into the atmosphere (PCBs, Chapter 8); put species together that had always been apart; and changed habitats through land clearing and other activities. If too many species become extinct, there is concern that the global ecosystem will reach a tipping point, collapse, and no longer provide vital ecosystem services that enable people to survive.

People will always impact biodiversity; and most species that people move elsewhere cannot be eliminated from their newfound habitats. However, many steps can be taken (Chapter 16) to ensure that such new ("novel") ecosystems continue to function so that people can survive. It is currently difficult to identify which species are critical to maintain ecosystems suitable for people; and it is probably the accumulation of many species. Consequently, it is probably prudent to maintain all species– to ensure that no species inadvertently becomes extinct.

13.2 Classifications and Variations of Organisms

Evolution can be considered analogous to a tree (Figure 13.1) in which each bud is a species and its increasingly distant ancestors are represented by moving along limbs toward the tree base. Closely related species are closely connected on branches near the tree's periphery. Distantly related species are on branches that diverged on different stems closer to the tree's base. Two of the major stems would be plants and animals. The animal branch would contain a subbranch of "chordates" (animals with a central nerve down their back) with

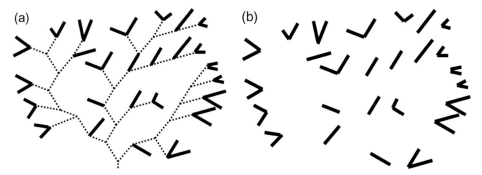

Figure 13.1 The analogy of evolution to a tree; (a) solid lines at tips represent species with dashed "stems" being evolutionary pathways, (b) branches ("evolutionary pathways") are hidden [22].

a subbranch for insects below that. The chordate branch would later contain a subbranch for vertebrates that would then have further subbranches for reptiles, birds, and mammals. The branch for plants would similarly have subbranches.

13.2.1 Classifications of Organisms

Scientists place existing and extinct species in their relative positions on this evolutionary tree through "taxonomic classifications." The challenge to classifying is that only existing species at the "tips" of the branches and some fossils currently exist (Figure 13.1b). Plants and animals are currently classified in the "Linnaean system" [23, 24]. It is a hierarchy from individual Varieties that are grouped into Species and then Genera (singular, "Genus"), Families, Orders, Classes, Phyla (singular "Phylum"), and Kingdoms (such as Animal and Plant Kingdoms). The system is continually being improved, and confusion exists in definitions. It is unclear, for example, whether a species should be distinguished as a group that cannot interbreed with similar species – an issue because some populations classified as different species of the same genus do interbreed, while others do not [25].

The assignment of organism groups to classes is constantly changing. Sometimes a variety is upgraded to a distinct species, or the opposite occurs. At various times, two (animals and plants), three (animals, plants, and microorganisms), and up to six kingdoms have been identified (animals, plants, fungi, protista, bacteria, and archaebacteria). Fungi, protista, bacteria, and archaebacteria will collectively be referred to as "microbes" in this book.

A scientific project known as the "Tree of Life" [26, 27] is developing a new classification by decoding DNA. The Tree of Life concept refers to related populations as "clades." Currently, a clade can be as small as a single population that is traditionally considered a species; or, it can be as large as – or larger than – a whole genus, such as all oak (*Quercus*) species.

Plants and animals can also be grouped by morphological, niche, and trophic levels (to be discussed) and other characteristics [28].

Eventually, a general change to the Tree of Life or a different approach may be beneficial [29]; however, incorporating the changes into national and international laws and programs could prove difficult. The Linnaean system is used in this book.

13.2.2 Numbers and Kinds of Species

Between 11 and 13.4 million species, including both known ones and educated extrapolations, are estimated to exist [30–32]. Differences seem to be in the numbers of invertebrate animals (primarily insects and other arthropods) and algae, protozoa, bacteria, viruses. One authority estimates [31] about 2 million fewer invertebrate animals, another [32] estimates about 2 million fewer algae, protozoa, bacteria, and viruses. Table 13.1 shows

Table 13.1 *Estimated thousands of species in the world by "creature type," based on one assessment [30]*

Creature type	Described species	Estimated number	Working figure	Accuracy
Viruses	4	50–1,000	400	Very poor
Bacteria	4	50–3,000	1,000	Very poor
Fungi	72	200–2,700	1,500	Moderate
Protozoa	40	60–200	200	Very poor
Algae	40	150–1,000	400	Very poor
Plants	270	300–500	320	Good
Nematodes	25	100–1,000	400	Poor
Crustaceans (Arthropods)	40	75–200	150	Moderate
Arachnids (Arthropods)	75	300–1,000	750	Moderate
Insects (Arthropods)	950	2,000–100,000	8,000	Moderate
Molluscs	70	100–200	200	Moderate
Chordates	45	50–55	50	Good
Other	115	200–800	250	Moderate
Totals	1750	3,635–111,655	13,620	Very poor

one estimate [30]. Of all species, 14 percent live in the ocean [32] – half of the protozoa and 25% of the algae – but very few fungi, plants, and prokaryotes.

Different species, genera, and families are found in different parts of the world, with patterns to their locations (Chapter 14).

13.2.3 Long-Term Changes in Species

At periodic intervals of millions of years, the Earth undergoes massive extinctions of species (Figure 2.5). The Permian event was the largest; an estimated 95% of existing species were eliminated. Less drastic events occur between these major ones at estimated 30-million-year intervals. Causes of these extinctions are suggested as being combinations of asteroid impacts [33, 34] perhaps generated when the solar system crosses certain parts of the Milky Way [35], climate changes (Chapter 9), and other factors [36].

Following these dramatic extinctions, the remaining species reorganize and again expand in number through evolution – mutations, genetic selection, colonization, and genetic drift in the new environment. Both the evolved species and the environment after the disturbance are inevitably radically different than before (Figure 2.4). For example, the world was dominated by dinosaurs before the Cretaceous extinction of 65 million years ago; mammals have emerged and dominated since [37]. This recent success of mammals is not because of their innate superiority; rather, fortuitous (for people) initial conditions after the Cretaceous extinction simply gave mammals an advantage (Chapter 2).

13.3 Niche, Food Web, Island Biogeography

The remainder of this chapter will address some questions that underlie the basic theories of ecology. Why are there different kinds of species? Why do some genetic groups contain many species and others contain only a few? Why do the numbers of species change? Why are different species found in different places? The different environments and genetic differentiation have led to the "niche," "food web," and "island biogeography" concepts.

13.3.1 Species Niches

Organisms expand in number and size as they occupy and exploit resources within an ecosystem. The environment within an ecosystem is not uniform and resources needed to survive – energy, elements, water, and others – are not uniformly accessible. Mutations and genetic differentiations lead to variations that enable each species to survive and compete successfully in a different range of environments within the ecosystem [38]. The environment can be placed within a multiple-axis matrix (theoretical volume or state space) with each axis being a different environmental variable (state variable) [39]. Figure 13.2 shows a two-dimensional matrix that depicts three species' "potential niches" – where they can live along the gradients of soil moisture and initial disturbance type.

A disturbance can be an environmental gradient when it sets the initial conditions of competitiveness of different species. For example, plant species vary in their regeneration mechanisms – windblown seeds that enter after disturbances, buried seeds that can endure some disturbances; and sprouts, tubers, rhizomes, and others that endure other kinds of

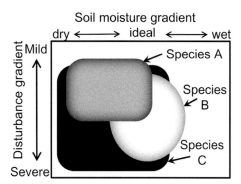

Figure 13.2 Three species arrayed within a two-dimensional matrix with the "realized niche" showing for each species. Dominating species are in the forefront.

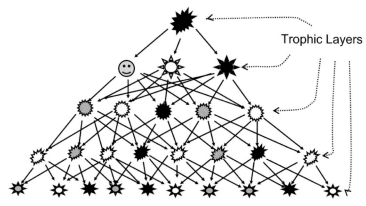

Figure 13.3 The "food web" concept, unique to animals [13].

disturbances. Consequently the disturbance type helps determine which species will first be present, begin growing, have an advantage, and dominate the new community [40].

Species living together can have overlapping niches. If one species does not completely exclude another, each will gain a competitive advantage and dominate part of the overlap, creating each species' "realized niche" (Figure 13.2) [41, 42]. If not for species A and B (Figure 13.2), species C could grow in a larger area; it is relegated to a realized niche in the fringe environments because it cannot compete in many places with the other species.

Animals and plants behave differently in such gradients. Nearly all plants grow best under the same, ideal conditions of moisture, sunlight, and temperature (Figure 13.2); but they are often found in suboptimal circumstances where they can compete successfully. Animals sometimes remain in physiologically suboptimum environments because of evolved or learned mental behaviors [43] even if more optimum environments become available.

13.3.2 Food Web

The niche concept is applicable to plants – and to animals if their behaviors are also considered. A similar concept to the niche is the "food chain," "food pyramid," "food web," or "food network" (Figure 13.3) formed in animal communities [13, 44]. Except for being on the bottom and eaten, green plants are excluded from this food web. Herbivores eat only plants. Other animals may eat one or several species of these herbivores; and other animals may eat these. Some animals are "omnivores," capable of eating both plants and animals. Hierarchical strata within the food web that show which animals are prey or predators to a species are "trophic levels."

If predator numbers at one trophic level decline, their prey at the immediately lower level increase and eat more animals at the level below them. Alternatively, this decline means fewer animals are eaten at the next lower level – and the fluctuation reverberates downward through the system. Changes also reverberate upward for similar reasons. Decline of some animals at a level can lead to increases of others at the same level since they would have less competition for prey; or it can lead to declines because there would be more pressure from higher levels to eat the remaining animals. The myriad of interactions within and between hierarchies leads to many, changing conditions within the communities, rather than a deterministic, stable condition [20, 21, 45–47].

Food webs are less stable if few species exist at any level, since the decline or extirpation of this one species can threaten the predator population's food supply [48]. Conceptually, an extremely efficient member of the food web could eat its prey to extinction, and lead to its own eventual starvation. Animals (especially insects) and predatory microbes appear to have evolved mechanisms so they do not eat their prey to extinction. Predators introduced from other parts of the world do not have this evolved behavior with local prey and often cause extinction of their new prey – such as the chestnut blight (*Cryphonectria parasitica*), Asian long-horned beetle (*Anoplophora glabripennis*), and the emerald ash borer (*Agrilus planipennis*) [49] introduced to eastern North America from Asia.

13.3.3 Island Biogeography, Competitiveness, Colonization

Small islands and similarly isolated areas commonly contain relatively few species for several reasons [50–52]. Potential immigrant species have difficulty reaching the area. The small area and concomitant small populations of existing species make them vulnerable to extinctions with perturbations that would only kill part of a population in a larger area.

An organism filling a niche or trophic position in a food web in a small, isolated area is often not very competitive compared to a species occupying a similar niche or trophic position on a larger area because the large-area species was selected after competing more vigorously with more species – the "island biogeography" concept [50, 51]. Consequently, introduced species from larger areas commonly displace and often drive to extinction native species that occupied the same niche or trophic position in smaller, previously isolated areas. Similarly, species from small, isolated areas are generally not very competitive with native species when introduced to large areas.

13.4 Species Pump, Carrying Capacity, Exotics

A species previously occupying a large area may only be able to survive in several, isolated areas known as "refugia" if a more aggressive competitor invades and dominates; the environment changes; or the area is fragmented by flooding, drought, glaciers, or other means. Populations in the different refugia may be subject to genetic drift, described earlier. They also may be exposed to different environmental pressures or the founder effect that causes divergent genetic selections. When the environment again changes and populations converge, they may have evolved differently enough to be different species. This process of divergence, isolation, genetic drift, and then convergence as different species is known as a "species pump" [53].

Previously isolated populations can also converge when they are still capable of breeding. They create distinctive hybrids. Places of convergence and hybridization of many animal and plant genotypes are referred to as "suture zones" [54, 55].

13.4.1 Species Extinctions

Extinctions always have occurred and will occur to some species, and new species will also always emerge. Extinctions of large, charismatic plants and animals are highly noticeable and they will be missed for aesthetic reasons. Some animals are key to maintaining complex food webs (Figure 13.3), and plants such as trees provide key structures for other plants and animals. Equally important is the diversity of vertebrates and arthropods, small plants, and microbes that ensure that ecosystems remain complex and therefore vibrant.

All species currently existing have survived tough, changing environments and competition. They are not fragile entities that easily become extinct; but they can become extinct for several reasons. They may be unable to compete successfully with introduced or mutated native species. They may be subjected to a condition to which they are not adapted, such as a rapid change in climate or toxic air or water. Or, they may be unlucky and a disturbance destroys the entire population. Species are especially vulnerable to extinctions when their populations are low, such as when confined to refugia or during shifts of other trophic levels in the food web. Inbreeding can then prove fatal. Disappearance of a few species from most communities seems to allow remaining species to rearrange their niches; however, communities reorganize quite dramatically and often more simplistically when many species become extinct [56].

Many species have become extinct because of periodic asteroid impacts (Figure 2.5) [33, 34] described earlier. A mass extinction of about 30 mammal species occurred in North America about 10 million years ago, associated with climate changes as North and South America joined and changed the ocean currents [37]. People apparently eliminated the giant emu (*Casuariidae* family) and other large animals [57] 50 thousand years ago when they first entered Australia. An estimated 33 families of birds and mammals became extinct during the last glaciation, at least partly attributed to human hunting. People apparently extirpated many species when they first colonized the Americas about 15 to 10 thousand years ago [58, 59]. It is unclear if extinctions of American species such as the mastodon (*Mammut americanum;* extinct 12,800 BC), North American camel (*Camelops hesternus*), giant beaver (*Castoroides leiseyorum* and *C. ohioensis*), short-faced bear (*Arctodus simus;* extinct 11,800 BC), and saber-toothed cat (*Smilodon* genus; extinct 6,000 BC) were a result of direct hunting [60], human-transported diseases, changes in habitat, a combination of the three, or other reasons [60–63]. The woolly rhinoceros (*Coelodonta antiquitatis*) and Irish elk (*Megaloceros giganteus*) became extinct when people returned to northern Eurasia after the recent glaciers left. An intense era of extinctions occurred around the Mediterranean Sea during the Roman Empire. An estimated 2,000 bird species were driven to extinction when Southeast Asian people first colonized the Pacific Islands [64, 65] where the dominant native species, often birds, were easy prey because of their weak competitive abilities (island biogeography).

Estimating extinction rates is difficult and controversial, marred by difficulties in counting, sampling, and otherwise estimating and extrapolating [11, 66]. It is accepted that people have recently increased extinction rates. An estimated background extinction rate is two species per year. Estimates of current extinction rates for all species vary from about 2,000 to 40,000 species per year [11, 66–69]. If recently estimated rates continued and new species were only formed at the background extinction rate, all species, including people, would become extinct between 350 and 8,000 years from now.

13.4.2 Species Discoveries and Rediscoveries

No perfect inventory of the world's species exists, or even of the "charismatic" megafauna and megaflora such as magnificent or beautiful mammals, birds, trees, or flowers. New species are still being discovered, including mammals and trees; and species once believed extinct are being found alive – although not at the current rate of extinction [70]. Many microbes, insects, plant species, and other life forms are still being identified, named, and systematically catalogued.

Even accessible places can harbor species that are unclassified or presumed extinct for several reasons: untrained people may not realize the species is unknown or presumed extinct and not report it; the remaining individuals hide in the most secure places as a species becomes rare; a species may disappear from a known niche but adapt to a new niche unbeknownst to biologists [71]; or, biologists and/or local people may hide an endangered species' presence to avoid attracting poachers, sightseers, or perceived counterproductive regulations.

Evolution of new species is not well studied but seems to occur at high numbers periodically for the many reasons described above. Evolution of a new species may not compensate for the biological diversity lost by another species' extinction. New species that evolve will probably be close relatives of existing species – known as "cheap species" [18]. For example, loss of the only two species of elephants would not be compensated in their contribution to biodiversity by the evolution of two new species of the abundant deer.

13.4.3 Size, Carrying Capacity, Fecundity, r vs K Selection

Many more species of arthropods and microbes exist than trees, mammals, and birds for logical reasons (Table 13.1).

Each area has a finite amount of resources that can be utilized by plants based on the quality of soils, climate, and moisture. The resources are collectively referred to as "growing space" [42]. Each area has an upper limit to the amount of living plant matter that it can

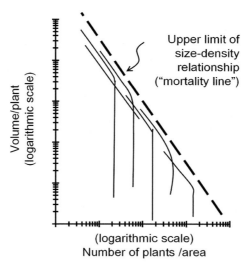

Figure 13.4 The maximum number of plants that an area can support ("upper limit") is related to the plant sizes by a logarithmic relationship [72–76]. Curved lines represent repeat measurements of plant communities.

support, with more productive areas supporting more living matter (Chapter 7). This living matter can be many small plants or a few large ones; and the upper limit has a slope of "–3/2" on a double-logarithmic axis (Figure 13.4) [72–76].

The finite amount of plant food can limit the animal food web size (Figure 13.3). An area's ability to support animals is referred to as its "carrying capacity" [43], and its upper limit is expressed as "animal unit equivalents (or months)" (AUMs) – the number of "animal units" (a specified single or collection of animals) that a unit area (such as a hectare) can support for a month [77]. An area can support many small animals or fewer large ones (Table 13.2).

Species that remain small, such as insects or herbaceous plants, have many more individuals in an area than large species. Consequently, there are many more opportunities for small species to breed with a variety of partners and produce a wider diversity of genetic combinations in the many offspring.

Both the time between reproductive cycles and the number of offspring produced at each cycle govern the total number of offspring that an individual produces. Plants and animals that grow to large sizes spend much energy and a long time growing before they reproduce, leading to a slower reproduction cycle and so a smaller opportunity for creating variations that lead to speciation. Often only a part of the population of large trees and animals contributes to reproduction, thus further limiting biological diversity. Trees in dominant positions of full sunlight set most flowers and seeds. Often dominant male animals primarily mate and contribute to the species' future genotypes.

Some plants produce many, small seeds with little food resources in each; consequently, the probability of survival of each is quite low but the probability that at least some will

Table 13.2 *Animal unit equivalents (ratio compared to "cow with calf") for different species used to determine the carrying capacity of rangelands [78]*

Class of animal	Animal unit equivalent	Class of animal	Animal unit equivalent
cow, 2,200 kg, dry	0.92	goat, mature	0.15
cow, 2,200 kg, with calf	1	kid, 1 year old	0.1
bull, mature	1.35	deer, white tailed, mature	0.15
cattle, 1 year old	0.6	deer, mule, mature	0.2
cattle, 2 years old	0.8	elk, mature	0.6
horse, mature	1.25	antelope, mature	0.2
sheep, mature	0.2	bison, mature	1
lamb, 1 year old	0.15	sheep, bighorn, mature	0.2

survive is high. Other plants produce fewer seeds but put considerable resources into each, so the probability of each seed becoming a plant is high but the total number is low. Animal species also have these two behaviors – producing many offspring or few but spending energy nursing and teaching them.

The term "K selection" is used to describe species that produce few offspring in each reproduction cycle but endow each with many resources. "r selection" describes the opposite – species that produce many offspring in each reproduction cycle with little endowment of resources [50, 79].

13.4.4 Exotic Species

A species sometimes enters a community where it had never been, with or without people's assistance. Known as an "exotic" species, the new arriver had not competed or coevolved with its new surroundings and can behave in any of several ways [80, 81].

• The invader can become immediately excluded from the new community by aggressive resident competitors or predators, especially if the exotic species enters a large community from a small, isolated one.
• Other, resident species can shift their niches or food webs to accommodate the new species, not eliminating any species but allowing the newcomer to become a member of the community.
• The invader can begin rapidly increasing in numbers, sometimes followed by a decline to a more stable population size as the niches of resident species or the food web shift to favor strong predators or competitors of the exotic species.
• The invader can begin as an infrequent, benign member of the community but expand exponentially as enough individuals become established to breed vigorously, as something changes in the environment, or as a mutation occurs in the exotic species. It can eventually displace the niches or food webs of native species.

• Some introduced species can eat their prey so aggressively that they drive the prey to extinction and then possibly go extinct themselves, as described earlier.
• Invading species that eliminate existing species can lead to further extinctions. For example, birds that pollinate were eliminated from New Zealand and so changed plant populations that depended on the pollinating birds [82].

13.5 Plant, Animal, and Microbial Behaviors

Behaviors observed in plants or animals are sometimes extrapolated to the other. Plants and animals have many very different behaviors, so such extrapolations need to be tested carefully. Some differences are:

1. Animals depend very strongly on plants for survival; but plants need animals much less. Plants are ultimately the source of all animal energy. They also alter the physical condition where animals live by providing hiding cover and nesting structures such as trees. Some animals are useful or necessary to plants for pollinating specific plants and for destroying competitors, such as through grazing.
2. Once they germinate, plants do not move. Consequently, an individual plant cannot "seek" a better place to live or escape from something hostile. The direct influence of a plant is only within its canopy and root zone, plus any shade beyond that. By contrast, animals move, often within set areas ("territories") or over set migration routes as individuals or groups. The territories are often limited by the territories of others of the same species.
3. Photosynthesizing plants generally do not eat each other; and a predator/prey hierarchy (food web) does not exist among plants.
4. Plants do not partition niches by behavioral mechanisms. Rather, they commonly grow under suboptimal conditions where they have a survival advantage (Figure 13.2). If more optimum conditions are available, a plant will grow in these conditions as well. The demise of one plant species in a community is usually an advantage to the others.
5. Different animal species need different complex organic nutrients that they cannot manufacture (Chapter 20). Consequently, they eat plants and/or each other and so exist in a hierarchy or web (Figure 13.3). Elimination of another animal species can destabilize a predator species. Conversely, plants do not take in nutrients in organic forms.
6. Unlike plants, animals commonly have feedback behaviors to help minimize overcrowding, such as territories and hierarchies within social groups.
7. Many plant species can survive as dormant seeds for decades, to germinate and grow much later; but very few animals can remain dormant for more than a few months.

Arthropods – insects and others – have many behaviors of other animals but can reproduce very rapidly and so quickly change genetically. This rapid change allows them to respond to the environment and develop mutualistic behaviors – such as pollination – with "higher" plants and animals. Many can stay dormant for months or years, and others metamorphose from one appearance to another.

Plants and animals also have many commonalities, such as the need for water and energy; variations between r- and K-behaviors in reproducing; similar element needs; and similar genetic recombination methods.

Fungi, bacteria, protistas (protozoans and algae), eubacteria, and archaebacteria act similarly to plants in their lack of mobility; however, they get food (energy) by digesting organic matter from living and dead plants, other microbes, and animals and sometimes by reducing minerals (for example, by converting iron from ferric to ferrous ions). They exist as single- and multiple-cell organisms. They reproduce rapidly both sexually and asexually and so can genetically adapt to new conditions rapidly. They become quiescent or grow as conditions are appropriate. They live throughout the world and in aerobic or anaerobic environments, including in guts of animals where they mutualistically help digest organic molecules. Some also split atmospheric nitrogen (N_2) molecules and pass the nitrogen to their host, higher plants.

References

1. T. E. Lovejoy. Biodiversity: What Is It? in Reaka-Kudla M. L., Wilson D. E., Wilson E. O., editors, *Biodiversity II Understanding and Protecting Our Biological Resources*. (Joseph Henry Press, 1997): pp. 7–14.
2. I. R. Swingland. Biodiversity. *Encyclopedia of Biodiversity*. 2001;1:377–91.
3. E. O. Wilson, F. M. Peter, editors. *Biodiversity*. (National Academy of Science (US) and Smithsonian Institutions, 1988).
4. M. A. Huston. *Biological Diversity: The Coexistence of Species*. (Cambridge University Press, 1994).
5. S. A. Levin. Ecosystems and the Biosphere as Complex Adaptive Systems. *Ecosystems*. 1998;1 (5):431–6.
6. M. L. Hunter Jr. *Coping with Ignorance: The Coarse-Filter Strategy for Maintaining Biodiversity*. Kohm K. A., editor. (Island Press, 1991).
7. The Nature Conservancy. *Natural Heritage Program Operations Manual*. Arlington, VA: 1982.
8. International Union for Conservation of Nature. *The IUCN Red List of Threatened Species. Version 2016–3*. 2016 (Accessed 28 February, 2016). Available from: www.iucnredlist.org.
9. CITES. *The Cites Appendices*. (Convention on International Trade in Endangered Species of Wild Fauna and Flora, 2013 [Accessed August 21, 2016]). Available from: https://cites.org/eng/app/index.php.
10. UN FAO. *Global Forest Resources Assessment 2000: Main Report*. (United Nations Food and Agriculture Organization, 2001).
11. S. L. Pimm, C. N. Jenkins, R. Abell, et al. The Biodiversity of Species and Their Rates of Extinction, Distribution, and Protection. *Science*. 2014;344(6187):1246752.
12. J. H. Tallis. *Plant Community History: Long-Term Changes in Plant Distribution and Diversity*. (Chapman and Hall, 1991).
13. M. Begon, C. R. Townsend, J. L. Harper. *Ecology: From Individuals to Ecosystems*. (Blackwell: Publishing, 2006).
14. P. A. Abrams. On Classifying Interactions between Populations. *Oecologia*. 1987;73(2):272–81.
15. J. F. Bruno, J. J. Stachowicz, M. D. Bertness. Inclusion of Facilitation into Ecological Theory. *Trends in Ecology & Evolution*. 2003;18(3):119–25.

16. P. R. Ehrlich, E. O. Wilson. Biodiversity Studies: Science and Policy. *Science*. 1991;253 (5021):758.

17. M. L. Oldfield. *The Value of Conserving Genetic Resources*. (Sinauer Associates Inc., 1984).

18. E. O. Wilson. *The Diversity of Life*. (WW Norton & Company, 1999).

19. J. Diamond. *Collapse: How Societies Choose to Fail or Succeed*. (Penguin, 2005).

20. D. Fortin, H. L. Beyer, M. S. Boyce, Wolves Influence Elk Movements: Behavior Shapes a Trophic Cascade in Yellowstone National Park. *Ecology*. 2005;86(5):1320–30.

21. W. J. Ripple, R. L. Beschta. Trophic Cascades in Yellowstone: The First 15 Years after Wolf Reintroduction. *Biological Conservation*. 2012;145(1):205–13.

22. P. Marshall, G. P. Berlyn, C. D. Oliver. The Relevance of Species Concepts in Sustainable Forestry. *Journal of Sustainable Forestry*. 2014;33:195–2010.

23. E. Mayr, P. D. Ashlock. *Principles of Systematic Zoology*, second edition. (McGraw-Hill, Inc., 1969).

24. E. Wiley. An Annotated Linnaean Hierarchy, with Comments on Natural Taxa and Competing Systems. *Systematic Biology*. 1979;28(3):308–37.

25. P. Marshall, G. P. Berlyn, C. D. Oliver. The Relevance of Species Concepts in Sustainable Forestry. *Journal of Sustainable Forestry*. 2014;33(2):195–210.

26. W. Judd, C. Campbell, E. Kellogg, A. Stevens, P. Donoghue. *Plant Systematics: A Phylogenetic Approach*. (Sinauer Associates, Inc., 2008).

27. D. R. Maddison, K.-S. Schulz, W. P. Maddison. The Tree of Life Web Project. In *Linnaeus Tercentenary: Progress in Invertebrate Taxonomy*. (Zootaxa, 2007): pp. 19–40.

28. B. D. Mishler. The Phylogenetic Species Concept (Sensu Mishler and Theriot): Monophyly, Apomorphy, and Phylogenetic Species Concepts, in Wheeler Q. D., Meier R., editors, *Species Concepts and Phylogenetic Theory: A Debate*. (Columbia University Press: 2000), pp. 119–32.

29. E. Mayr. Biological Classification: Toward a Synthesis of Opposing Methodologies. *Essential Readings in Evolutionary Biology*. 2014:354.

30. R. T. Watson, V. H. Heywood, I. Baste, et al. *Global Biodiversity Assessment*. (Cambridge University Press, 1995).

31. A. D. Chapman. *Numbers of Living Species in Australia and the World, 2nd Edition*. (Australian Biodiversity Information Services, 2009).

32. C. Mora, D. P. Tittensor, S. Adl, A. G. Simpson, B. Worm. How Many Species Are There on Earth and in the Ocean? *PLOS Biology*. 2011;9(8):e1001127.

33. L. W. Alvarez. Experimental Evidence That an Asteroid Impact Led to the Extinction of Many Species 65 Million Years Ago. *Proceedings of the National Academy of Sciences*. 1983;80(2): 627–42.

34. K. O. Pope, S. L. D'Hondt, C. R. Marshall. Meteorite Impact and the Mass Extinction of Species at the Cretaceous/Tertiary Boundary. *Proceedings of the National Academy of Sciences*. 1998;95 (19):11028–9.

35. D. M. Raup, J. J. Sepkoski. Periodicity of Extinctions in the Geologic Past. *Proceedings of the National Academy of Sciences*. 1984;81(3):801–5.

36. R. J. Twitchett. The Palaeoclimatology, Palaeoecology and Palaeoenvironmental Analysis of Mass Extinction Events. *Palaeogeography, Palaeoclimatology, Palaeoecology*. 2006;232 (2):190–213.

37. T. Flannery. *The Eternal Frontier: An Ecological History of North America and Its Peoples*. (Atlantic Monthly Press, 2001).

38. H. Walter, S.-W. Breckle. *Walter's Vegetation of the Earth: The Ecological Systems of the Geo-Biosphere*. (Springer, 2002).

39. G. E. Hutchinson. Concluding Remarks, in Hutchinson G. E., editor, *Cold Spring Harbor Symposium on Quantitative Biology*. 22. (Yale University, 1957): pp. 415–27.

40. C. D. Oliver. Forest Development in North America Following Major Disturbances. *Forest Ecology and Management*. 1980;3:153–68.

41. J. Kimmins. *Forest Ecology*. (Macmillan Publishing Company, 1987).

42. C. D. Oliver, B. C. Larson. *Forest Stand Dynamics*, update edition: (John Wiley, 1996).

43. C. S. Elton. *Animal Ecology*. (University of Chicago Press, 2001).

44. M. Newman. *Networks: An Introduction*. (Oxford University Press, 2013).
45. O. J. Schmitz. Predator Diversity and Trophic Interactions. *Ecology*. 2007;88(10):2415–26.
46. O. J. Schmitz. *Resolving Ecosystem Complexity (Mpb-47)*. (Princeton University Press, 2010).
47. J. R. Smith, O. J. Schmitz. Cascading Ecological Effects of Landscape Moderated Arthropod Diversity. *Oikos*. 2016;125(9):1261–72.
48. G. E. Hutchinson. Homage to Santa Rosalia or Why Are There So Many Kinds of Animals? *The American Naturalist*. 1959;93(870):145–59.
49. W. K. Moser, E. L. Barnard, R. F. Billings, et al. Impacts of Nonnative Invasive Species on US Forests and Recommendations for Policy and Management. *Journal of Forestry*. 2009;107 (6):320–7.
50. R. H. MacArthur, E. O. Wilson. *Theory of Island Biogeography*. (Princeton University Press, 1967).
51. L. D. Harris. *The Fragmented Forest: Island Biogeography Theory and the Preservation of Biotic Diversity*. (University of Chicago Press, 1984).
52. Y. Haila. A Conceptual Genealogy of Fragmentation Research: From Island Biogeography to Landscape Ecology. *Ecological Applications*. 2002;12(2):321–34.
53. J. Flenley. The Origins of Diversity in Tropical Rain Forests. *Trends in Ecology & Evolution*. 1993;8(4):119–20.
54. C. L. Remington. Suture-Zones of Hybrid Interaction between Recently Joined Biotas. in *Evolutionary Biology*. (Springer, 1968): pp. 321–428.
55. N. G. Swenson, D. J. Howard, R. Harrison. Do Suture Zones Exist? *Evolution*. 2004;58(11): 2391–7.
56. R. Gastaldo, W. DiMichele, H. Pfefferkorn. Out of the Icehouse into the Greenhouse: A Late Paleozoic Analog for Modern Global Vegetational Change. *GSA TODAY*. 1996;6(10).
57. T. Flannery. *The Future Eaters: An Ecological History of the Australasian Lands and People*. (Grove Press, 2002).
58. B. Fagan, N. Durrani. *People of the Earth: An Introduction to World Prehistory*. (Routledge, 2015).
59. R. D. MacPhee, H.-D. SUES. *Extinctions in near Time: Causes, Contexts, and Consequences*. (Springer Science and Business Media, 2013).
60. D. J. Meltzer, J. I. Mead. The Timing of Late Pleistocene Mammalian Extinctions in North America. *Quaternary Research*. 1983;19(1):130–5.
61. A. D. Barnosky. "Big Game" Extinction Caused by Late Pleistocene Climatic Change: Irish Elk (Megaloceros Giganteus) in Ireland. *Quaternary Research*. 1986;25(1):128–35.
62. A. D. Barnosky, P. L. Koch, R. S. Feranec, S. L. Wing, A. B. Shabel. Assessing the Causes of Late Pleistocene Extinctions on the Continents. *Science*. 2004;306(5693):70–5.
63. J. L. Gill, J. W. Williams, S. T. Jackson, K. B. Lininger, G. S. Robinson. Pleistocene Megafaunal Collapse, Novel Plant Communities, and Enhanced Fire Regimes in North America. *Science*. 2009;326(5956):1100–3.
64. M. McGlone. The Polynesian Settlement of New Zealand in Relation to Environmental and Biotic Changes. *New Zealand Journal of Ecology*. 1989:115–29.
65. P. J. Bellingham, D. R. Towns, E. K. Cameron, et al. New Zealand Island Restoration: Seabirds, Predators, and the Importance of History. *New Zealand Journal of Ecology*. 2010;34(1):115.
66. B. Lomborg. *The Skeptical Environmentalist: Measuring the Real State of the World*. (Cambridge University Press, 2003).
67. T. E. Lovejoy. A Projection of Species Extinctions, in Barney G. O., editor, *The Global 2000 Report to the President of the United States: Entering the Twenty-First Century*. II. (Pergamon Press, 1980): pp. 328–31.
68. N. E. Stork. *Measuring Global Biodiversity and Its Decline*. (Joseph Henry Press, 1997).
69. F. D. Smith, R. M. May. Estimating Extinction Rates. *Nature*. 1993;364:494–6.
70. P. H. J. Maas. Rediscovered Species and Subspecies, 2013 (Accessed July 13, 2016). Available from: www.petermaas.nl/extinct/lists/rediscovered.htm.
71. T. Low. *The New Nature: Winners and Losers in Wild Australia*. (Penguin Books Australia Ltd., 2002).

72. K. Yoda, T. Kira, H. Ogawa, K. Hozumi. Self-Thinning in Overcrowded Pure Stands under Cultivated and Natural Conditions (Intraspecific Competition among Higher Plants. Xi). *Journal of Biology, Osaka City University.* 1963;14:107–29.

73. J. White. The Thinning Rule and Its Application to Mixtures of Plant Populations, in White J., editor, *Studies in Plant Demography: A Festschrift for John Harper.* (Academic Press, 1985): pp. 291–309.

74. J. White, J. L. Harper. Correlated Changes in Plant Size and Number in Plant Populations. *Journal of Ecology.* 1970;58:467–85.

75. T. J. Drew, J. W. Flewelling. Some Recent Japanese Theories of Yield-Density Relationships and Their Application to Monterey Pine Plantations. *Forest Science.* 1977;23:517–34.

76. T. J. Drew, J. W. Flewelling. Stand Density Management: An Alternative Approach and Its Application to Douglas-Fir Plantations. *Forest Science.* 1979;25(3):518–32.

77. D. L. Scarnecchia, M. Kothmann. A Dynamic Approach to Grazing Management Terminology. *Journal of Range Management.* 1982:262–4.

78. M. Pratt, G. A. Rasmussen. *Determining Your Stock Rating.* (Utah State University Cooperative Extension, 2001 [Accessed May 22, 2016]). Available from: http://extension.usu.edu/files/pub lications/publication/NR_RM_04.pdf.

79. L. B. Slobodkin. *Growth and Regulation of Animal Populations.* (Dover Publications, 1980).

80. F. W. Allendorf, L. L. Lundquist. Introduction: Population Biology, Evolution, and Control of Invasive Species. *Conservation Biology.* 2003;17(1):24–30.

81. D. Simberloff. *Invasive Species: What Everyone Needs to Know.* (Oxford University Press, 2013).

82. D. Kelly, J. J. Ladley, A. W. Robertson, et al. Mutualisms with the Wreckage of an Avifauna: The Status of Bird Pollination and Fruit-Dispersal in New Zealand. *New Zealand Journal of Ecology.* 2010;34(1):66.

14

Biodiversity: Communities and Landscapes

14.1 Ecosystems and a Paradigm Shift

The term "ecosystem" was proposed by Tansley in 1935 to study the interactions among multiple organisms and their environment as systems [1]. Ecosystems can be as small as the microbes in a Petri dish or as large as the Earth [2]. This chapter will first discuss the community as an ecosystem.

14.1.1 Plant Communities (Stand, Grassland, Field, Steppe)

Plant communities – contiguous, uniform areas (Chapter 13) – are usually between 2 and 200 hectares in size and can be subdivided, combined, or delineated differently as conditions change. Discipline-specific names for plant communities are used in agriculture, animal husbandry, forestry, hydrology, and urban management (stand, field, pasture; Figure 2.6). The name differences sometimes lead to unnecessary confusion. "Plant community" will generally be used in this book to refer to any and all such areas [3, 4].

Plant communities can be used to delineate ecosystems at one scale; and landscapes and larger areas can delineate ecosystems at larger scales (Figure 2.6). Animal communities often change locations and so are difficult to delineate as ecosystems, although they are part of an ecosystem when present.

14.1.2 Paradigm Shift

The study of ecosystems at one time emphasized the stability and species interdependencies as a coevolved, deterministic community with no inputs beyond its borders [5, 6]. It was presumably so interconnected that it would collapse if any component species were lost. Barring disruptions by people, ecosystems were believed to be approaching a steady-state, or "a beautiful, mature 'climax' stage that becomes its naturally permanent condition" [7]. At an extreme, ecosystems were considered "superorganisms" in which the species in a community could not live without each other [8].

Although rejected by ecologists by the late twentieth century [7], the "stable" and "climax" paradigms linger among nonscientists and scientists in fields peripheral to

211

ecology, even as late as 2012 [9, 10]. People sometimes still implicitly base resource decisions on the concept that "nature knows best and that human intervention is bad by definition" [7] without explicitly admitting it. Some influences of this outdated paradigm have probably helped biodiversity – the establishment of "natural reserves," for example. Others, such as not intervening to prevent overcrowding in elephant herds or not managing overly crowded forests, have led to endangered species, catastrophic forest fires, and human displacements and hardships [11].

Scientists now recognize that ecosystems are quite dynamic and resilient – a distinctly different perspective. Competition and redundancy are major influences. Plant and animal communities are complex, dynamic systems that develop, change, decline, and reorganize in new ways.

A fundamental principle is that interactions of organisms within communities are not necessarily obligate relationships; that is, a species rarely depends on other unique species within a community to survive. Rather, different species that were previously in separate communities commonly form new communities together – often after disturbances. And species that were previously together can usually thrive apart in new communities.

Most ecosystems can be better understood by using competition to explain observed interactions rather than coevolution, even though some coevolution and other interactions occur [12, 13]. Coevolution prevents predators from eating their prey to extinction and enables some animals to pollinate certain plants, as discussed before. However, competition creates overlaps among plant niches as well as animal redundancies in the food web that make ecosystems vibrant and resilient.

14.2 Plant Community Dynamics

Plant community dynamics are very similar whether the area is dominated by annual plants, perennial non-woody herbaceous plants, shrubs, trees, or a mixture [14–17]. The pattern of plant changes following a disturbance is often referred to as "succession." The changes have been divided into two types. "Primary succession" occurs when the plants invade bare rock and slowly build soil. "Secondary succession" occurs when plants invade a disturbed area where the soil is largely intact [18]. There are gradients between these.

14.2.1 *Initial and Relay Floristics*

Secondary succession was formerly believed to occur by "pioneer" species first invading and preparing the environment so later arriving "mid-successional" species could survive, grow, and prepare the way for even later-arriving "climax" species in a "relay floristics" pattern (Figure 14.1a) [14]. Climax species were presumed to succeed themselves and form a stable forest that predominated before civilized people disrupted it [19, 20].

During the past few decades, it has become accepted that nearly all plants invade during distinct intervals shortly after disturbances ("initial floristics") (Figure 14.1b) [14]. They

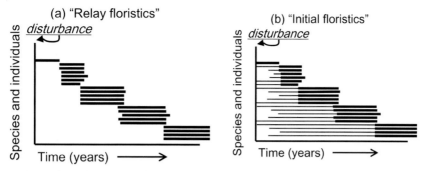

Figure 14.1 Schematic of alternative paradigms of plant community development [14]. Bold lines represent times of dominance; normal lines represent living, but not dominant. Reproduced with permission of Springer.

compete for growing space (Chapter 13) [21] relinquished by plants killed by the disturbance. Different species dominate at different times following the disturbance, giving the appearances that they invaded at different times (relay floristics; Figure 14.1a) and are different ages, even though they are not.

14.2.2 Disturbances and Plant Community Interactions

Disturbances kill vegetation, release growing space, and so enable other plants to invade (Figure 14.15a). They also give some invading species a competitive advantage by the regeneration mechanisms favored by the disturbance – the initial condition (Figure 13.2) [21]. Beginning in the first growing season following the disturbance, many herbaceous and woody plants germinate or otherwise emerge from preexisting seeds, tubers, or sprouts, while seeds of other plants fall into the area and germinate (Figure 14.9c). Species diversity is high at first because many short-lived and long-lived species become established together shortly after a disturbance.

Each new season, the growing space is reoccupied by new plants and the existing, continuously expanding perennials until these perennials permanently occupy all growing space [15, 22–24]. Species dominance can change over many decades even though all plants invade shortly after a disturbance (Figure 14.2) [15, 17, 21]. Less competitive individuals either die or survive but grow little in less favorable environments, such as in shade (Figure 14.2b) [25].

Disturbances known as "vegetation-replacing disturbances" can eliminate all preexisting vegetation (Figure 14.15a). Others, known as "partial disturbances" (Figures 14.3, 15c) kill some plants but leave others alive [27]. Some partial disturbances release so little growing space that surviving plants expand, but new individuals do not enter – or do not survive if they do enter [28]. Others allow surviving plants to expand and new ones to begin [29–31]. Some species usually seen in the understory will not reproduce without a disturbance that removes shading plants and exposes them to full sunlight [6]. Other species may not survive

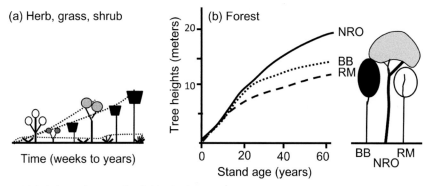

Figure 14.2 Schematic growth of (a) shrub [15] and (b) forest [21, 26] community following a disturbance.

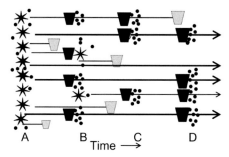

Figure 14.3 Schematic development of a plant community with partial disturbances [15]. Black stars are disturbances and trapezoids are living individuals of different species; dots are seeds.

without large, vegetation-replacing disturbances that give them enough sunlight to germinate, grow, and dominate [25, 32, 33].

With some variations, the process described above occurs in deserts, grasslands[1] [34], pastures [15], shrublands [23, 35, 36], and forests [21]. Desert communities can be quite diverse because it is rare that a few, competitive species grow enough to exclude the many other species. Grasslands and shrublands can be quite high or low in plant diversity, depending on whether a few perennials have excluded other plants.

Grassland and pasture communities subjected to frequent burnings or light grazing can become dominated by a few grass species; or they can be more diverse if subjected to a different disturbance – heavy grazing that removes grass and allows a diversity of light-seeded non-grasses to invade [34, 37–39]. A forest community (stand) can vary between conifer-dominated to mixed-broadleaf depending on the initiating disturbance (Figure 14.4). Forests with partial disturbances can also favor different species to dominate (Figure 14.3) [21].

---

[1] "Grassland," "steppe," "prairie," and "pasture" have overlapping definitions. "Grassland" will be used except in special circumstances.

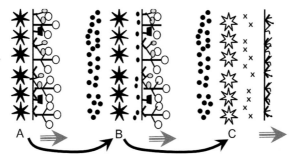

Figure 14.4 Schematic plant community subjected to same type of vegetation-replacing disturbances at times A and B, but a different type at time C [15].

In many places, the same area can exist as forest, grassland, shrubland, or desert – in different "Vegetation Covers" (states) – depending on the initial conditions following a disturbance. If woody species are not present at first, perennial non-woody species can exclude new species, including woody shrubs and trees, after they have occupied all growing space. Consequently, an area can exist as a grassland or brushland [15, 21, 23] for many decades. If perennial species are not present, annuals occupy the area continuously. If trees are present, they eventually dominate the growing space and exclude other species from entering for many decades or centuries.

14.3 Emergent Patterns: Plant Community Structures

Plant communities have emergent patterns – structures or states – that can be recognized and anticipated in many communities, despite differences in species and evolutionary backgrounds. Structures are recognizable even in plant communities where large species numbers make individual species behaviors difficult to predict [40]. The many ways that changes can occur are referred to as community "development pathways" [41].

Shrub/grasslands can change structures (states; Figure 14.5a) from nearly uniform grass structures (Figure 14.8a) to areas dominated by annual and perennial non-grass herbaceous plants through disturbances that release soil growing space (Figure 14.8b) [34, 37–39]. Or, they can remain as grasslands if subjected to light disturbances that promote the grasses' rhizomes. In some soils and climates, non-grass herbaceous structures can change to shrub or forest structures barring disturbances.

Figure 14.5b shows a general pattern found in the world's forests [21, 42, 43]. A disturbance that eliminates all previous trees and large shrubs creates the "open structure" (Figures 14.9a, c). Without another disturbance, the open structure could grow to a grassland, shrub, or forest community. If enough trees are present, a forest would emerge and grow to the "dense structure" (Figures 14.12a, 15b). Later, trees and other vegetation grow close to the forest floor and create the "understory structure." Very small disturbances can

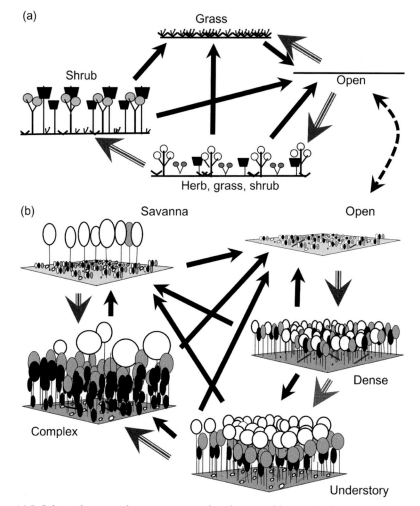

Figure 14.5 Schematic vegetation structures and pathways with growth (large, gray arrows) and disturbances (smaller black arrows) (a) grass/shrubland, (b) forest [41]. Dashed line shows that many communities can change between forest and shrub/grasslands when in open structures.

also cause small trees and other vegetation to invade the dense forest and form the understory structure more quickly. With time or partial disturbances, the understory structure changes to the "complex structure" (Figure 14.16).

If a disturbance leaves a few standing trees, the result is the "savanna structure," which can regrow to the complex structure over time or remain as a savanna with frequent partial disturbances such as light fires or grazing (Figures 14.5, 11, 12b). Disturbances can change most structures back to the open or savanna structure.

Grass and shrub communities seem to have similar changes in structures, although they have not been as well studied as forests.

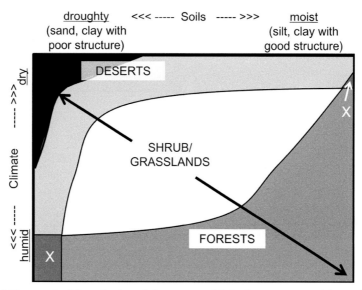

Figure 14.6 Desert, shrub/grassland, and forest potential land cover types niches schematically distributed in a two-dimensional matrix. (Arrow indicates extent of shrubs/grasslands. Areas shown as "X" can be desert, shrub/grassland, or forest.)

14.3.1 Vegetation Cover Types

The open, mineral soil of shrub/grasslands and the open and savanna structures of forests are times when new plants become established. Depending on the conditions at this time, grass, shrub, or tree species can invade, follow different pathways [8], and so become shrub/grasslands, forests dominated by different species, or even remain devoid of vegetation as deserts (Figure 14.7b–d).

The niches (state-spaces) of land covers can be partly described by soil conditions, climates (Figure 14.6), and disturbance regimes (not shown in Figure 14.6). Forests are only found in moist soils or climates and deserts are common in dry areas (Figure 14.7a); however, a community can sometimes be either desert or forest where either the soil or climate is dry and the other moist ("Xs" in Figures 14.6, 7b–d). A community can also be a desert where the soil is moist but disturbances prevent vegetation growth (Figures 14.7c, 8b). Shrub/grasslands can occupy all but the driest environments, and thus are found in many places where either deserts (Figures 14.7–8) or forests (Figure 14.9) could also occupy the plant community.

Land areas are commonly classified by their currently predominant or "expected" vegetation cover (Figure 7.1). The "expected" cover is useful in trying to assess what vegetation has been displaced by agriculture or urban areas; but its use may sometimes be a relic of the climax concept discussed earlier (Chapter 13). Since areas change among vegetation covers, experts may not agree on what is expected (Figure 9.5). For example, another vegetation cover map of South America shows much less forest cover and more

Figure 14.7 (a) Sand dunes where insufficient water prevents vegetation (Atacama Desert, Chile). (b) "Sand dune fixation" to prevent blowing sand dunes from creating deserts where forests and shrubs now exist (Mediterranean area, Turkey). Courtesy of Professor Melih Boydak. (c) Desert being created by overgrazing (Wyoming, USA). (d) Sand dunes stabilized by planted forest (Ukraine).

grasslands [44] than Figure 7.1. Vegetation cover type maps such as Figure 7.1 are extremely useful if their context is understood.

The ability to change among cover types explains confusion in plant community classifications where "expected" vegetation is assigned to an area. For example, a forest stand in the open structure could also be classified as a field or shrub/grassland. The many reasons that a community changes its land cover leads to further confusion. It is probable that the cause of desertification in the African Sahel (south of the Sahara Desert) [45–47] is overgrazing and can be prevented; however, it is sometimes assumed to be caused by the changing climate and to be irreversible.

14.3.2 Animal and Plant Dependencies on Structures

Different animal species generally need one or several specific structures to obtain their food, hiding cover, temperature regulation, and nests or dens (Figures 14.5, 10). These structures are often used synonymously with "habitats." Some animals exist within a single

Figure 14.8 Shrub/grassland in (a) Wyoming, USA, and (b) Tibetan Plateau, People's Republic of China in places too dry to support forests; overgrazing is changing this area of Tibetan grassland to shrub or desert.

structure. Other species move among several structures [48]. Species that are commonly found together are referred to as "guilds" [49], a helpful term in coarse filter conservation.

Plants also become established in some structures but not others; once established, some plants such as trees can continue growing as the structure changes. Most plants need open or savanna structures with their unoccupied growing space to begin growing. Species such as bigleaf mahogany (*Swietenia macrophylla*) of Central and South America and Nuttall and cherrybark oaks (*Quercus pagoda* and *nuttallii*) of the southeastern United States can only

(a)

(b)

(c)

Figure 14.9 Shrub/grasslands in (a) Russian Far East and (b) Scotland, where forests are excluded by fires and grazing, respectively. (c) Open structure in Ukraine, created by clear-cutting that could grow into forest, shrubland, or grassland. (A black and white version of this figure will appear in some formats. For the color version, please refer to the plate section. Color plate 7 (Figure 14.9c only).)

survive if they begin in the full sunlight of the "open" structure [25, 32]. However, they emerge to dominance in the dense and understory structures.

Many short-lived species complete their lifecycles in the open or savanna structures; the full sunlight generates high rates of photosynthesis giving high nutrition to these many plants' flowers, fruits, and leaves. The plants are close to the ground and available to animals (Figure 4.9c). Many pollinators and other butterflies, insects, and birds inhabit these structures. Mammals and their predators are commonly found in openings where the succulent plants are accessible as well as nutritious. The open structure offers no "thermal cover" relief from extreme heat and cold and can exclude browsing animals when snow accumulates.

If a few grass species reoccupy the growing space after a disturbance and exclude other species, they can limit the biodiversity to just these grasses and a few insects and birds even though they provide nutritious vegetation and seeds often harvested or grazed by domestic livestock and wild ruminants. If woody shrubs become established without trees following the disturbance, they can eventually exclude non-woody annual and perennial plants, change the biodiversity, and remain a relatively stable vegetation cover until another disturbance occurs. The area can grow to a forest if many trees become established following the disturbance.

Savannas contain most of the annual and longer-lived plant and animal species found in open structures (Figure 14.10). However, the slight shade precludes some plant species and causes others not to grow, compete, or produce sugar and/or flowers as vigorously [25]. And the browse is not as nutritious for animals. Some tree-nesting animals utilize the savannas exclusively, while other open-dwelling animals avoid savannas.

Annual herbs and newly initiating plants are generally excluded from dense structures; existing herbs and newly initiating ones commonly die. And as a stand grows to the dense structure the resident animal species also change [51]. Succulent vegetation is generally in the upper canopy and inaccessible to ground-dwelling animals. Consequently, many animals and plants found in open structures do not live well here. Dense structures provide hiding cover for predators and prey as well as thermal cover. Some arthropods feed on dying plants here, and some birds feed on these arthropods.

Understory and complex structures harbor some annual and perennial herbaceous plants and germinating woody plants near the forest floor, but these plants contain little nutrition in the low sunlight. Some generalist animals and a unique group of specialists live in these structures and eat decaying wood or the fungi and arthropods that live on this dead wood. Other animals prey on these animals. These complex forests provide much vertical diversity of limbs, branches, and leaves as well as diverse features such as cavities in trees, large branches, and large snags and logs that animals use for hiding, nesting, and thermal cover (for example, Figure 14.16).

14.4 Animals' Habitats, Movements, and Behaviors

Animals both utilize and affect plant communities [52]. They may depend on a single stand structure, need different ones for different functions – hiding and feeding – or utilize different structures as seasons change (Figure 14.10) [53]. Most feeding is done in open

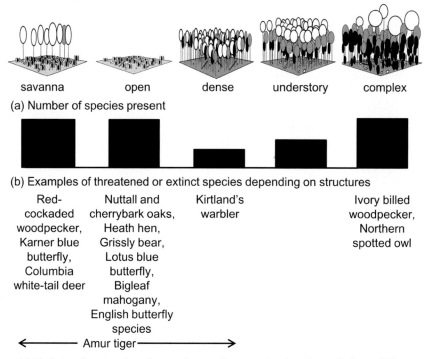

savanna   open   dense   understory   complex

(a) Number of species present

(b) Examples of threatened or extinct species depending on structures

| Red-cockaded woodpecker, Karner blue butterfly, Columbia white-tail deer | Nuttall and cherrybark oaks, Heath hen, Grissly bear, Lotus blue butterfly, Bigleaf mahogany, English butterfly species | Kirtland's warbler | | Ivory billed woodpecker, Northern spotted owl |

← ——— Amur tiger ——————— →

Figure 14.10 Some threatened, endangered, or extinct species that depend(ed) on different stand structures (habitats) that became limited in size [50].

or savanna areas where a diversity of nutritious plant and animal species are accessible near the ground.

Animal reproduction cycles have evolved to times of different food availabilities [53, 54]. They commonly produce young during late spring and early summer when new flowers, growing shoots, and leaves contain much protein, sugar, and starch but little fiber. During midsummer, grasses, leaves, and forbs (non-grass, herbaceous broadleaf plants) are available; and in late summer when animals need to store energy for winter, fruits and seeds are commonly abundant and eaten. Foods become increasingly scarce in winter and are generally sought in evergreen or dry annual herbs and grasses, woody twigs and buds, and starch-filled roots.

14.4.1 Movements

Some animals migrate relatively short distances up and down slopes to escape snow or to reach water [55]. Others migrate hundreds or thousands of kilometers during different seasons, such as neotropical migrant birds and monarch butterflies (*Danaus plexippus*). Others do not migrate.

Figure 14.11 (a) Pine savanna kept open by grazing, Montana, USA. (b) Oak savanna kept open by fire, Russian Far East. (A black and white version of this figure will appear in some formats. For the color version, please refer to the plate section. Color plate 9.)

Carnivorous animals follow the food cycles of herbivores and eat many during a year – young in spring and summer and old and feeble in winter. Predators higher on the food web (Figure 13.3) need larger areas to survive because they eat many animals at lower trophic levels. The size of area needed depends on the "carrying capacity" of prey, which is based on the amounts of favorable structures (Figure 14.10–11, Table 15.2) [56].

Rodents, some birds, and a few other animals feed on seeds, insects, or fungi in many structures. The dense structure is often used by both prey and predators for hiding when adjacent to open stands; and some prey avoid such edges. The dense structure is also used for nests, birthing, hiding, and feeding young. Savanna structures are used by birds that nest in cavities of large trees.

14.4.2 Social Groups

Animals live in specific locations, in groups with others of their species, in both locations and groups, or in locations for some seasons and groups for others [57]. Some groups ("herds," "packs," or "flocks") form seasonally for different activities: mating, migrating, protection, or sharing resources. Social hierarchies often form among males and among females within mammal and bird groups. "Alpha" (dominant) males commonly do much of the breeding.

Individual animals or groups can also occupy distinct "territories" that may or may not overlap with territories of others of their species [54, 58]. These territories are used for mating and feeding. Their size varies with animal size, trophic level, and feeding habits. And, territories can be a variety of shapes and even discontinuous to encompass certain habitats.

14.4.3 Feeding Patterns

Animal feeding habits differ based on the energy efficiencies of the foods they eat (Chapter 20). Grasses are difficult to digest, so ruminants and other animals such as horses graze for much of their time to stay alive; they graze especially intensively in winter to stay warm. At the other extreme, carnivores get concentrated energy in fats and proteins. Carnivores such as cats only eat every few days but sleep much of the time. Animals such as rodents, birds, and deer that feed on nuts receive concentrated energy in oils, while those feeding on flower nectar have less concentrated energy in sugars. A few animals, such as rodents, bats, and bears, slow their metabolisms in winter in a sleeplike condition that saves energy. Trophic levels have been discussed earlier (Figure 13.3).

14.4.4 Behaviors

Animals display different behaviors depending on the occasion [54, 58]. They often have ritualized "competitions" with others of their species when protecting their territory, courting, or establishing social hierarchy in a group. These displays rarely lead to deaths. They have much angrier, injurious behaviors when protecting their young, social group, or food. Carnivores inflict intentionally fatal injuries when killing other animals for food, but these are not associated with anger.

Some behaviors of higher animals are proving to be learned behaviors that previously were assumed to be genetically inherited. Cat species and wolves that were assumed shy by instinct had actually learned to fear people; they are now becoming less fearful and more aggressive as fewer people hunt them [59]. Similarly, many animals are preferring environments heavily occupied by people – fruit bats in cities (*Pteropus* species) [60] and American bald eagles (*Haliaeetus leucociphalus*) and polar bears (*Ursus maritimus*) around garbage dumps.

14.5 Landscapes of Multiple Plant Communities

A forest stand or shrub/grassland community can support only one structure (Figures 14.11) at a time and so provide habitat for only some species (Figure 14.10). Consequently, biodiversity needs to be considered and managed at larger scales of many communities – the landscape (Figure 14.13).

A landscape (Figure 2.6) has patterns (states) just as individual communities have structures [61, 62]. Contiguous communities in the savanna and open structures are "open forest," shrub, or grassland landscapes. Contiguous dense, understory, and/or complex communities are "closed forest" landscapes. The borders between open structures and closed forests are "edges." Open areas or closed forests away from edges are "interiors;" some species are only found in interiors of open areas [63] or closed forests [64]. Areas with both open and closed structures are diverse or "fragmented" forest landscapes.

At a given time, a landscape can have predominantly open, closed forest, or fragmented characteristics depending on what structures predominate. The terms "alpha," "beta," and "gamma" diversity refer to within-community, between-community, and total diversity [65]. Different aspects of biodiversity have been approximated by various indexes [66]: species richness, species diversity, species evenness, taxonomic and phylogenetic measures, and abundance/biomass comparisons.

All community structures are necessary for the greatest landscape biodiversity. Grasslands were maintained by human-caused, mild fires in grasslands until it was realized that interspersed, intensive disturbance "pulses" actually created more plant and animal diversity and nutrition [67, 68]. A landscape can change from one pattern – open, closed, or fragmented – to another as communities grow or are disturbed. As a result, plant and animal populations change (Figure 14.12).

14.5.1 Changes at the Landscape Level

Disturbances immediately kill individual plants and animals, but their greatest impacts are to set in motion changes in landscape patterns.

In fire-prone areas before many roads or fire control infrastructures existed, fires often traveled for weeks or months over large areas [21]. At times the fires "crept" along the forest floor in openings and beneath large trees in savanna and some closed structures, killing small trees and sometimes weakening big trees' stems and roots. Occasionally, they would "flare up" into the overstory as destructive "crown fires" and travel rapidly and very destructively for a few meters or many kilometers before again becoming "ground fires." The result was a mosaic of open areas with burned snags from crown fires (Figure 14.9a); savanna forests where the smaller trees were killed (Figures 14.11, 12b); and dense understory and complex structures in places the fire missed or burned lightly (Figures 14.5, 12a, 16). This diversity of structures caused subsequent fires also to burn variably as a ground or crown fire and maintain the mosaic.

In other places, disturbances occurred less frequently but were more catastrophic. Most landscapes contained both a variable distribution of structures and an irregular topography even if one structure predominated. Consequently, a large disturbance would act differently as the topography and stand structures varied. It would leave a mixture of structure patterns and growth rates but predominantly open and savanna structures (Figure 14.13a–b) [69]. Following such disturbances, undisturbed areas would be "refugia" for species that need closed forests; and species depending on openings could disburse across most of the landscape [70]. As the forest regrew (Figure 14.13c–f), open-dependent species would become confined to refugia that remained open because of poor soils or frequent, small disturbances; and closed forest species could disburse across most of the landscape. A landscape thus exhibited fluctuations and occasional isolations in species resembling the species pump (Chapter 13). Different guilds were able to migrate across the landscape at different times, and so corridors may not have been necessary for species survival but may have been supplementary. Edges, water locations, and seasonal cycles also affected species presences and migrations.

14.5.2 Recent Changes in Landscapes and Species Declines

Quite large disruptions and fluctuations in vegetation cover have occurred during the existence of most currently living species (Figure 9.5). Vegetation covers that were common 18 thousand

Figure 14.12 (a) Continuous closed forest that supports little tiger prey (and tigers) in Northeast China [56]. (b) Savanna/open/dense mosaic in Jim Corbett National Park (Uttarakhand, India) maintained by managed burning that supports tiger prey and tigers; note deer in mid-ground. (A black and white version of this figure will appear in some formats. For the color version, please refer to the plate section. Color plate 10.)

years ago are now much more restricted, such as shrub/grasslands in South America and Africa (Figure 9.5). Forests are now much more widespread in South America, Europe, and Africa than at that time. No plants or animals existed in most of Canada and the Nordic countries because of the glaciers. The disruptions changed structures at all scales and probably caused losses of species; however, some remnants of many structures probably existed continuously during the fluctuations because many species depending on the various structures still survive.

Endangered species are predominantly those living in forests (Figure 14.14). Farms and urban areas presently occupy about 20% of the area that would otherwise be predominantly forests [71]. This change in forest area should not account for the recent increase in species extinctions (Chapter 13) given past, larger forest area changes (Figure 9.5).

There is concern that forest harvests threaten species. To assess the relationship between harvest and species threat, this book compared three measures of forest harvest with nine measures of species endangeredness using the UNFRA species groups data [72]. Forest harvest measures were:

1. percentage annual change in forest area,
2. harvest as a percentage of standing volume, and
3. average harvest per ha of forest.

The nine species measures were (UNFRA species groups):

- percentage endangered species of all species;
- percentage country-endemic species of the endangered species; and
- percentage of forest species that are both country-endemic and endangered for each of the seven UNFRA species groups (Chapter 13).

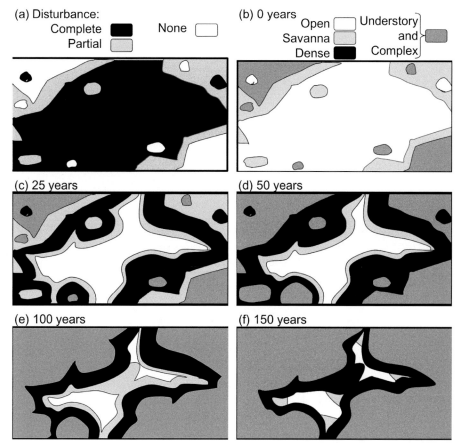

Figure 14.13 Simulation of changes in stand structure distribution across a landscape for 150 years following a fire [70]. Reproduced with permission by Taylor and Francis Group Ltd: www.informa world.com

Only countries with the forest activity and species measures different from zero were included. Up to 215 countries were included in each analysis. Linear, polynomial, and other regressions were run and the best fit (highest R^2) for each comparison was used.

No regressions had a coefficient of determination (R^2) greater than 0.01, indicating that there is almost no relationship (less than 1%) between forest harvest and species endangeredness. Some countries with high harvest rates or high forest area declines have high concentrations of endangered forest species while other countries in the same continent such as Africa with similar harvest or forest area changes have few or no endangered species. And, some countries with little or no forest harvest or forest area decline – and sometimes an increase in forest area – have many endangered forest species while others do not.

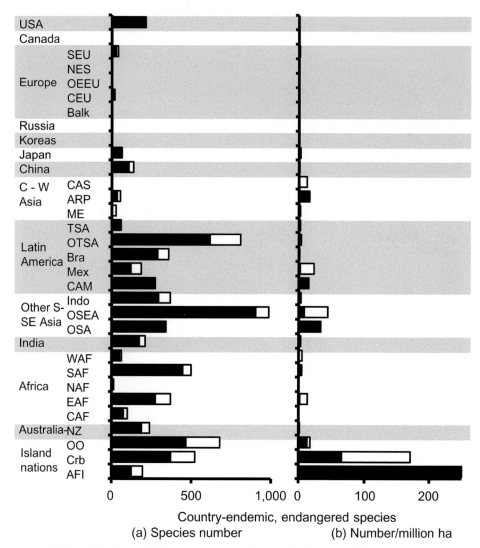

Figure 14.14 (a) Number and (b) concentration of forest (black = per-million-ha forest) and non-forest (white = per-million-ha non-forest) country-endemic endangered species by country groups [72]. Data from UNFRA groups.

Although forest harvest amounts apparently do not influence the high extinction rates in forests, two other factors may be important: the uniformity of recent human-induced disturbances and the interruption of species movement patterns.

Before the recent uniformity of human disturbances, mixtures of refugia and corridors probably allowed species to survive that could not to use the predominant habitats. Many human-induced disturbances such as clear-cuts, selection harvests, and controlled burns of grasslands are uniformly done over large areas. Only a small proportion of the

Figure 14.15 (a) Large area in western Washington, USA, uniformly clear-cut. (b) Large area in western Washington, uniformly grown from open to dense structure 60 years after being uniformly clear-cut [77]. (c) Recent selective harvesting in Russian Far East. (d) Forest many years after being selectively harvested in Tennessee, USA [21]. Figure 14.15b reproduced with permission of Taylor and Francis Group.

"historical range of variability" [73] of initial conditions in residual structures, regeneration propagules [21, 74], habitats as refugia, and "geodiversity" (variations in "geosites" – surface materials, forms, and physical features [75, 76]) are left (Figure 14.15a). The uniformity is carried forward as clear-cut or "high graded" (degraded) forests regrow for many decades as species-depauperate forests.

Much of the world's current forest areas – and possibly shrub and grasslands – are not only lacking needed structures over large areas, but are also interspersed with agriculture fields, reservoirs, and urban areas as well as crossed by highways, railroads, canals, pipelines, transmission lines, and other features. These structures fragment the habitat, limiting the ability of species to move to suitable habitats and breed over large enough areas to avoid inbreeding. This fragmentation risks species extinctions [78].

14.6 Species Movements with Climate Change

Both plant and animal species have migrated over hundreds of miles in the past; animal species generally move more rapidly. The species found in an area at a given time may not be those most suited for the various niches. Rather, they may be the species that were able to move to the location at a given time and may actually be moving to an even more suitable climate, albeit at an unnoticeably slow pace.

With past climate changes, plant and animal species migrated to more favorable areas, became extinct [79], or remained behind in favorable microclimates as isolated "relict" species. Some species are still migrating both poleward in response to the last deglaciation and possibly toward the equator as the weather changes approaching the next glaciation (Chapter 9) [80]. Contrary to earlier assumptions [81], species do not migrate as intact communities [82]. Consequently, a plant community defined by species exists at only one time; and the predominant vegetation type in an area can change with the climate [83]. "We cannot . . . see the current Amazon rainforest as the area's 'normal' condition, because for the last million years the Earth's climate has spent about 80 percent of the time in glacial conditions" [84].

Plants are very sensitive to the local climate during germination and early establishment, but can persist in a less optimal climate once established (Chapter 13). It is possible to find communities with tree species living together that could not have become established in the same climate (Figure 14.16) [85].

Plants and animals do not need to migrate as far up and down mountains as they do horizontally across flat areas to find suitable microsites for germination (Hopkins' Bioclimatic Law, Chapter 7). Species often concentrate in mountains [87], probably partly for the above reason but also because mountains have more barriers and micro-topographic, microclimatic, and geosite variations that lead to refugia and speciation (Chapter 11).

Migrations following glacial retreats seem to have occurred much more rapidly than migrations after continents met. Possibly, the climate changed quite rapidly following glacial retreats; and/or migrating species were moving into lands that had no or few already established species to compete with. Plant species migrate by germinating in the available growing space of open and savanna structures (Figure 14.5) in the directions of favorable climate. Young plants of each species contain the genetic characteristics most adapted to the recent climate, so their progeny are probably best suited to the local climate. These two factors suggest that disturbed areas and young trees will facilitate plant adaptation and migration with climate change. Structures with older trees will be appropriate as habitats for other migrating species, especially animals.

Figure 14.16 Large Douglas firs (person touching; *Pseudotsuga menziesii*) became established during the warm period 1,000 years ago; mountain hemlocks (dark crowns to right of person; *Tsuga mertensiana*) established during the Little Ice Age (approximately 300 years ago) and Pacific silver fir (small, foreground; *Abies amabalis*) established 100 years ago [21, 86]. Courtesy of University Press of Colorado.

With changing climates, various species will amalgamate in diverse geosites as they migrate. It may be prudent to ensure that all geosites are maintained to aid survival of as many species as possible [75, 76].

Species can be temporarily or permanently prevented from migrating with changing climates by broad, east/west mountains and valleys, river floodplains, or similar barriers. Migrating species can also diverge as they migrate, with one population moving up or down mountains and another moving to different latitudes [82]. Different genotypes are selected as species migrate and/or climates change, enabling populations isolated in refugia to survive for centuries and divergent populations to evolve into different species.

Species whose niche has shrunk can persist for many generations in fortuitous places across the landscape. From a biodiversity conservation perspective, it may be appropriate to ensure that these individuals survive by providing all habitats – such as shrub/grasslands in the Amazon (Figure 9.5), not just the habitats of the currently predominant vegetation.

The rate of global warming during the past century – and the upward and poleward movement of suitable climates for species – may have exceeded the rates of warming during melting of the last continental glaciers (Chapter 9). There is considerable concern that the rapid change will lead to dramatic species extinctions [88]. Even if species are not exterminated by their migrations being outpaced by the changing climates, insect and other animal species migrate more rapidly than plant species and attack plant species that are not adapted to withstand the newly arriving animals [9, 89].

References

1. A. G. Tansley. The Use and Abuse of Vegetational Concepts and Terms. *Ecology.* 1935;16 (3):284–307.
2. J. Kimmins. *Forest Ecology.* (Macmillan Publishing Company, 1987).
3. M. J. Crawley. The Structure of Plant Communities. *Plant Ecology.* 2009:475–531.
4. R. Ornduff, P. M. Faber, T. Keeler-Wolf. *Introduction to California Plant Life.* (University of California Press, 2003).
5. S. G. Boyce, C. D. Oliver. The History of Research in Forest Ecology and Silviculture. *Forest and Wildlife Science in America: A History.* (Forest History Society, 1999), pp. 414–53.
6. M. A. Huston. *Biological Diversity: The Coexistence of Species.* (Cambridge University Press, 1994).
7. W. K. Stevens. New Eye on Nature: The Real Constant Is Eternal Turmoil. *New York Times.* 1990 July 31, 1990;Sect. B.
8. S. A. Levin. Self-Organization and the Emergence of Complexity in Ecological Systems. *BioScience.* 2005;55(12):1075–9.
9. D. Huddart, T. Stott. *Earth Environments: Present, Past, and Future.* (Wiley-Blackwell, 2012).
10. M. C. Urban, J. J. Tewksbury, K. S. Sheldon. On a Collision Course: Competition and Dispersal Differences Create No-Analogue Communities and Cause Extinctions During Climate Change. *Proceedings of the Royal Society of London B: Biological Sciences.* 2012;279(1735):2072–80.
11. C. A. Spinage. Too Many Elephants: A Continent-Wide Problem: Part I. *African Ecology.* (Springer, 2012), pp. 675–713.
12. C. J. Lortie, R. W. Brooker, P. Choler, et al. Rethinking Plant Community Theory. *Oikos.* 2004;107(2):433–8.
13. S. Báez, D. A. Donoso, S. A. Queenborough, et al. Ant Mutualism Increases Long-Term Growth and Survival of a Common Amazonian Tree. *The American Naturalist.* 2016;188(5):567–75.
14. F. E. Egler. Vegetation Science Concepts I. Initial Floristic Composition, a Factor in Old-Field Vegetation Development with 2 Figs. *Vegetatio.* 1954;4(6):412–17.
15. J. L. Harper. *Population Biology of Plants.* (Academic Press, 1977).
16. W. A. Niering. Vegetation Dynamics (Succession and Climax) in Relation to Plant Community Management. *Conservation Biology.* 1987;1(4):287–95.
17. C. D. Oliver. Forest Development in North America Following Major Disturbances. *Forest Ecology and Management.* 1980;3:153–68.
18. E. P. Odum. *Fundamentals of Ecology.* (Saunders, 1971).
19. F. E. Clements. *Plant Succession: An Analysis of the Development of Vegetation.* (Carnegie Institution of Washington, 1916).
20. F. E. Clements. Nature and Structure of the Climax. *Ecology.* 1936;24(1):252–84.
21. C. D. Oliver, B. C. Larson. *Forest Stand Dynamics,* update edition. (John Wiley, 1996).
22. W. H. Drury, I. C. Nisbet. Succession. *Journal of the Arnold Arboretum.* 1973;54(3):331–68.

23. W. Niering, G. Dreyer, F. Egler, J. Anderson Jr. Stability of Viburnum Lentago Shrub Community after 30 Years. *Bulletin of the Torrey Botanical Club*. 1986:23–7.
24. W. A. Niering, F. E. Egler. A Shrub Community of Viburnum Lentago, Stable for Twenty-Five Years. *Ecology*. 1955;36(2):356–60.
25. C. D. Oliver, E. C. Burkhardt, D. A. Skojac. The Increasing Scarcity of Red Oaks in Mississippi River Floodplain Forests: Influence of the Residual Overstory. *Forest Ecology and Management*. 2005;210(1):393–414.
26. C. D. Oliver. Development of Northern Red Oak in Mixed Species Stands in Central New England. *Yale University School of Forestry and Environmental Studies Bulletin*. 1978;91:63.
27. K. R. Covey, A. L. Barrett, M. S. Ashton. Ice Storms as a Successional Pathway for Fagus Grandifolia Advancement in Quercus Rubra Dominated Forests of Southern New England. *Canadian Journal of Forest Research*. 2015;45(11):1628–35.
28. C. D. Oliver, E. P. Stephens. Reconstruction of a Mixed-Species Forest in Central New England. *Ecology*. 1977;58(3):562–72.
29. K. L. O'Hara. *Multiaged Silviculture: Managing for Complex Forest Stand Structures*. (Oxford University Press, 2014).
30. E. Rossi, I. Granzow-de la Cerda, C. Oliver, D. Kulakowski. Wind Effects and Regeneration in Broadleaf and Pine Stands after Hurricane Felix (2007) in Northern Nicaragua. *Forest Ecology and Management*. 2017;400:199–207.
31. U. M. Goodale, G. P. Berlyn, T. G. Gregoire, K. U. Tennakoon, M. S. Ashton. Differences in Survival and Growth among Tropical Rain Forest Pioneer Tree Seedlings in Relation to Canopy Openness and Herbivory. *Biotropica*. 2014;46(2):183–93.
32. L. K. Snook. Catastrophic Disturbance, Logging and the Ecology of Mahogany (Swietenia Macrophylla King): Grounds for Listing a Major Tropical Timber Species in CITES. *Botanical Journal of the Linnean Society*. 1996;122(1):35–46.
33. J. Fernandez-Vega, K. R. Covey, M. S. Ashton. Tamm Review: Large-Scale Infrequent Disturbances and Their Role in Regenerating Shade-Intolerant Tree Species in Mesoamerican Rainforests: Implications for Sustainable Forest Management. *Forest Ecology and Management*. 2017;395:48–68.
34. S. L. Collins. Fire Frequency and Community Heterogeneity in Tallgrass Prairie Vegetation. *Ecology*. 1992;73(6):2001–6.
35. E. Van Der Maarel. Pattern and Process in the Plant Community: Fifty Years after as Watt. *Journal of Vegetation Science*. 1996;7(1):19–28.
36. A. S. Watt. Pattern and Process in the Plant Community. *Journal of Ecology*. 1947;35(1/2):1–22.
37. S. L. Collins, S. M. Glenn, D. J. Gibson. Experimental Analysis of Intermediate Disturbance and Initial Floristic Composition: Decoupling Cause and Effect. *Ecology*. 1995;76(2):486–92.
38. S. Fuhlendorf, D. Engle. Application of the Fire–Grazing Interaction to Restore a Shifting Mosaic on Tallgrass Prairie. *Journal of Applied Ecology*. 2004;41(4):604–14.
39. S. D. Fuhlendorf, D. M. Engle. Restoring Heterogeneity on Rangelands: Ecosystem Management Based on Evolutionary Grazing Patterns. *BioScience*. 2001;51(8):625–32.
40. S. P. Hubbell, R. B. Foster. Biology, Chance, and History and the Structure of Tropical Rain Forest Tree Communities. *Community Ecology*. 1986;19:314–29.
41. C. D. Oliver, K. L. O'Hara. Effects of Restoration at the Stand Level, in Stanturf J. A., Madsen P., editors, *Restoration of Boreal and Temperate Forests*. (CRC Press, 2004): pp. 31–59.
42. J. F. Franklin, T. A. Spies, R. Van Pelt, et al. Disturbances and Structural Development of Natural Forest Ecosystems with Silvicultural Implications, Using Douglas-Fir Forests as an Example. *Forest Ecology and Management*. 2002;155(1):399–423.
43. P. S. Park, C. D. Oliver. Variability of Stand Structures and Development in Old-Growth Forests in the Pacific Northwest, USA. *Forests*. 2015;6(9):3177–96.
44. GifeX. *South America Vegetation*, 2017 (Accessed July 1, 2017). Available from: www.gifex.com/detail-en/2009–11-19–11222/South-America-vegetation.html.
45. S. M. Herrmann, A. Anyamba, C. J. Tucker. Recent Trends in Vegetation Dynamics in the African Sahel and Their Relationship to Climate. *Global Environmental Change*. 2005;15(4):394–404.

46. P. Hiernaux, C. L. Bielders, C. Valentin, A. Bationo, S. Fernandez-Rivera. Effects of Livestock Grazing on Physical and Chemical Properties of Sandy Soils in Sahelian Rangelands. *Journal of Arid Environments*. 1999;41(3):231–45.

47. M. Mortimore, B. Turner. Does the Sahelian Smallholder's Management of Woodland, Farm Trees, Rangeland Support the Hypothesis of Human-Induced Desertification? *Journal of Arid Environments*. 2005;63(3):567–95.

48. M. L. Hunter Jr. *Wildlife, Forests, and Forestry. Principles of Managing Forests for Biological Diversity*. (Prentice Hall, 1990).

49. D. Simberloff, T. Dayan. The Guild Concept and the Structure of Ecological Communities. *Annual Review of Ecology and Systematics*. 1991;22:115–43.

50. C. Oliver, D. Adams, T. Bonnicksen, et al. *Report on Forest Health of the United States by the Forest Health Science Panel*. (Center for International Trade in Forest Products, University of Washington, 1997).

51. M. C. Duguid, E. H. Morrell, E. Goodale, M. S. Ashton. Changes in Breeding Bird Abundance and Species Composition over a 20 year Chronosequence Following Shelterwood Harvests in Oak-Hardwood Forests. *Forest Ecology and Management*. 2016;376:221–30.

52. O. J. Schmitz. *Resolving Ecosystem Complexity (Mpb-47)*. (Princeton University Press, 2010).

53. M. L. Hunter. *Wildlife, Forests and Forestry*. (Prentice Hall, 1998).

54. J. H. Shaw. *Introduction to Wildlife Management*. (McGraw-Hill, 1985).

55. S. A. Gauthreauz Jr. *Animal Migration, Orientation and Navigation*. (Academic Press, 2012).

56. X. Han, C. D. Oliver, J. Ge, Q. Guo, X. Kou. Managing Forest Stand Structures to Enhance Conservation of the Amur Tiger (Panthera Tigris Altaica). in *A Goal-Oriented Approach to Forest Landscape Restoration*. (Springer, 2012): pp. 93–128.

57. D. J. Sumpter. *Collective Animal Behavior*. (Princeton University Press, 2010).

58. L. Boitani, T. Fuller, editors. *Research Techniques in Animal Ecology: Controversies and Questions*. (Columbia University Press, 2000).

59. P. Canby. The Cat Came Back: Alpha Predators and the New Wilderness. *Harper's Magazine*. 2005:95–102.

60. T. Low. *The New Nature: Winners and Losers in Wild Australia*. (Penguin Australia Ltd, 2002).

61. M. Krishnadas, A. Kumar, L. S. Comita. Environmental Gradients Structure Tropical Tree Assemblages at the Regional Scale. *Journal of Vegetation Science*. 2016;27(6):1117–28.

62. L. D. Audino, S. J. Murphy, L. Zambaldi, J. Louzada, L. S. Comita. Drivers of Community Assembly in Tropical Forest Restoration Sites: Role of Local Environment, Landscape and Space. *Ecological Applications*. 2017.

63. M. A. Van Horn, R. M. Gentry, J. Faaborg. Patterns of Ovenbird (*Seiurus aurocapillus*) Pairing Success in Missouri Forest Tracts. *The Auk*. 1995:98–106.

64. C. Richter. Deep-Forest Birds and Hostile Edges. In *Restoring North America's Birds: Lessons from Landscape Ecology*. (Yale University Press, 2002): p. 99.

65. R. H. Whittaker. Evolution and Measurement of Species Diversity. *Taxon*. 1972:213–51.

66. S. A. Khan. *Methodology for Assessing Biodiversity*. (Centre of Advanced Study in Marine Biology, 2006).

67. R. Voleti, S. L. Winter, S. Leis. Patch Burn-Grazing: An Annotated Bibliography. *Papers in Natural Resources*. 2014 (Paper 462):12.

68. J. Henning, G. Lacefield, M. Rasnake, et al. *Rotational Grazing*. (University of Kentucky, College of Agriculture, 2000).

69. A. C. Camp, C. D. Oliver, P. Hessburg, R. Everett. Predicting Late-Successional Fire Refugia Pre-Dating European Settlement in the Wenatchee Mountains. Forest Ecology and Management 95: 63–77. *Forest Ecology and Management*. 1997;95:63–77.

70. C. D. Oliver, A. Osawa, A. Camp. Forest Dynamics and Resulting Animal and Plant Population Changes at the Stand and Landscape Levels. *Journal of Sustainable Forestry*. 1997;6(3–4):281–312.

71. J. H. Tallis. *Plant Community History: Long-Term Changes in Plant Distribution and Diversity*. (Chapman and Hall, 1991).

72. UN FAO. *Global Forest Resources Assessment 2000: Main Report*. (United Nations Food and Agriculture Organization, 2001).

73. P. Morgan, G. H. Aplet, J. B. Haufler, et al. Historical Range of Variability: A Useful Tool for Evaluating Ecosystem Change. *Journal of Sustainable Forestry*. 1994;2(1–2):87–111.

74. D. Piotto, F. Montagnini, W. Thomas, M. Ashton, C. Oliver. Forest Recovery after Swidden Cultivation across a 40-Year Chronosequence in the Atlantic Forest of Southern Bahia, Brazil. *Plant Ecology*. 2009;205(2):261–72.

75. J. J. Lawler, D. D. Ackerly, C. M. Albano, et al. The Theory Behind, and the Challenges of, Conserving Nature's Stage in a Time of Rapid Change. *Conservation Biology*. 2015;29 (3):618–29.

76. J. Hjort, J. Gordon, M. Gray, M. Hunter Jr. Valuing the Stage: Why Geodiversity Matters. *Conservation Biology DOI*. 2015;10.

77. C. D. Oliver, K. L. O'Hara, P. J. Baker. Effects of Restoration at the Stand Level, in Stanturf J. A., Madsen P., editors, *Restoration of Boreal and Temperate Forests*. (Taylor and Francis, 2015).

78. L. Fahrig. Effects of Habitat Fragmentation on Biodiversity. *Annual Review Of Ecological Evolutionary Systems*. 2003(34):487–515.

79. T. Flannery. *The Eternal Frontier: An Ecological History of North America and Its Peoples*. (Atlantic Monthly Press, 2001).

80. E. C. Pielou. *After the Ice Age: The Return of Life to Glaciated North America*. (University of Chicago Press, 2008).

81. E. L. Braun. *Deciduous Forests of Eastern United States*. (Blakiston Company, 1950).

82. M. B. Davis. Late-Glacial Climate in Northern United States: A Comparison of New England and the Great Lakes Region, in Cushing E. J., Wright H. E., editors, *Proceedings of the Vii Congress of the International Association for Quaternary Research*. 7. (Yale University Press, 1967): pp. 11–43.

83. R. Gastaldo, W. DiMichele, H. Pfefferkorn. Out of the Icehouse into the Greenhouse: A Late Paleozoic Analog for Modern Global Vegetational Change. *GSA TODAY*. 1996;6(10).

84. M. Maslin. The Climatic Rollercoaster, in: Fagan B., editor, *The Complete Ice Age*. (Thames and Hudson, 2009): pp. 62–91.

85. C. D. Oliver, A. Adams, R. J. Zasoski. Disturbance Patterns and Forest Development in a Recently Deglaciated Valley in the Northwestern Cascade Range of Washington, USA. *Canadian Journal of Forest Research*. 1985;15(1):221–32.

86. C. D. Oliver. Mitigating Anthropocene Influences in Forests in the United States, in Sample V. A., Bixler R. P., Miller C., editors, *Forest Conservation in the Anthropocene*. (University Press of Colorado, 2016): pp. 99–111.

87. S. L. Pimm, C. N. Jenkins, R. Abell, et al. The Biodiversity of Species and Their Rates of Extinction, Distribution, and Protection. *Science*. 2014;344(6187):1246752.

88. V. A. Sample, R. P. Bixler, C. Miller. *Forest Conservation in the Anthropocene*. (University Press of Colorado, 2016).

89. A. L. Carroll, S. W. Taylor, J. Régnière, L. Safranyik, editors. Effect of Climate Change on Range Expansion by the Mountain Pine Beetle in British Columbia. In *Mountain Pine Beetle Symposium: Challenges and Solutions*, October 30–31, 2003; (Natural Resources Canada, 2003).

15

Robust and Threatened Communities and Species

15.1 Coarse Filter Conservation

Species' ranges and the sizes and geographic distributions of species can give insights about their vulnerability to species extinctions from a coarse filter approach (Chapter 13) [1–3]. Species with large ranges are probably least vulnerable to extinction because a relatively small disturbance cannot impact much of the population, because there is ample room for becoming resilient by diversifying genetically, and because species occupying large areas are most competitive.

Different classifications have been developed for delineating ranges or groups of ranges, often for different purposes [4–6] and with different inadvertent biases (Chapter 14). This book uses the following scales to discuss concepts, but does not advocate a specific classification:

- Floristic Realms
- Ecological Zones (Figure 7.1)
- Potential Vegetation Types
- Global Land Covers
- Global Vegetation Covers

15.2 Floristic Realms

The Earth's terrestrial vegetation can be divided into five "floristic realms" (Figure 15.1) [4, 7]. Each realm contains genera that are usually absent elsewhere. The Holarctic Realm is home to pines (*Pinus*), firs (*Abies*), spruces (*Picea*), oaks (*Quercus*), beech (*Fagus*), maples (*Acer*), birches (*Betula*), and others. The Australian Realm is home to *Eucalyptus* and other genera. These plants are now found elsewhere because people moved them or occasionally for other reasons. The realms largely result from plate tectonic changes of the past 250 million years (Figure 9.1) [8]. As flora became isolated, they differentiated genetically, with areas that separated earliest having the greatest differences.

Unlike plants, many animals migrated rapidly as continents met – or sooner in the cases of birds. Consequently, many animal genera are widely distributed, although others still reflect their geographic origin.

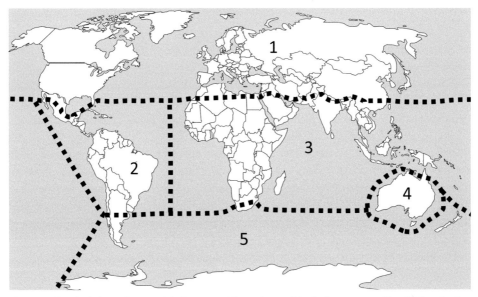

Figure 15.1 Floristic realms – global geographic regions with distinct genera (1 = Holarctic, 2 = Neotropic, 3 = Paleotropic, 4 = Australian, 5 = Antarctic) [4, 7].

The Holarctic Realm (1, Figure 15.1) – temperate and boreal North America, Europe, and Asia – has long been together but separate from other continents and subcontinents until quite recently (Figure 9.1). The Neotropic Realm (2, Figure 15.1) consists of tropical South America and central America up through the lowland, tropical forests of southern Mexico. It was once connected to Africa but then separated, and has connected to North America in geologically recent times (Figure 9.1). The Paleotropic Realm (3, Figure 15.1) extends from Africa south of the Sahara Desert, through tropical India and into tropical southeast Asia, including some larger Pacific islands. Apparently, the species migrated laterally throughout this realm once India, Africa, and Southeast Asia connected with Eurasia. The Australian Realm (4, Figure 15.1) is concentrated around Australia. The Antarctic Realm (5, Figure 15.1) consists of New Zealand, some other South Pacific Islands, the southern tip of Africa, and temperate South America, which were once joined with Antarctica. The species migrated along a common ecoregion before Antarctica became isolated and cold (Chapter 9).

Distantly related plant species can still be found among different floristic realms. Most current flowering plant families existed when the dinosaurs were decimated about 65 million years ago. Animals evolved much more since then with variations among floristic realms [9].

Species from one floristic realm occasionally becomes introduced to another, non-touching realm. Monkeys arrived in isolated South America from Africa, probably by accidentally floating across the Atlantic Ocean on trees [10]. When continents joined, slow intermingling occurred among the floristic realms' species (Figure 9.1) [11, 12].

Further species differences occur in geographically distant places within a floristic realm [8]. Such isolated species of the same genera often occupy similar climates, landforms, and niches; and, they appear and behave similarly. They can sometimes interbreed if contact is renewed.

Plants and animals are also distributed unevenly within floristic realms. The advancing ice sheets generally forced plants and animals southward in North America (Figure 9.5a). Amazon plants were isolated in refugia between glaciers advancing downward from the Andes and drier climates forcing plants upward from the Amazon into the Andes (Chapter 9). The Amazon forest that diverged into different refuges later expanded with the warmer climate of 8,000 years ago (Figure 9.5b); closely related species from different refugia are now separated by rivers, low mountains, or other geographic barriers [13]. The humid climates of 8,000 years ago changed many deserts into grasslands and enabled flora and fauna to move there. Warmer, drier times since then have stimulated species to migrate uphill and away from the equator [14, 15].

The floristic realms are not equal in size and contain different amounts of each ecological zone (Figure 15.2, Table 15.1). The realms also contain different concentrations of species and endangered species.

The Paleotropic and Neotropic Realms contain the greatest species concentrations, possibly because they contain most tropical forests. Even though much larger than the Antarctic Realm, North America, Temperate Asia, and Australia contain lower species concentrations. It is possible that the Antarctic Realm area has shrunk more than the species numbers with the isolation and freezing of Antarctica, which eliminated its area from the realm. Australia's low species concentration but high proportion endangered is probably because it is dry, highly leached of elements that plants and animals need (Chapter 11), and small with a lot of competitive exotic species threatening native ones. Europe/North Africa and Africa contain the smallest proportion of endangered species – possibly because people of the same culture have been occupying them for a long time or because they are large areas (island biogeography, Chapter 13). Tropical Asia, North America, and Australia contain the greatest proportion of endangered species, probably for independent reasons.

The lowest concentrations of species are generally found in cold (high latitude) and dry (desert) climates in all floristic realms (Figure 15.3), with increasing concentrations on the less extreme peripheries of these climates.

Higher plant species concentrations are found in warm, humid regions with very high concentrations throughout most tropical areas.

15.3 Vegetation, Land Covers, Biodiversity Hotspots

Ecological zones have been delineated and described in Chapter 8. Ecosystems can be delineated to finer resolutions such as vegetation types, potential vegetation types, and land covers.

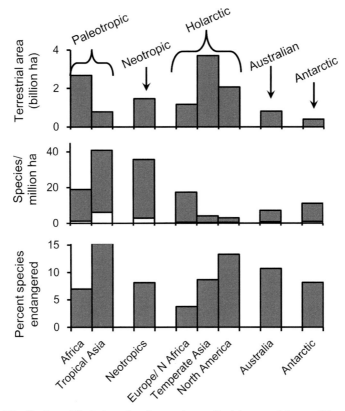

Figure 15.2 Distribution of floristic realms by continents for (a) terrestrial area, (b) species concentration (weighted average; Data from UNFRA groups; white = endangered [16] and black = other [17]), and (c) percent of endangered species (weighted average; UNFRA groups). (Russia is grouped with Asia.) [18].

15.3.1 Potential Vegetation Types

Subgroups of similar vegetation have been created for this book by subdividing floristic realms (Figure 15.1) into groups of similar potential vegetation types (Figure 7.1) – moist forests, dry forests, grasslands, deserts, and others (Table 15.1). The result is 56 unique terrestrial subdivisions that can serve as "Potential Vegetation Types" for coarse filter biodiversity conservation. Potential Vegetation Types describe the vegetation as if agriculture and urban areas were in pre-anthropogenic vegetation types and probably show a forest bias (Chapter 14) [18].

The sizes of these unique types vary dramatically – from 10 million ha for Neotropical shrublands to over 1.3 billion hectares for neotropical wet forests (Table 15.1). Tundra, polar, shrub, and dry forest ecological zones do not occur in all realms. Tundra, Polar, and Desert vegetation types occupy 25% of the terrestrial Earth. They provide limited ability for species to live and are considered uninhabitable for most people. Half of the

Table 15.1 *Distribution of the world's Potential Vegetation Types (millions of ha and percent) [18, 19]*

Region[A]	Wet Forest[B]	Dry Forest[C]	Shrub[D]	Grassland[E]	Desert[F]	Mountain[G]	Tundra[H]	Polar
Paleotropic								
Africa	890	405	601	48	898	189	0	0
	7%	*3%*	*5%*	*0%*	*7%*	*1%*	*0%*	*0%*
Tropical	652	159	121	116	430	439	0	0
Asia	*5%*	*1%*	*1%*	*1%*	*3%*	*3%*	*0%*	*0%*
Neotropic								
Neotropics	1,330	202	10	94	77	260	0	0
	10%	*2%*	*0%*	*1%*	*1%*	*2%*	*0%*	*0%*
Holarctic								
Europe/	501	76	0	122	9	102	0	0
N Africa	*4%*	*1%*	*0%*	*1%*	*0%*	*1%*	*0%*	*0%*
Temperate	770	0	0	210	468	932	141	206
Asia	*6%*	*0%*	*0%*	*2%*	*4%*	*7%*	*1%*	*2%*
North	562	9	0	297	120	354	266	358
America	*4%*	*0%*	*0%*	*2%*	*1%*	*3%*	*2%*	*3%*
Australia/	99	59	107	147	416	27	0	0
Antarctic	*1%*	*0%*	*1%*	*1%*	*3%*	*0%*	*0%*	*0%*
(not including Antarctica)								
Total	4,804	910	839	1,034	2,418	2,303	407	564
World	*36%*	*7%*	*6%*	*8%*	*18%*	*17%*	*3%*	*4%*

[A] See Figure 7.1.
[B] Wet forest = Tropical Rain and Moist Deciduous; Subtropical Humid; Temperate Oceanic and Continental; and Boreal Coniferous Forests.
[C] Dry forest = Tropical and Subtropical Dry Forests.
[D] Shrubland = Tropical Shrubland.
[E] Grassland = Subtropical and Temperate Steppes.
[F] Desert = Tropical, Subtropical, and Temperate Deserts.
[G] Mountain = Tropical, Subtropical, Temperate, and Boreal Mountain Systems.
[H] Tundra = Boreal Tundra Woodlands.

remaining area is potentially wet forests, although very little occurs in Australia. Some of these wet forests have been converted to urban and agriculture land [20]. Mountains comprise 17% of the area, and shrubs and grasslands together comprise 14%. Other classifications may assign more area to shrubs or grasslands and less to forests. Within nearly all groups are non-tidal wetlands (Figure 10.9). They comprise 4% of the terrestrial Earth's area [21, 22].

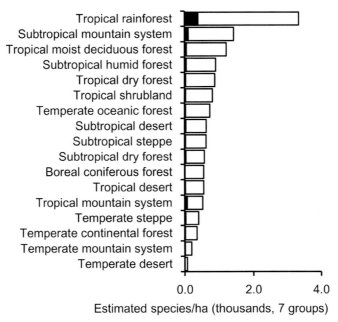

Figure 15.3 Estimate of endangered (black) and other (white) species richness by ecoregion [18]. Data from UNFRA groups.

15.3.2 Land Covers

"Global Land Covers" (Figure 4.2, Table15.2) contrast with Potential Vegetation Types (Figure 7.1, Table 15.1) because they reflect current vegetation and include human-dominated landscapes. They were inferred globally from several sources [18, 19, 23, 24]. Accessible land covers (Table 15.2) are shown in Figure 4.2, Table 15.2.

Grasslands (Table 15.2, Figure 15.4) are a mixture of cultivated pastures and semi-managed grazing land.

The smaller closed forest area and larger grassland area in Table 15.2 compared to 15.1 reinforces other studies showing most agriculture areas result from clearing forests [20]. Despite over 7 billion people living in the world, forests and shrublands still cover 36% of the world's land area and a greater proportion of the accessible land. Grasslands cover only slightly less than closed and open forests.

15.3.3 Past and Current Changes in Land Covers

The world's areas in different land covers have changed dramatically over the past 20 thousand years with changing climates (Figure 9.5, Chapter 14).

In addition, forests have been cleared for agriculture in China, India, Europe, and parts of Africa and the America's over the past few thousand years. Some areas have regrown to forests (Chapter 5). Forests currently occupy about two-thirds of their potential area under the

Table 15.2 *Global distribution of total and accessible land covers [18, 19, 23, 24]*

Land cover	All land cover		Accessible land cover	
	(%)	(million ha)	(%)	(million ha)
Closed forests	Total: 23%		Total: 28%	
Closed forests	22%	3,332	27%	3,006
Forest plantations	1%	113	1%	113
Open	Total: 35%		Total: 47%	
Open forests	3%	445	4%	445
Shrubs/trees	9%	1,302	12%	1,302
Forests fallow	1%	127	1%	127
Grasslands	23%	3,368	30%	3,368
Other, not human-dominated	Total: 28%		Total: 7%	
Deserts	16%	2,418	0%	0
Polar	4%	564	0%	0
Tundra	3%	407	0%	0
Other & uncertain	3%	408	4%	408
Inland water	2%	349	3%	349
Urban	Total: 3%		Total: 4%	
Impermeable	0.45%	67	0.60%	67
Built environment	0.20%	30	0.27%	30
Other urban	2.35%	349	3.14%	349
Agriculture	Total: 11%		Total: 14%	
Perennial crops	1%	142	1%	142
Annual crops	10%	1,410	13%	1,410
Total land	100%	14,830	100%	11,115

present climate. They will be further reduced with urban expansion and possible agriculture expansion (Chapters 5, 22). Forest clearing for agriculture has not been equal in all landforms. Many alluvial floodplain landforms have been nearly completely converted to agriculture, leaving little forest area and the biodiversity it once supported.

In much of the world, past grazing and human-generated fires in croplands and grasslands associated with agriculture activities commonly moved into adjacent forests, probably leaving the forests quite open. Evidence of past savanna structures are found in many places where the forests are now quite dense. When people began using kilns, smelters, fireplaces, and steam engines, they cleared the forest further for wood fuel [26].

The land cover shifted from agriculture and openings to dense forests in large parts of eastern North America – and probably South America – about 450 years ago when European colonists' diseases decimated Native American peoples [27–29]. Dense forests grew and encroached on former farmlands, savannas, and grasslands. As the European-American population increased, they again cleared and burned forests and recreated more open and savanna structures.

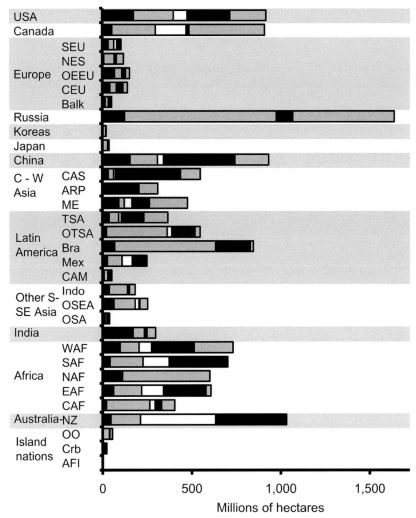

Figure 15.4 Distribution of major vegetation covers by country groups. (From left: black = agriculture, gray = forests, white = shrub, second black = pastures/grassland, right gray = other.) [19, 24, 25].

The world's forests began changing again dramatically during the past century. Older forests (complex and understory structures) were harvested for timber – and are still harvested in some areas. The forest area and volume are decreasing in the developing world but increasing in the developed world (Figures 30.3, 5). Other conservation efforts have promoted fire prevention, so the forests are growing denser; this fire prevention probably explains the paucity of open forests in the world (Table 15.2).

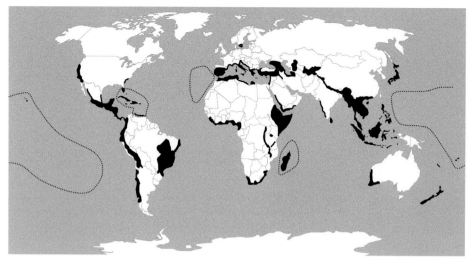

Figure 15.5 Biodiversity hotspots identified as areas where the survival of many species are of concern [30–32].

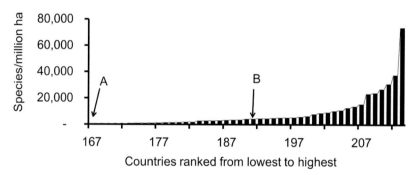

Figure 15.6 The 20% (right of A) and 10% (right of B) of world countries with highest species concentration in ascending order of concentration [18]. Data from UNFRA groups.

From a coarse filter perspective – the conservation of all habitats – the world now probably has an excess of dense forests and a shortage of open, savanna, and complex forests, despite the overharvesting in the developing world (Chapter 30).

15.3.4 Biodiversity Hotspots, Species Concentrations

Because species, threatened species, and biodiversity are not uniformly distributed, efforts to avoid extinctions can be focused on those areas of high biodiversity and/or threatened species, sometimes known as "biodiversity hotspots" (Figure 15.5) [30–32].

Relatively few, small countries contain the highest concentrations of species in the world (Figure 15.6) [18] (UNFRA groups).

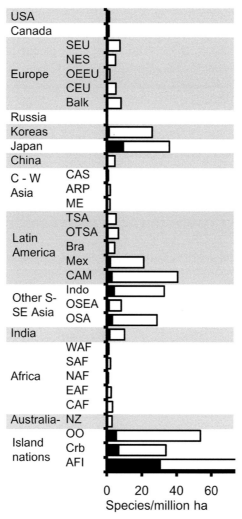

Figure 15.7 Concentration of endangered (black) and other (white) species number by country groups [18]. Data from UNFRA groups.

Twenty percent of the countries (Figure 15.6) have more than 500 targeted (UNFRA groups, Chapter 13) species per million ha; together, these countries contain 0.1% of the world's land area and 0.4% of the human population. And, 10% of the countries (right of "B") have more than 4,500 species per million ha, but contain 0.005% of the land area and 0.12% of the human population. The countries to the right of "A" are primarily islands (island biogeography, Chapter 13), with some Central American nations.

Figure 15.7 shows concentrations of the same species as Figure 15.6, except by geographic regions and also shows endangered species. Not all tropical rain forests contain high species concentrations. Brazil and other South American countries, and others in

Southeast Asia, India, and Africa have relatively low species concentrations, possibly because of their large areas of uniform topography and climate. Confined spaces have high concentrations of species, such as island regions and peninsulas – African, Caribbean, and Oceanic Islands; Japan, Central America, and the Koreas. A combination of island biogeography principles and rising sea level during the past several thousand years may have concentrated species into these smaller areas. Many tropical areas of high species concentrations contain mountains – Indonesia, Malaysia, Columbia, and Peru – which further harbor many species (Chapter 14). Mexico also has a high concentration of species with great topographic relief, is on a recently joined suture zone of two floristic realms, and contains both tropical and temperate climates.

Figures 15.6–7 suggest concerted efforts to restore habitats and conserve biodiversity in a quite small proportion of the world could dramatically reduce threats of species extinctions. Research by Zeydanli [33] further suggests that key topographic features exist within a landscape that are critical to many endangered species; consequently, it may be possible to be very efficient and effective in conserving biodiversity through a general plan of habitat management in larger areas and targeting key areas for preservation or restoration.

15.4 Fine Filter Conservation

Once areas that may be sensitive to species threats have been identified and addressed through coarse filter conservation, individual species that are still threatened can be efficiently pinpointed through fine filter approaches.

15.4.1 Species Threatened with Extinction

The proportion of threatened species within different taxonomic groups varies. Comparing Tables 15.3 and 13.1 shows that 16% of the mammals, birds, reptiles, amphibians, and fish are threatened and 4% of the plants. Less than 1% of other groups are listed as threatened – because the assessment is incomplete, the other species are more resilient, or they are less identifiable and verifiable as threatened. The 24 thousand identified, threatened species are within the range of estimated 2- to 40 thousand species that become extinct each year (Chapter 13).

The flowering plant group seems to have the greatest number of threatened species (Table 15.3, Figure 15.8), especially in tropical areas, partly because the total number of plant species is large. The proportion of plant species that are threatened is not high. Fish seem to be threatened in most coastal, fish-abundant regions. Mammal species are most threatened in the tropics and birds seem threatened in all places.

Threats to "non-charismatic" species – amphibians, molluscs, reptiles, and insects – suggest that people may be poisoning or otherwise depleting the environment more than expected. These species are building blocks needed for complex ecological communities.

Table 15.3 *Numbers of threatened species by major taxonomic groups [34].*
(a) Taxonomic groups that are mostly assessed worldwide; (b) groups not
completely assessed, so may be under-recorded

Taxonomic group	Species numbers	Taxonomic group	Species numbers	Taxonomic group	Species numbers
(a) *Good estimates*		(b) *Incomplete estimates*			
Amphibians	2,068	All algae	9	Mollusks	1,984
Birds	1,460	Fish	2,359	Reptiles	1,079
Mammals	1,194	Insects	1,268	Fungi and protists	34
Gymnosperms	400	Flowering plants	10,941		
		Mosses and ferns (and fern allies)	293	Crustaceans, coral, and other invertebrates	1,218

The same poisons could also concentrate in humans and large charismatic animals in the upper food pyramid (Figure 13.3)

Vertebrate animal species distributions are somewhat congruent with plant species, but different groups vary [35]. Both mammal and bird species are least concentrated ("species rich") in cold and dry areas, moderately rich in temperate moist climates, and most rich in moist tropical environments. Bird species are richest in the northern Andes of South America, but not very rich in parts of southeastern North America and Asia north of the Himalayas. Mammal species are richest in eastern Africa and generally very rich in equatorial Africa, the Andes, and northern South America.

Several reasons why more species live in the tropics have been put forward. The longer growing season in warm climates allows animals and plants to breed more often each year, thus creating more genetically different individuals to form species. The climate and long growing season also increases the productivity, and therefore may increase the numbers of niches for species to exploit. Also, the tropics have not undergone extremely cold cycles during glacial maxima, so resident species were not eliminated by cold, ice, or forced migrations.

Some genetic groups are concentrated outside of the tropics [13]. Marine benthic organisms tend to be more diverse at temperate latitudes, probably because colder water contains more oxygen and because of upwellings on the west sides of continents (Chapter 7). Sea birds follow the same pattern, probably because they are attracted to fish, which are attracted to these other marine organisms. Lichen diversity tends to be greatest in cold and/or dry climates because they are apparently outcompeted by vascular plants in other places. Parasitic wasps tend to reach peak diversity at mid-latitudes as do

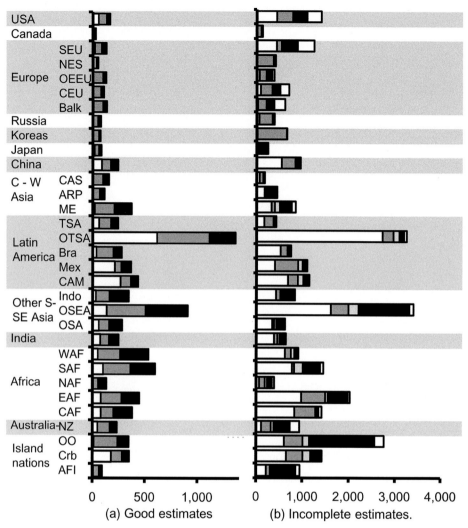

Figure 15.8 Numbers of threatened species by country groups and type [34]. (a) Good estimates: white = amphibians, gray = birds, black = mammals. Note scale change. (b) Data based on incomplete estimates. From left: first white = plants, dark gray = fishes, light gray = reptiles, black = other invertebrates, right white = mollusks. Data used with permission of IUCN.

nematodes, also because there are fewer competitors and predators to check their populations [13].

15.4.2 Local Variations in Species Distributions

The numbers of species found together within a genus vary dramatically by geographic location. Oak (*Quercus*) species numbers are high in Mexico and eastern North America but low in northern Europe. One native species of pine – Scotts pine (*Pinus sylvestris*) – predominates from the Scandinavian countries to southern France and the Pyrenees–Alps–Carpathian Mountains and extends into Asia. By contrast, Asia Minor contains at least four native pine species; and southeastern North America contains at least thirteen. Other species are concentrated elsewhere, with many fir (*Abies*) species in western North America, although some firs are found in many Holarctic forests. Similarly, a few evergreen *Rhododendron* species are found in North America and Asia Minor, but many are found in the Himalayas. Only three species of *Liquidambar* are found in the world – one in southeastern North America, one in southwestern Asia Minor, and one in southeastern Asia. Causes of these differences in species distributions are unclear and probably varied.

Speciation is occurring at present. Balsam fir (*Abies balsamea*) began returning north from southeastern North America after the last glaciation. One population moved to the tops of the Appalachian Mountains, while another moved farther north at low elevations and into Canada. The Appalachian population is now isolated and considered a separate species – Fraser fir (*Abies fraseri*) [14].

References

1. M. L. Hunter Jr. *Coping with Ignorance: The Coarse-Filter Strategy for Maintaining Biodiversity.* Kohm K. A., editor. (Island Press, 1991).
2. M. L. Hunter, G. L. Jacobson, T. Webb. Paleoecology and the Coarse-Filter Approach to Maintaining Biological Diversity. *Conservation Biology.* 1988;2(4):375–85.
3. The Nature Conservancy. *Natural Heritage Program Operations Manual.* (The Nature Conservancy, 1982).
4. H. Walter, S.-W. Breckle. *Walter's Vegetation of the Earth: The Ecological Systems of the Geo-Biosphere.* (Springer, 2002).
5. T. H. Snelder, B. J. Biggs. *Multiscale River Environment Classification for Water Resources Management.* (Wiley Online Library, 2002 [Accessed May 21, 2017]). Available from: http://onlinelibrary.wiley.com/doi/10.1111/j.1752-1688.2002.tb04344.x/abstract.
6. R. G. Bailey. Delineation of Ecosystem Regions. *Environmental Management.* 1983;7:365–73.
7. H. Walter. *Vegetation of the Earth in Relation to Climate and the Eco-Physiological Conditions.* (English Universities Press, 1973).
8. T. Flannery. *The Eternal Frontier: An Ecological History of North America and Its Peoples.* (Atlantic Monthly Press, 2001).
9. T. Flannery. *The Future Eaters: An Ecological History of the Australasian Lands and People.* (Grove Press, 2002).
10. L. Wade. Monkey Ancestors Rafted across the Sea to North America. *Science.* 2016 (April 20, 2016).
11. S. A. Jansa, F. K. Barker, R. S. Voss. The Early Diversification History of Didelphid Marsupials: A Window into South America's "Splendid Isolation." *Evolution.* 2014;68(3):684–95.
12. G. G. Simpson. *Splendid Isolation: The Curious History of South American Mammals.* (Yale University Press, 1980).
13. M. A. Huston. *Biological Diversity: The Coexistence of Species.* (Cambridge University Press, 1994).
14. M. B. Davis. Late-Glacial Climate in Northern United States: A Comparison of New England and the Great Lakes Region, in Cushing E. J., Wright H. E., editors, *Proceedings of the VII*

Congress of the International Association for Quaternary Research. 7. (Yale University Press, 1967), pp. 11–43.

15. J. E. McCormack, H. Huang, L. L. Knowles, et al. *Encyclopedia of Islands*. 2009;4:841-3.
16. C. White. *Revolution on the Range: The Rise of a New Ranch in the American West*. (Island Press, 2012).
17. X. Lee, M. L. Goulden, D. Y. Hollinger, et al. Observed Increase in Local Cooling Effect of Deforestation at Higher Latitudes. *Nature*. 2011;479(7373):384–7.
18. UN FAO. *Global Forest Resources Assessment 2000: Main Report*. (United Nations Food and Agriculture Organization, 2001).
19. UN FAO. *State of the World's Forests 2011*. (United Nations Food and Agriculture Organization, 2011).
20. J. H. Tallis. *Plant Community History: Long-Term Changes in Plant Distribution and Diversity*. (Chapman and Hall, 1991).
21. I. Aselmann, P. Crutzen. Global Distribution of Natural Freshwater Wetlands and Rice Paddies, Their Net Primary Productivity, Seasonality and Possible Methane Emissions. *Journal of Atmospheric Chemistry*. 1989;8(4):307–58.
22. C. Prigent, F. Papa, F. Aires, et al. Changes in Land Surface Water Dynamics since the 1990s and Relation to Population Pressure. *Geophysical Research Letters*. 2012;39(8).
23. UN FAO. *FAOSTAT: Download Data*. (United Nations Food and Agriculture Organization, 2009 [Accessed 2012 September 16, 2012]). Available from: http://faostat3.fao.org/download/.
24. Z. Liu, C. He, Y. Zhou, J. Wu. How Much of the World's Land Has Been Urbanized, Really? A Hierarchical Framework for Avoiding Confusion. *Landscape Ecology*. 2014;29(5):763–71.
25. UN FAO. *FLUDE: The Forest Land Use Data Explorer*. (United Nations Food and Agriculture Organization, 2015 [Accessed May 25, 2016]). Available from: www.fao.org/forest-resources-assessment/explore-data/en/.
26. J. Perlin. *A Forest Journey: The Story of Wood and Civilization*. (The Countryman Press, 2005).
27. J. E. Fickle. *Green Gold: Alabama's Forests and Forest Industries*. (University of Alabama Press, 2014).
28. C. C. Mann. *1491: New Revelations of the Americas before Columbus*. (Vintage Books, 2005).
29. H. Savage Jr. *Lost Heritage: Wilderness America through the Eyes of Seven Pre-Audubon Naturalists*. (William Morrow, 1970).
30. N. Myers, R. A. Mittermeier, C. G. Mittermeier, G. A. Da Fonseca, J. Kent. Biodiversity Hotspots for Conservation Priorities. *Nature*. 2000;403(6772):853–8.
31. K. J. Willis, M. B. Araújo, K. D. Bennett, et al. How Can a Knowledge of the Past Help to Conserve the Future? Biodiversity Conservation and the Relevance of Long-Term Ecological Studies. *Philosophical Transactions of the Royal Society B: Biological Sciences*. 2007;362 (1478):175–87.
32. R. Mittermeier, R. Gil, M. Hoffman, et al. *Hotspots Revisited: Earth's Biologically Richest and Most Endangered Terrestrial Ecoregions*. (University of Chicago Press, 2005).
33. U. Zeydanli. *Mapping Areas for Species Protection in Turkey*. Personal Communication, based on a Seminar Presentation. Dr. Zeydanli was a Fulbright Senior Fellow to Yale University, from Ankara, Turkey. (Global Institute of Sustainable Forestry, Yale University, 2016).
34. International Union for Conservation of Nature. *The IUCN Red List of Threatened Species. Version 2016–3*. 2016 (Accessed February 28, 2016). Available from: www.iucnredlist.org.
35. R. Grenyer, C. D. L. Orme, S. F. Jackson, et al. Global Distribution and Conservation of Rare and Threatened Vertebrates. *Nature*. 2006;444(7115):93–6.

16

Managing Biodiversity

16.1 People's Effects on Biodiversity

Few people were concerned with preventing species extinctions or conserving biodiversity until about one hundred years ago. Animals were sometimes hunted to extinction and plants were allowed to become extinct if not immediately useful. Plants and animals were moved throughout the world with little regard for what they did to native species.

People have affected species by changing their genetics, where they live on the Earth, their abundance, their habitats, their behaviors, and even their existence [1].

People have changed both the genotypes and behaviors of plants and animals by domesticating them. Except for the dog, domesticated about 30 thousand years ago, most domestications are believed to have begun in the past 12 thousand years – pigs, goats, sheep, cows, and horses [2, 3]. Plants were largely domesticated with the change from hunter-gatherer to agrarian cultures [4]. Genetic changes with domestication through breeding were first intuitive, but have become systematic with Mendelian genetics and genetic engineering. Domestication is discussed in more detail in Chapter 21.

The genotypes of non-domesticated species have also changed because of people's activities. Black pines (*Pinus nigra*) growing around Saros Bay in the Aegean Sea are genetically very crooked because millennia of shipbuilders harvested the straight trees (Figure 16.1a). Intensive goat, sheep, and pig grazing and burning in the coastal Mediterranean vegetation for thousands of years caused evolution of many shrubby, often thorny species ("maquis" vegetation) that compete well and depend on these open and constantly disturbed conditions. Recent declines in the grazing and burning culture have actually endangered these "human-induced" maquis species (Figure 16.1b) [5, 6].

The amounts and distributions of habitats have changed (Table 15.2) with agriculture expansion and abandonment, endangering some species [7] and favoring others. Diversions of water have affected lakes and wetlands – most notably the Aral Sea [8, 9] – and so led species to extinction. Dams, levees, fences, roads, and other infrastructures have also displaced or changed habitats and altered migration routes.

Figure 16.1 (a) Genetically crooked black pine trees beside Saros Bay, Aegean Sea, Turkey. (b) Maquis vegetation along the northeastern Mediterranean Sea, Turkey [5, 6].

Exotic species introductions have accelerated, creating a recent rise in native species extinctions as people colonized new areas or introduced foreign cultures. Goats were intentionally introduced to Pacific islands, grassland plant species to North America from Mongolia, pigs and others species to North America, and many species to Australia [10, 11]. Invasive exotic "hitchhiker" species such as insects, microorganisms, and mollusks inadvertently accompany imported goods and attack, displace, and/or extirpate native species.

People also change the behavior of plants and animals without necessarily changing their genetic makeup. A desired agricultural or ornamental plant is put in an artificially dominant position through "weed" control. Animals also change their behavior as people's behaviors change [12, 13].

People extirpated large animal and bird species when they first entered Australia, North and South America, and the Pacific Islands, and elsewhere (Chapter 13). By contrast, many large animals still exist in Africa, although people have been living there longer than anywhere else. Chemical uses and air and water pollution seem to be inadvertently endangering species. Non-agriculture encroachments into tropical forests is leading to "bushmeat hunting" – the hunting of wild animals, endangered or otherwise, for food whose origin is disguised (Chapter 30) [14]. Modern demands for animal parts – rhinoceros horns, elephant tusks, and tiger parts – are also promoting hunting of endangered species which, although illegal, is very lucrative.

16.2 Approaches to Promoting Biodiversity

Specific ideas for promoting biodiversity are suggested below that may be implemented together or individually. Three important things about promoting biodiversity are:

• Primary ways to conserve biodiversity are to maintain unpolluted air and water; to limit spreading exotic species; and to maintain all vegetation types, habitats, vegetation structures, and connectivity among them throughout the world.
• Conserving biodiversity is not an absolute preservation of the present condition. Species have always moved to new locations, found new habitats, gone extinct, and emerged anew. Consequently, the objective is to slow the rate of change to the background level (Chapter 14).
• Biodiversity is a long-term value, and so is best conserved when people are wealthy and educated enough to be concerned about the long term. Consequently, biodiversity can best be conserved when complex societies are maintained and peaceful (Chapter 6) [15].

Biodiversity can be promoted through a combination of policies and activities. Specific ideas are suggested below that may be implemented together or individually.

16.3 Policies for Promoting Biodiversity

Regional, national, and international laws have been developed beginning in the 1970s (US Endangered Species Act [16]; international Convention on the International Trade of Endangered Species [17]) that identify, mandate protection, regulate trade, or otherwise promote the protection and recovery of endangered species.

Maintaining all habitats has been promoted under the concepts of "rewilding," "landscape management," and "restoration" [18–25]. The International Union for the Conservation of Nature (IUCN), the Nature Conservancy (TNC), and the World Wildlife Fund (WWF) all try to conserve habitats throughout the world – with slightly different focuses [26].

All species and habitats will probably best be conserved if representative ecosystems are in "reserves" (Figure 16.2). The world contains over 15% of its forests in reserves and more reserves in other vegetation types. These reserves could be better distributed to conserve needed ecosystems appropriately.

The forest and other lands not in agriculture, urban areas, or other human dwellings could complement the reserves by people proactively restoring them to needed structures [18, 19]

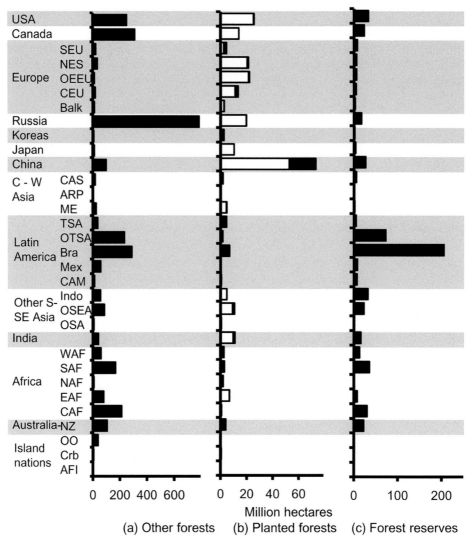

Figure 16.2 Forest land uses by country groups [37] (b) white = planted with native species, black = planted with introduced species.

and integrating management for biodiversity with other values (Chapters 28–30) [27–30]. These activities would ensure all habitats, structures (Figure 14.5), corridors [31, 32] and other features, and ecosystem services [33, 34] are restored, maintained, and enhanced while providing other resource raw materials and economic returns as well as rural employment [35, 36]. (See "Kristianstads Vattenrike" case study, Chapter 3.)

    Other parts of each area could be intensively managed for agriculture crops, cities, mines, and similar places [18, 35].

16.3.1 Species Models and Inventories

The public inventories of all species and explicit identification of endangered species (Chapter 13) are important steps in systematically addressing biodiversity. Enough knowledge and data now exists to begin understanding and identifying which, where, and why species are most endangered and to begin avoiding extinctions. Modeling technologies can help identify threatened species groups and individuals and develop ways to save them.

16.3.2 Targeting Areas of High Biodiversity Sensitivity

Figures 15.5–7 indicate that a significant part of the biodiversity concern could be alleviated by concentrating on relatively small parts of the terrestrial world. Such areas have been identified by several groups with slightly different emphases (Figure 15.3) [38–49] and a "Key Biodiversity Area Database" is being developed [50].

16.3.3 Biodiversity Mitigation Bank

An area of high species concentration does not necessarily mean that the species are endangered. Regression analyses for this book showed that only 31% ($R^2 = 0.31$) of the concentration of endangered species was correlated to the concentration of all species in the world's countries, indicating that high species concentrations can occur with viable communities.

   Those countries with greatest biodiversity, however, should not bear the cost of conserving species and thus be punished for owning a global treasure. A mechanism is needed to share conservation costs among all people, such as by requiring a country to own "biodiversity credits" in order to participate in global trade. The credits could be "owned" either by a country with high biodiversity maintaining well-conserved biodiversity or a country with little biodiversity purchasing excess credits from countries with high biodiversity through a "biodiversity mitigation bank."

16.3.4 Maximizing Land Area for Habitats

As much as 19% of terrestrial land may be in agriculture by 2050, with a concomitant reduction in forest area from 27% to 19% (Table 22.2). Also, urban expansion may increase from currently 3% to 9% of the total land by 2050 [51], further encroaching on land that harbors biodiversity.

   Considering the wide fluctuations in land cover during the past 20,000 years (Figure 9.5), the current reductions will not be a crisis; however, more potential habitat to support biodiversity could be provided and less aquifer drawdown would occur by reducing the area in cropland and minimizing the area of human structures. Crop plant production has become more efficient in the past century; however, total food production could become less intrusive by several activities discussed in Chapters 22–23.

16.3.5 Quarantines and Economic Development

A global policy could disallow unprocessed organic materials from being transported across oceans or similar barriers in order to prevent exotic hitchhiker pests from accompanying the raw material and becoming harmful in the importing country [52]. Food, wood (including pallets), pulp, flower, horticulture, and similar industries could establish their processing facilities and make finished products that are pest-free in the region of raw material origin. The process is similar to current quarantine actions and may require amending many current trade agreements. The policy would have the added benefit of increasing secondary manufacture and thus economic development in countries with currently "extractive economies" (Chapter 6) [53].

16.4 Activities to Promote Biodiversity

Policies need to be complemented with proactive field activities. Examples of field activities are suggested below.

16.4.1 Landcare

An innovation is emerging in Australia and possibly elsewhere termed "landcare" [54–56]. Technically trained people are employed to restore degraded areas, including forests, wetlands, and shrublands, proactively by a combination of removing aggressive, exotic species and replanting native species. It is a commitment to biodiversity that has other implications.

   Land care could become a social driver in rural areas, thus reversing the current urban migration (Chapter 5) while creating ecosystem services, commodities, aesthetically pleasing rural landscapes, secondary employment, and other social capital [57–59].

16.4.2 Moving Species Elsewhere

Escaped animals and plants have become indigenous in the wild in other parts of the world – usually inadvertently. On the one hand, introduced animals and plants may be displacing native species in their new territories; on the other, they may be kept from extinction by living in these new homes – and in some cases may be impossible to eradicate. So, these species may become accepted as parts of "novel" ecosystems [13, 60].

16.4.3 Reintroducing Species and Species Analogies

Animals and plants are sometimes reintroduced where they have become extinct (or nearly so). The California condor (*Gymnogyps californianus*) numbers were increased by captive breeding before the animals were released to the wild [61].

Proactive measures to "conserve" or "recover" endangered species are creating species with novel genetic makeups [20, 62] such as the "new" American chestnut (*Castanea dentata*) which was bred with the Chinese chestnut (*Castanea mollissima*) for blight resistance [63]. The Florida cougar (*Puma concolor coryi*) was saved from inbreeding and extinction by breeding it with cougars from other regions [64]. The wild horse (*Equus ferus*) in central Asia is recovering by breeding individuals found in zoos and reintroducing them to the wild. And, domestic cattle (*Bos Taurus*) are being bred to recover the wild Auroch (*Bos taurus primigenius*) in Europe [65]. Such efforts save a large proportion of the genetic makeup that would have been completely lost if the species became extinct.

"Pleistocene rewilding" is a suggestion to replace species in niches and trophic levels (Figures 13.2–3) that existed before human-related extinctions [21, 22, 25]. A closely related animal (or plant) would be introduced in the wild to replace extinct species. At an extreme, the African elephant could replace the extinct mastodon in North America; an Asian camel could replace the extinct North American camel; and an Asian tiger could replace the extinct saber-toothed cat of North America. The large areas needed for viable populations of large animals may limit the feasibility of some rewilding.

Whether introductions are continued will depend on their scientific and social merit as well as their cost. Introducing similar tree species to North America, such as the ash trees (*Fraxinus* species) of the Russian Far East, may be economically viable because of its lumber value in case the introduced emerald ash borer insect (*Agrilus planipennis*) drives the American white ash (*Fraxinus americana*) to extinction.

16.4.4 Chemical Issues and Integrated Pest Management

Many chemicals intentionally or inadvertently introduced to the atmosphere or soil have had negative consequences on biodiversity (Chapter 7) [66], and we may discover others also have negative consequences. The protocols of Integrated Pest Management (IPM; Chapter 21) are probably the best way to address the chemical issues, including fossil fuels [67]. And, agriculture using fewer chemicals is becoming promising for other reasons as well (Chapters 22–23).

16.4.5 Hunting to Promote Biodiversity

Hunting can benefit or harm biodiversity depending how it is done [68, 69]. Unplanned hunting that eliminates the largest or most healthy animals or reduces a population below the critical size needed for survival can drive the species to extinction. On the other hand, planned hunting can cull unhealthy or physically deformed animals that would have been most subject to predators, regulate the population size to avoid exceeding its habitat's carrying capacity, regulate the male:female ratio, instill a fear of humans in such aggressive animals as wolves and the cat family [12], provide incomes and thus incentives to conserve the animal populations – ivory and horn hunters, for example [70–73] – and provide income to fund conservation of threatened species [74].

References

1. C. Rammel, S. Stagl, H. Wilfing. Managing Complex Adaptive Systems – A Co-Evolutionary Perspective on Natural Resource Management. *Ecological Economics*. 2007;63(1):9–21.
2. N. D. Ovodov, S. J. Crockford, Y. V. Kuzmin, et al. A 33,000-Year-Old Incipient Dog from the Altai Mountains of Siberia: Evidence of the Earliest Domestication Disrupted by the Last Glacial Maximum. *PLoS One*. 2011;6(7):e22821.
3. D. W. Anthony. *The Horse, the Wheel, and Language: How Bronze-Age Riders from the Eurasian Steppes Shaped the Modern World*. (Princeton University Press, 2010).
4. D. Rindos. *The Origins of Agriculture: An Evolutionary Perspective*. (Academic Press, 2013).
5. A. Grove, O. Rackham. Threatened Landscapes in the Mediterranean: Examples from Crete. *Landscape and Urban Planning*. 1993;24(1–4):279–92.
6. R. Tomaselli. The Degradation of the Mediterranean Maquis. *Ambio*. 1977:356–62.
7. K. S. Thomson. Marginalia: Benjamin Franklin's Lost Tree. *American Scientist*. 1990;78(3): 203–6.
8. B. Malmqvist, S. Rundle. Threats to the Running Water Ecosystems of the World. *Environmental Conservation*. 2002;29(02):134–53.
9. P. P. Micklin, W. D. Williams. *The Aral Sea Basin*. (Springer, 1996).
10. P. Salo, E. Korpimäki, P. B. Banks, M. Nordström, C. R. Dickman. Alien Predators Are More Dangerous Than Native Predators to Prey Populations. *Proceedings of the Royal Society of London B: Biological Sciences*. 2007;274(1615):1237–43.
11. A. A. Burbidge, N. McKenzie. Patterns in the Modern Decline of Western Australia's Vertebrate Fauna: Causes and Conservation Implications. *Biological Conservation*. 1989;50(1):143–98.
12. P. Canby. The Cat Came Back: Alpha Predators and the New Wilderness. *Harper's Magazine*. 2005:95–102.
13. T. Low. *The New Nature: Winners and Losers in Wild Australia*. (Penguin Australia Ltd, 2002).
14. P. Lindsey, G. Balme, M. Becker, et al. *Illegal Hunting & the Bushmeat Trade in Savanna Africa: Drivers, Impacts & Solutions to Address the Problem*. (Panthera, Zoological Society of London, Wildlife Conservation Society, 2013).
15. K. Conca, A. Carius, G. D. Dabelko. *Building Peace through Environmental Cooperation*. (Worldwatch Institute, 2005).
16. O. A. Houck. The Endangered Species Act and Its Implementation by the US Departments of Interior and Commerce. *University of Colorado Law Review*. 1993;64(2):277.
17. CITES. *The Cites Appendices*. (Convention on International Trade in Endangered Species of Wild Fauna and Flora, 2013 [Accessed August 21, 2016]). Available from: https://cites.org/eng/app/index.php.
18. J. A. Stanturf, B. J. Palik, R. K. Dumroese. Contemporary Forest Restoration: A Review Emphasizing Function. *Forest Ecology and Management*. 2014;331:292–323.
19. D. C. Dey, E. S. Gardiner, J. M. Kabrick, J. A. Stanturf, D. F. Jacobs. Innovations in Afforestation of Agricultural Bottomlands to Restore Native Forests in the Eastern USA. *Scandinavian Journal of Forest Research*. 2010;25(S8):31–42.
20. C. J. Donlan. Restoring America's Big, Wild Animals. *Scientific American*. 2007;296(6): 70–7.
21. J. Donlon, J. Berger, C. E. Bock, et al. Pleistocene Rewilding: An Optimistic Agenda for Twenty-First Century Conservation. *The American Naturalist*. 2006;168:660–81.
22. T. Flannery. *The Eternal Frontier: An Ecological History of North America and Its Peoples*. (Atlantic Monthly Press, 2001).
23. J. H. Cissel, F. J. Swanson, P. J. Weisberg. Landscape Management Using Historical Fire Regimes: Blue River, Oregon. *Ecological Applications*. 1999;9(4):1217–31.

24. J. M. C. Silva, C. Uhl, G. Murray. Plant Succession, Landscape Management, and the Ecology of Frugivorous Birds in Abandoned Amazonian Pastures. *Conservation Biology.* 1996;10 (2):491–503.

25. A. C. Kitchener. Re-Wilding Ireland: Restoring Mammalian Diversity or Developing New Mammalian Communities? *Irish Naturalists' Journal.* 2012:4–13.

26. E. A. Gordon, O. E. Franco, M. L. Tyrrell. *Protecting Biodiversity: A Guide to Criteria Used by Global Conservation Organizations.* (Yale School of Forestry & Environmental Studies, 2005).

27. S. G. Letcher, R. L. Chazdon. Rapid Recovery of Biomass, Species Richness, and Species Composition in a Forest Chronosequence in Northeastern Costa Rica. *Biotropica.* 2009;41(5): 608–17.

28. D. Piotto, F. Montagnini, W. Thomas, M. Ashton, C. Oliver. Forest Recovery after Swidden Cultivation across a 40-Year Chronosequence in the Atlantic Forest of Southern Bahia, Brazil. *Plant Ecology.* 2009;205(2):261–72.

29. P. Kareiva, S. Watts, R. McDonald, T. Boucher. Domesticated Nature: Shaping Landscapes and Ecosystems for Human Welfare. *Science.* 2007;316(5833):1866–9.

30. O. J. Schmitz. *The New Ecology: Rethinking a Science for the Anthropocene.* (Princeton University Press, 2016).

31. R. L. Chazdon, C. A. Harvey, O. Komar, et al. Beyond Reserves: A Research Agenda for Conserving Biodiversity in Human-Modified Tropical Landscapes. *Biotropica.* 2009;41(2): 142–53.

32. C. A. Harvey, W. A. Haber. Remnant Trees and the Conservation of Biodiversity in Costa Rican Pastures. *Agroforestry Systems.* 1998;44(1):37–68.

33. M. L. Oldfield. *The Value of Conserving Genetic Resources.* (Sinauer Associates Inc., 1984).

34. N. Myers. Environmental Services of Biodiversity. *Proceedings of the National Academy of Sciences.* 1996;93(7):2764–9.

35. R. S. Seymour, M. L. Hunter, Jr. Principles of Ecological Forestry, in Hunter M. L., Jr., editor, *Maintaining Biodiversity in Forest Ecosystems.* (Cambridge University Press, 1999): pp. 22–61.

36. C. R. Margules, R. L. Pressey. Systematic Conservation Planning. *Nature.* 2000;405(6783): 243–53.

37. UN FAO. *FLUDE: The Forest Land Use Data Explorer.* (United Nations Food and Agriculture Organization, 2015 [Accessed May 25, 2016]). Available from: www.fao.org/forest-resources-assessment/explore-data/en/.

38. S. Jennings, R. Nussbaum, N. Judd, et al. The High Conservation Value Forest Toolkit. *Proforest.* 2003.

39. N. Myers, R. A. Mittermeier, C. G. Mittermeier, G. A. Da Fonseca, J. Kent. Biodiversity Hotspots for Conservation Priorities. *Nature.* 2000;403(6772):853–8.

40. Conservation International. Hotspots 2016 (Accessed September 16, 2016). Available from: www.conservation.org/How/Pages/Hotspots.aspx.

41. D. M. Olson, E. Dinerstein. The Global 200: Priority Ecoregions for Global Conservation. *Annals of the Missouri Botanical Garden.* 2002:199–224.

42. World Wildlife Fund. *Global 200.* (World Wildlife Fund, 2012 [Accessed September 16, 2016]). Available from: www.worldwildlife.org/publications/global-200.

43. World Wildlife Fund. *About Global Ecoregions.* (World Wildlife Fund, 2016 [Accessed September 16, 2016]). Available from: http://wwf.panda.org/about_our_earth/ecoregions/about/.

44. Ramsar. *Ramsar Sites around the World.* (The Ramsar Convention Secretariat, 2014 [Accessed September 16, 2016]). Available from: www.ramsar.org/sites-countries/ramsar-sites-around-the-world.

45. Bird Life International. *Important Bird and Biodiversity Areas (IBAS).* (Bird Life International, 2016 [Accessed September 16, 2016]). Available from: www.birdlife.org/worldwide/programmes/sites-habitats-ibas.

46. Alliance for Zero Extinction. *Aze Map.* (American Bird Conservancy, 2013 [Accessed September 16, 2016]). Available from: www.zeroextinction.org/maps/AZE_map_12022010s .pdf.

47. World Conservation Monitoring Center IUCN. *The World Database on Protected Areas (WDPA)*. (ProtectedPlanet, 2010 [Accessed September 16, 2016]). Available from: www.protectedplanet.net.

48. UNESCO. *World Heritage List*. (U.N. Educational, Scientific, and Cultural Organization, World Heritage Center, 2016 [Accessed September 16, 2016]). Available from: http://whc.unesco.org/en/list/.

49. NatureServe. *The Biodiversity Indicators Dashboard*. ([Accessed September 16, 2016]). Available from: http://dashboard.natureserve.org/.

50. Key Biodiversity Areas Partnership. *World Database of Key Biodiversity Areas* 2016 (Accessed September 16, 2016). Available from: http://birdlaa2.memset.net/kba/home.

51. K. C. Seto, B. Güneralp, L. R. Hutyra. Global Forecasts of Urban Expansion to 2030 and Direct Impacts on Biodiversity and Carbon Pools. *Proceedings of the National Academy of Sciences*. 2012.

52. T. A. Crowl, T. O. Crist, R. R. Parmenter, G. Belovsky, A. E. Lugo. The Spread of Invasive Species and Infectious Disease as Drivers of Ecosystem Change. *Frontiers in Ecology and the Environment*. 2008;6(5):238–46.

53. D. Acemoglu, J. Robinson. *Why Nations Fail: The Origins of Power, Prosperity, and Poverty*. (Crown Business, 2012).

54. NRM Group. *An Evaluation of Investments in Landcare Support Projects*. (URS Australia and Griffin, 2001).

55. A. Campbell, G. Siepen. *Landcare: Communities Shaping the Land and the Future*. (ICON Group International, 1994).

56. R. Youl, S. Marriott, T. Nabben. *Landcare in Australia*. (SILC and Rob Youl Consulting Pty Ltd, 2006).

57. J. Cary, T. Webb. Landcare in Australia: Community Participation and Land Management. *Journal of Soil and Water Conservation*. 2001;56(4):274–8.

58. E. Compton, R. B. Beeton. An Accidental Outcome: Social Capital and Its Implications for Landcare and the "Status Quo." *Journal of Rural Studies*. 2012;28(2):149–60.

59. A. Curtis, M. Lockwood. Landcare and Catchment Management in Australia: Lessons for State-Sponsored Community Participation. *Society and Natural Resources*. 2000;13(1):61–73.

60. R. J. Hobbs, S. Arico, J. Aronson, et al. Novel Ecosystems: Theoretical and Management Aspects of the New Ecological World Order. *Global ecology and biogeography*. 2006;15(1):1–7.

61. N. F. Snyder, H. Snyder. *The California Condor: A Saga of Natural History and Conservation*. (Academic Press San Diego, 2000).

62. E. O'Rourke. The Reintroduction and Reinterpretation of the Wild. *Journal of Agricultural and Environmental Ethics*. 2000;13(1–2):145–65.

63. D. F. Jacobs. Toward Development of Silvical Strategies for Forest Restoration of American Chestnut (Castanea Dentata) Using Blight-Resistant Hybrids. *Biological Conservation*. 2007;137(4):497–506.

64. D. S. Maehr, R. C. Lacy, E. D. Land, O. L. Bass Jr., T. S. Hoctor. *Evolution of Population Viability Assessments for the Florida Panther: A Multiperspective Approach*. (University of Chicago Press, 2002).

65. P. J. Seddon, A. Moehrenschlager, J. Ewen. Reintroducing Resurrected Species: Selecting Deextinction Candidates. *Trends in Ecology and Evolution*. 2014;29(3):140–7.

66. R. Carson. *Silent Spring*. (Houghton Mifflin, 1962).

67. M. L. Flint, R. van den Bosch. The Philosophy of Integrated Pest Management. *Introduction to Integrated Pest Management*. (Springer, 1981): pp. 107–19.

68. M. Festa-Bianchet, M. Apollonio. *Animal Behavior and Wildlife Conservation*. (Island Press, 2003).

69. R. Steneck. *An Ecological Context for the Role of Large Carnivorous Animals in Conserving Biodiversity*. (Island Press, 2005).

70. R. Barnes. The Conflict between Humans and Elephants in the Central African Forests. *Mammal Review*. 1996;26(2–3):67–80.

71. N. Leader-Williams, S. Milledge, K. Adcock, et al. Trophy Hunting of Black Rhino Diceros Bicornis: Proposals to Ensure Its Future Sustainability. *Journal of International Wildlife Law and Policy*. 2005;8(1):1–11.
72. E. Milner-Gulland, N. Leader-Williams. A Model of Incentives for the Illegal Exploitation of Black Rhinos and Elephants: Poaching Pays in Luangwa Valley, Zambia. *Journal of Applied Ecology*. 1992:388–401.
73. J. G. Goldman. Can Trophy Hunting Actually Help Conservation? *Conservation: The Source for Environmental Intelligence*. 2014.
74. M. Recio. Quandary on Texas Ranch: Can You Protect Rare Species by Hunting It? *McClatchy Newspapers*. 2012.

Part VI

Water

17

Hydrologic Cycle

17.1 Distribution and Properties of Water

Water is a global temperature moderator, is necessary for life, and is used by people for growing and preparing food, hydration, energy, manufacturing, transport, cleaning, and recreation.

About 2% of the Earth's water is in the solid state (ice) in glaciers, permanent snowfields, and permafrost (Table 17.1). Only 0.1% is in the atmosphere as gaseous vapor and liquid and ice in clouds. Of the remaining approximately 97.5% in the liquid state, 99% is saline – contains many dissolved cations – and is found in the oceans, salt water lakes and seas, and saline aquifers.

Less than 1% of the water is in terrestrial freshwater lakes, rivers, groundwater and freshwater aquifers. This "freshwater" (not saline) is most useful. Like energy and mineral elements, the amount of freshwater in the world is roughly constant; freshwater's abundance is largely based on its circulation. The atmospheric water evaporates as fresh water from the oceans and circulates rapidly to the terrestrial land and oceans as precipitation. The terrestrial precipitation then passes through forests, croplands, steppes, plants, animals, soil waters, streams, rivers, floodplains, and estuaries moderately rapidly and often repeatedly. It passes through aquifers slowly. The freshwater then returns to the salty oceans where it circulates very slowly. Although non-salty, the glacier, snowfields, and permafrost have little interchange with the other water stocks.

17.1.1 Thermal Properties

Water has high heat capacity, latent heat, turbulent, and conductive properties; and the large amount of water helps distribute and moderate global temperature (Chapters 7–8). Water vapor is also a greenhouse gas and recirculates heat energy back to the Earth instead of its being lost to space (Chapter 8). On the other hand, clouds, glaciers, and snowfields reflect incoming radiation, thus making the Earth cooler.

During the last glacial period, much more water was held in glaciers, ice, snowfields, and permafrost than at present (Figure 9.5a). Less was held in the atmosphere and oceans, and ocean sizes were slightly smaller.

Table 17.1 *Global water stocks [1]*

	Proportion of global water (%)	Proportion of fresh-water (%)	Renewal time (hours, days, years)
Oceans and Saltwater Lakes	96.546	0	?
Oceans	96.54	0	2,500 years
Saltwater Lakes	0.006	0	?
Atmospheric Water	0.001	0.04	8 days
Surface Terrestrial Water	0.01	0.3	15 days–10,000 years
Rivers	0.0002	0.006	15 days
Freshwater Lakes	0.007	0.26	15 years
Glaciers and Permanent Snow	1.74	68.7	1,500–10,000 years
Subsurface Terrestrial Water	1.71	?	?
Soil Water and Wetlands	0.002	0.08	1–5 years
Groundwater	0.76	30.06	1,400 years
Saline and Brackish Groundwater	0.93	?	?
Ice and Permafrost	0.022	0.86	10,000 years
Terrestrial Biological Water			
Water Inside Plants	0.0001	0.003	Several hours
Transformed to Organic Molecules			Hour–millions of years

17.1.2 Chemical Properties

Water is a polarized molecule with an "elbow" shape. Ionic and polarized covalent compounds dissolve in it. Consequently, it characteristically transports anions and cations needed for plant growth in soil, streams, and rivers (Figure 10.2). Pure water is neutral between acidic and alkaline, but dissolved substances can create extremely acidic and basic water. It also dissolves some minerals and rocks and redeposits them in different locations and concentrations where they crystallize as the water evaporates, changes temperature, or becomes otherwise saturated (Chapter 26).

Atmospheric, gaseous water molecules generally condense around a polarized particle (natural or artificial "cloud seeding") and produce clouds, ice crystal "snowflakes," rain, fog, or smog as more water molecules attach. Atmospheric substances can dissolve in the raindrops or snowflakes and be brought to earth, such as acid rain (Chapter 8).

Non-polarized substances such as oils do not dissolve well in water, but can form emulsions or invert emulsions if mixed appropriately. Or, they can separate into distinct layers of liquids stratified by their density.

Gases can also dissolve in water, with greater pressure and cooler temperatures enabling more gas to dissolve. Greater oxygen in cool oceans at higher latitudes increases fish abundance there, but very cold temperatures at extremely high latitudes can offset the advantages of greater oxygen. On the other hand, less carbon dioxide dissolved in warm waters at low latitudes means its concentration is more readily kept low by removal by carbonate-producing organisms (Figures 8.3–4). A concern is that increased carbon dioxide in the atmosphere is increasing carbon dioxide dissolved in the oceans, making oceans more acidic and possibly dissolving coral, calcareous shells of many sea creatures, and deposits of limestone and other carbonates. When these carbonates dissolve, they release even more carbon dioxide.

Organic matter or nutrients also become suspended in soil water, lakes, rivers, and streams. Too much organic matter or too many nutrients can cause organic matter to grow and lead to "eutrophication," in which microbes and animals consume this organic matter and grow, respire, and shade the water [2]. These activities deplete the water's dissolved oxygen and kill fish and other biota. On the other hand, some organic matter in water, and the nutrients to maintain it, are needed for fish and other aquatic life.

Different types of organic matter also dissolve in water. Tannins from oaks (*Quercus*), cedars (*Chamaecyparis*), and other trees can prevent poisonous aerobic and anaerobic microbes. Although safe for fish, swimming, and drinking, these waters have an ink-like appearance and are known as "black water" streams, lakes, or rivers (Chapter 12). Disease microorganisms can also live as organic matter in water, such as those causing dysentery and giardia [3, 4].

Water can be harmful for humans and biodiversity if polluted or eutrophicated by human-induced and other sediments, biological materials, or chemicals. Sediments that fill a lake can limit its ability to catch and store excess water for flood prevention and irrigation (Figure 19.2). Some waters are harmful because they contain chemicals such as arsenic and methane from surrounding geologic formations [5–8]. Others are harmful because industrial, agricultural, municipal, and home chemicals intentionally or otherwise get into them. Some chemicals are innately harmful while other harmful ones are produced by reactions of unrelated chemicals that happen to be in the water. Even though in very low concentrations in water, chemicals such as mercury can become concentrated in sediments, plants, and animals in wetlands and in fish and birds high in the food web (Figure 13.3) [9]. Some chemicals migrate from waste storage to other waters (Figure 17.3). Filters and other technologies help mitigate these pollutants and are helpful as long as society supports the mitigations; however, technologies are often not maintained during periods of social disarray and can result in toxic chemicals getting into the water. An ideal goal is to allow none of these pollutants in water; but extremely low concentrations are difficult to detect and remove.

17.1.3 Physical Properties

Water can move objects by floating or by "rafting" them on ice. If turbulent, water can also suspend light sediments and push others along the bottom as bed load (Chapter 10).

As rocks move and collide in water, they break and commonly become smaller and smoother, giving river and beach stones characteristically rounded and often polished appearances.

Liquid water contracts as it cools until about four degrees centigrade. When it cools further and its molecules slow down enough to form ice crystal lattices (Figure 26.3), the molecules are actually held farther apart and so the volume expands and density declines. Below zero degrees centigrade, essentially all water molecules are in the crystal structure of ice. This ice is lighter than liquid water and so floats instead of sinking to lake and river bottoms. Being at the water's surface, ice melts readily in the summer.

Carbon dioxide slowly diffuses to the bottoms of still lakes where the high pressure causes a large absorption of it and other anaerobic gases. Lakes in temperate zones have fall and spring "turnovers" as the differential water temperatures and densities between the top and bottom cause the water to rotate, releasing the gases [10]. This annual circulation apparently does not occur in tropical lakes; and dissolved carbon dioxide can accumulate at the depths, with disastrous effects if a rare turnover occurs ("killer lakes," Chapter 11) [11].

Boreal, temperate, and tropical lakes that accumulate organic matter can eventually have sphagnum mosses growing as floating organic matter (Chapter 12).

The expansion of freezing water also can also break rocks where water is in cracks; and it can lift boulders and fence posts within the soil ("frost heaving").

17.2 Hydrologic Cycle

Approximately 1.4 million cubic kilometers of water circulate between the oceans, atmosphere, groundwater, aquifers, rivers, lakes, and glaciers (Table 17.1). Water changes in availability in different places with seasons because of changing rates of melting, freezing, precipitation, evaporation, and evapotranspiration.

The circulation of water is known as the "hydrologic cycle" (Figure 17.1) [12–14]. The hydrologic cycle is studied through the renewal (residence) times of water in various "stocks" (Table 17.1) and "flows" [15].

Four characteristics of water stocks are important:

1. Total volume of water stored;
2. Time required for the water in a particular stock to be renewed (renewal or residence time) – replaced with different water as it cycles;
3. Rate at which water enters a stock (inflow rate); and
4. Rate at which water leaves a stock (discharge rate).

Water volume divided by renewal time gives the "sustainable flow rate" – the flow rate at which the stock's water volume will not change.

Most glaciers, frozen soils, and some aquifers hold water that was collected during some past climatic and/or geologic period and are not being replenished – at least not at a rapid rate. These are known as "fossil water stocks."

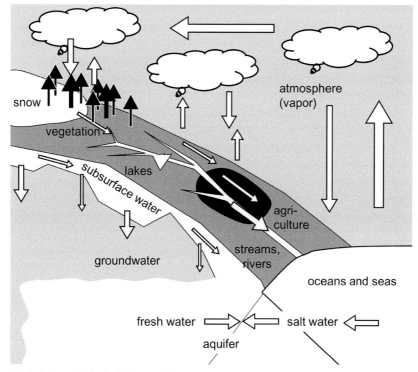

Figure 17.1 Schematic hydrologic cycle [12, 14].

17.3 Saltwater and Atmospheric Water Stocks

Streams or rivers eventually reach oceans or seas. Seas (lakes) differ from oceans in being partly or completely isolated from global ocean currents, being more or less salty than oceans, and having smaller tides.

Approximately two and one half percent of global runoff drains to seas without outlets such as the Aral Sea, Great Salt Lake, Dead Sea, Caspian Sea, Lake Chad, and others [1]. The levels of such landlocked waters commonly fluctuate with the climate. Salt concentrations in them generally increase and sometimes precipitate as crystals.

Atmospheric water vapor is an extremely small proportion of the Earth's freshwater with a rapid flow rate (Table 17.1). This water vapor changes from liquid as it evaporates from oceans, living organisms, lakes, rivers, and land – and returns to liquid or solid (ice) as it forms clouds and falls to the Earth as rain, snow, sleet, and hail. Water evaporates into the atmosphere primarily around the equator and to a lesser extent at approximately 60 degrees latitude (Chapter 7). Water vapor will spend an average of one week in the atmosphere before returning as precipitation (Table 17.1). The rate of water flow into and out of the atmosphere, the broad area covered, its redistribution of heat (Chapters 7–8), and its desalting ability make it extremely important.

More water is also held in the atmosphere in the warmer climates and thicker troposphere around the equator. Global atmospheric circulation sends water vapor to cooler climates, where much of it condenses as clouds. Precipitation is distributed even differently from evaporation and water vapor because of the further influence of global winds, mountains and other topographic features, and pressure systems (Chapter 8). Precipitation usually reaches the ground surface, but can sometimes evaporate while descending.

Much of the world's precipitation falls back into the oceans. That falling on land remains on various surfaces, runs off into streams and rivers, is evaporated, or infiltrates soils. In areas of limited precipitation and no vegetation, moisture remaining near the soil surfaces can evaporate and form a dry barrier within the soil. This barrier prevents water in wet soils at greater depths from migrating upward in a "wick-like" process ("capillary action") and both evaporating and allowing cations from below to flow to the surface soil water. Vegetation can dry soils to a greater depth through "evapotranspiration," as will be discussed.

17.4 Surface Terrestrial Water Stocks

Terrestrial stocks hold approximately three and one half percent of the Earth's total water [16]. They can be subdivided into surface, subsurface, and biological water.

17.4.1 Rivers and Streams

Most global precipitation not falling into oceans eventually drains to the oceans, infiltrates below ground, or evaporates. Precipitation that falls on land and moves as surface water and shallow subsurface water to lake, river, and stream networks is collectively referred to as surface storage. Although locally abundant, surface storage accounts for less than 1% of the global freshwater (Table 17.1).

Rivers and streams characteristically have a variety of slopes, flow rates, flow directions, rocks, sand and gravel, and other features; but they can be subdivided into contiguous areas referred to as "reaches" where they exhibit relatively uniform behaviors. Reaches serve to organize streams and rivers just as stands, pastures, and fields organize forests, grasslands, and farmlands.

Streams and rivers sometimes flow over relatively steep surfaces with irregularities that cause rapids. Such reaches generally pick up or push inorganic rocks, pebble, sand, silt, or clay. The result is an abrasive scouring of the stream or river bed, a moving of sediment downstream, and an increase of suspended inorganic sediment (Chapter 10).

Organic matter suspended or dissolved in the water suffers a different fate in these scouring reaches. Oxygen in water is replenished by contact with air; and rapidly moving rivers and streams throw water droplets into the air where their large surface area absorbs much oxygen. Aerobic microbes can then decompose organic matter in the water replenished with air. Consequently, rapids and waterfalls help keep water aerated, counteracting eutrophication [2]. Such river reaches are commonly referred to as "white water" because their bubbly water is white.

Water flowing overland sometimes creates new channels and can lead to severe erosion, especially in small-particle clay soils with little organic matter to bind the clay together (Chapter 10). Sudden, dramatic overland flows ("flash floods") are common in deserts where convective storms are common and little soil exists for infiltrating the water. Dramatic floods and slope failures also occur during "rain-on-snow" events, in which a warm rain melts snow and causes high water volumes from both rain and melting snow to flow overland on top of frozen soil or to infiltrate and oversaturate unfrozen soils.

17.4.2 Freshwater Lakes

Pools, ponds, lakes, and seas form where water accumulates until an outlet allows it to flow away. Such water bodies with outlets generally contain fresh water since dissolved cations and anions flow out with the water. Lakes are in geomorphologic depressions either with relatively impermeable bottoms or where the lakebed surface is below the surrounding water table.

Still water in lakes causes suspended sediment to drop, often in distinct annual layers of soil textures known as "varves" [17]. Water flowing out of lakes commonly contains less sediment than water entering, eventually causing lakes to fill with sediment and form alluvial floodplains or bogs.

Lakes can be valuable for fish and other aquatic organisms if there are sufficient nutrients and plants and animals for food and if oxygenated water enters a natural lake (from "white water" rivers, for example) or where oxygen dissolves into the lake.

17.4.3 Glaciers and Snow

Glaciers, permanent snowfields, ice, and permafrost are also large reservoirs of fresh water. They generally have low flow rates and so contribute relatively little directly to the hydrologic cycle except during times of rapid melting (Chapter 12). Indirectly, they contribute by their reflecting incoming solar radiation and the ice's heat capacity.

Temporary snow is not listed in Table 17.1. Snow is added in winter and released quite rapidly in spring – much later than when the snow fell. In evergreen forests, part of the snow accumulates on the foliage and sublimates – dissolves into the atmosphere directly as a gas. The complicated relationship between forests and precipitation will be discussed later.

Water from snow varies greatly with the snow's depth and water content. Snow can lose much of its water content through sublimation during prolonged cold, dry periods.

17.5 Below-Surface Terrestrial Water Stocks

Many terms are used for water entering the soil. As it saturates the soil and moves downward and laterally near the surface, it will be referred to in this book as "subsurface water."

As it moves farther down as "groundwater," it can reach an area of soil and bedrock saturated with water known as the "water table." The water table may be a "perched water table" formed by a less permeable or impermeable layer beneath, or it may be the surface of a much larger body of groundwater known as an "aquifer" that saturates the geologic structure. Caves that hold water are "cisterns." Some authorities use "subsurface water," "perched water table," and "aquifer" synonymously.

17.5.1 Soil Water and Wetlands

Water attaches to the soil as it flows into the Earth [13, 18, 19]. Once the soil water exceeds field capacity (Chapter 10), the excess water moves through the soil downward and downhill. The water can eventually emerge as a "spring" or pass into a stream, river, or lake; evapotranspire into the atmosphere; or move downward to subsurface stocks – perched water tables, aquifers, or cisterns.

Terrestrial wetlands commonly allow water to collect; remain; and eventually run off, evaporate, or infiltrate into the groundwater [20]. They slow the rapid runoff to streams, rivers, and oceans. They can recharge an aquifer (Chapters 12, 15) [21]. When streams and rivers reach oceans with tides, the fresh water often backs up quite far inland during high tides and causes a freshwater tidal zone rich in nutrients known as an "estuary." Downstream from this is a mixture of fresh and salt water ("brackish water") where the stream and ocean waters mix as tides ebb and flow. The continuous reversal of water flows creates very rich habitats with organic matter, aeration, and nutrients. These estuaries provide much of the lower end of the aquatic and ocean food pyramids (Figure 13.3).

17.5.2 Aquifers

Aquifers provide the largest amount of freshwater accessible for human use (Table 17.1), although lakes and rivers contain greater flows. The slow sustainable flow rates of aquifers mean they are sometimes overdrawn. More precipitation flows underground and to aquifers in some landforms than others (Figure 17.2) [13, 18, 19]. Landforms with relatively impermeable bedrock such as weathered shields, igneous bedrocks, and basalt flows commonly hold relatively little subsurface water (Chapter 10). Some water does accumulate in bedrock cracks, but usually an interconnected aquifer is not formed.

Most metamorphic and sedimentary landforms, including karsts and coastal plains, form interconnected areas of porous sediments or bedrock full of water. These allow much of their precipitation to flow into aquifers.

Deep, porous, overlying materials such as volcanic ash, sand, loess, alluvial floodplains, glacial outwash and till, and peat bogs allow much water to infiltrate. Depending on the structure of the underlying bedrock, water can form a water table within the deposits, flow to an underlying aquifer, and/or flow as subsurface water through the soil and into streams where it eventually emerges as surface water.

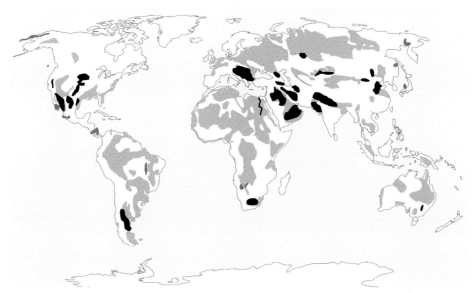

Figure 17.2 Major aquifers in the world (gray and black) [22–25] and aquifers of greatest stress (black) [26].

Water in aquifers is sometimes confined from above by sloping, impermeable layers so that water enters in one region, flows beneath the impermeable layer, and emerges through a break in the layer at a lower elevation nearby or many kilometers away (Figure 11.4). Such "confined aquifers" can have a "recharge" period of hundreds of years between water entering at the top and emerging at the bottom. Chemically polluted water entering the top of a confined aquifer may appear innocuous at first, but may eventually affect people far away and in the distant future as the polluted water finally emerges. For example (Figure 17.3), radioactive waste has been stored for decades near the Columbia River, Washington State, USA. Soon after its storage, the waste began to leak and the contamination is now slowly migrating through an aquifer to the Columbia River [27]. Present models suggest the waste's radioactivity may decay to a harmless level before reaching the river [28].

Aquifers near oceans and sea coasts can create a "halo" of non-salty water beneath the land (Figure 17.1). If excessive wells pump out too much fresh water, salt water can move into this aquifer, a process known as "salt water intrusion." It is difficult to return such aquifers to fresh water.

Some aquifers are being unsustainably depleted; water is leaving faster than it is recharging (Figure 17.2). This depletion has been attributed to excessive water withdrawal ("drawdown"), increased growth of trees whose evapotranspiration keeps water from infiltrating to the aquifer [30], and/or the "fossil" nature of the aquifer's water. For example, most water of the Ogallala Aquifer in the central United States is believed to have been

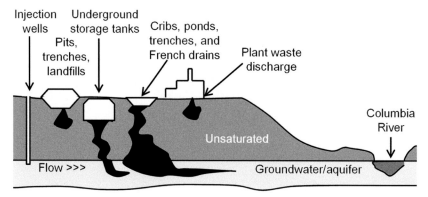

Figure 17.3 Flow of radioactively contaminated groundwater (black) from the Hanford nuclear facility into the groundwater and toward the Columbia River [27, 29]. Courtesy of Columbia Institute for Water Policy.

deposited when the glaciers melted (Chapter 9); and it is now being "mined" for water. Parts of it are running out of water, leaving the dependent cities and farms in desperate conditions [31].

17.5.3 Saline Subsurface Water and/or Groundwater

Irrigating with insufficient water only adds subsurface water to the rooting zone and allows it to evapotranspire, leaving behind the cations and anions that were dissolved in the irrigation water. The resulting soils become "saline" – contain too many cations and anions – and inhibit plant growth. Fertilizers also add cations and anions to the soil. Heavy irrigation is needed to flush these anions and cations downward beyond the rooting zone, especially if fertilizers and pesticides have been applied (Chapter 21).

An opposite situation occurs in areas with saline aquifers but nonsaline subsurface water (Figure 17.4). As long as a dry subsoil zone exists between the subsurface, fresh water in the plant rooting zone and the saline aquifer beneath, the rooting zone water remains non-salty and plants can grow. (See "Goulburn–Broken Catchment" Case Study, Chapter 3.) If a continuous column of wet soils connects the subsurface rooting zone and deeper saline zones, cations can migrate upward to the rooting zone and make it too saline for plant growth. In places, trees that evapotranspire the subsurface root zone water have kept it from infiltrating downward and forming a wet column to the saline groundwater that would enable cations from there to migrate to the root zone. Cutting trees in such places stopped the evapotranspiration, enabled the subsurface freshwater to flow downward and eliminate the dry barrier, and thus led to saline surface soils. Similarly, irrigating areas that had a previous dry subsoil barrier have created wetted columns that allow the cations to migrate upward and salinize the subsurface root zone water.

Ice and permafrost have been discussed in Chapter 12.

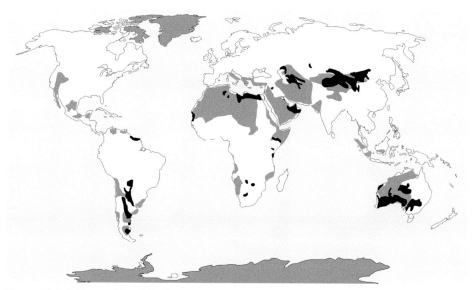

Figure 17.4 Saline aquifers (black) [23–25] and areas with no river discharge (gray) [24, 32, 33]. Saline areas are mapped differently by [34].

17.5.4 Terrestrial Biological Water

All living plants and animals hold about as much water as falls on the Earth as precipitation in one day – a very small fraction of the total global water. The total transformed water in all living and dead organic matter is about the equivalent of six days of global precipitation (Table 17.1). On the other hand, living plants rapidly circulate large quantities of water through evapotranspiration.

In addition to holding intact water, photosynthesis (Figure 20.1) is a rare case where the water molecule is broken apart. The component elements are incorporated into organic matter, but water molecules reform as the organic matter is destroyed through respiration or fires.

17.6 Water Flows

Water flows among the various stocks through the atmosphere as evaporation, evapotranspiration, and precipitation; across the terrestrial surface as overland flow or stream/river channel flow; and into and through landforms as infiltrating subsurface and groundwater flow. Rivers, oceans, atmosphere, and some other water-holding entities behave as both stocks and flows since they simultaneously hold and move water.

In most of the world, water used by a country originates from precipitation within the country and is found as surface, ground water, or an "overlap" as water moves above or below ground. Some water, usually surface water, originates from another country; and a minor amount leaves some countries (Figure 17.5). Treaties sometimes govern rights to water potentially moving across borders, as will be discussed in Chapter 18.

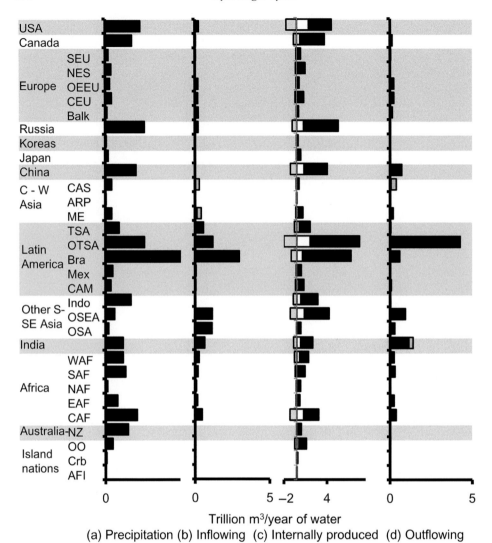

0　　　　　　0　　　　5 –2　　4　　　0　　　　5

Trillion m³/year of water

(a) Precipitation (b) Inflowing (c) Internally produced (d) Outflowing

Figure 17.5 (a) Precipitation and water, (b) inflowing into all countries in each country group, (c) internally produced from surface and groundwater, and (d) outflowing waters by country groups. (For all graphs: white = groundwater, black = surface water. For (b) and (d): gray = water committed to downstream country. For (c): negative gray = overlap between surface and groundwater) [35].

17.6.1 Evaporation and Evapotranspiration

Evaporation is the vaporization of water. Evapotranspiration is the vaporization of water coming from the plant's pores ("stomata"), usually found in the leaves. Evaporation can bring water from deep in the soil, through its roots and stem, and out of its leaves. Grasses in grasslands can become dormant and evapotranspire little during warm, dry summers; but forests continue to evapotranspire and so deplete soil moisture much more [36]. Evaporation usually only vaporizes water near the soil surface.

Water moves in the atmosphere as vapor or liquid mist or ice crystals in clouds in patterns described earlier (Chapter 7) and falls to earth as fog drip or precipitation – rain, snow, sleet, or hail.

17.6.2 Soil Infiltration and Groundwater Flow

Water infiltrates through soil pores at a rate based on the soil structure and texture (Chapter 10), with the un-infiltrated water remaining as puddles on the surface or flowing overland.

Some infiltrated water is held at each depth within the soil, while excess water ("gravity water") continues moving and creating a "wetted column" (Chapter 10). The gravity water accumulates as subsurface water and slowly flows vertically, horizontally, or diagonally through the soil depending on barriers, soil textures, and water gradients. Sometimes it can stop moving or move so slowly that it creates a "seasonal water table" that varies seasonally and between locations.

17.6.3 Vegetation, Water Flow, and Rainfall

The ability of forests and vegetation to increase rainfall has been proposed for over 150 years [37, 38]; and concern about Amazon deforestation has recently raised this issue again [39–43]. A similar proposal that "rainfall follows the plow," also generated about 150 years ago, led to disastrous results from farmers who tried to farm dry lands [44].

Absence of noticeable rainfall changes in eastern North America and central Europe forests following clearing and regrowth during the past three centuries has led many to attribute large-scale rainfall patterns to global climate patterns (Chapter 8) rather than forest evapotranspiration. Desertification generally does not occur with tree harvesting or over-grazing unless the soil is destroyed and was droughty to begin with (Chapter 14).

Trees can locally increase humidity through both evapotranspiration and shade; however, they dry deeper soils through evapotranspiration even though their shade keeps soil surfaces moist by limiting evaporation. They can also increase precipitation in foggy areas where fog condenses on trees and drips to the ground, as discussed below. The relationship between vegetation and water flow involves many opposing influences that vary in importance seasonally and spatially (Figure 17.6).

In both growing and dormant seasons, leafy vegetation will take up soil moisture from the rooting zone into its roots and evapotranspire it through leaves, drying the soil. When without leaves because the plants are deciduous (in winter) or the vegetation has been harvested, evapotranspiration does not occur, the soils are wetter, and more water infiltrates to streams and aquifers [30, 47–49]. Forests commonly deplete moisture more than grass-lands, as described above [36].

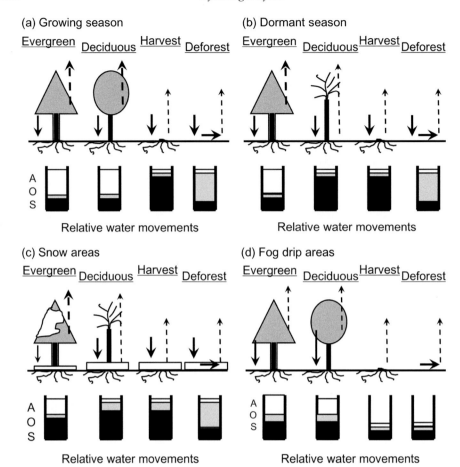

Figure 17.6 Vegetation and water flow. (Vertical arrows: solid = precipitation; dashed = evaporation/evapotranspiration, horizontal arrows = overland flow; A [white] = atmospheric water vapor; O [gray] = surface water; S[45] = subsurface water) [46].

Where a forest or shrubland is harvested and regrows, the dead and regrowing roots maintain a soil structure that allows water to infiltrate soil. Regrowing forests commonly reduce soil moisture to pre-harvest levels in less than two decades. If land conversion in fine-textured soils destroys the structure so that incoming water (precipitation or snowmelt) cannot infiltrate it (Chapter 10), the water flows overland as a rapid flood, followed by a drought (Figures 17.7–8). Consequently, vegetation – and especially trees – aids water's even flow but reduces the total amount of water. Harvesting vegetation increases the flow but only until vegetation regrows. Permanent removal of vegetation maintains a high flow but eventually ruins the soil structure and so gradually reduces infiltration and increases destructive overland flow.

Figure 17.7 Erosion of stream after soil structure and infiltration capacity were destroyed by over-grazing, poor tilling, and no streamside buffers, in Armenia. See also Figure 14.8b.

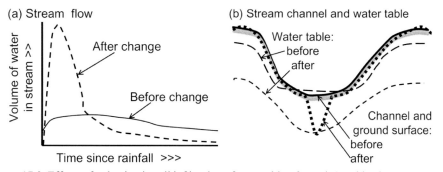

Figure 17.8 Effects of reduction in soil infiltration often resulting from changed land uses. (a) Water flow changes. (b) Water table, stream channel, and ground surface change.

Forests historically existed in a mosaic of stand structures that shifted across the landscape with disturbances and regrowth (Figures 14.5b, 14.12). The open and savanna structures lost less water to evapotranspiration (Figure 28.2).

Forests can increase soil moisture in topographic positions where fog collects on trees and drips into the soil, increasing the incoming precipitation and soil moisture. Without the trees, less fog drips and the soils are drier [50–53].

Forests can also change the amount and pattern of snowmelt and resultant water flow. Some snow landing on trees can sublimate as vapor and never reach the soil, but less snow sublimates that lands in nearby harvested or non-forest areas. Consequently, a forest can reduce soil moisture originating as snow; however, if the forest shade causes snow on the ground to melt slower than in clearings, the forest can help avoid floods and overland flow of rapidly melting snow [54–56].

The relative importance of each factor – evapotranspiration, soil structure, fog drip, snow sublimation, and shade – can create different relationships between vegetation and water flow amount and rate. These relationships can be determined empirically for each water catchment area and the vegetation can be managed accordingly.

17.6.4 Overland and Channel Flow

If precipitation or melting snow and ice produce more water than can infiltrate the soil, the remaining water flows on the surface. This overland flow occurs at times of heavy rain or rapid snow melt, when soils are frozen, and/or in places of poor soil structure. With much overland flow, a large volume of water rapidly reaches the streams and flows for a short time. After this initial surge, less water flows to the stream channel and to subsurface flow than if a better soil structure had allowed more water to infiltrate the soil and move slowly.

Paving, ditching, and collapse of the soil structure from deforestation, plowing or otherwise operating on wet soils, and heavy grazing (Figure 17.7) can lead to more overland flow and less subsurface flow. Overland flow has several effects (Figure 17.8).

- The rush of water erodes the stream and river channels, causing them to become deeper, with steeper banks – channel deformation;
- Less water infiltrating the soils means the water table lowers and, where present, aquifers are not recharged;
- The small water volume of the lowered water table flows slowly through the ground. It can come to the surface in the deeply incised channel and sustain a low flow between surges following rains.

References

1. I. A. Shiklomanov, J. C. Rodda. *World Water Resources at the Beginning of the Twenty-First Century.* (Cambridge University Press, 2004).
2. S. R. Carpenter, D. Ludwig, W. A. Brock. Management of Eutrophication for Lakes Subject to Potentially Irreversible Change. *Ecological Applications.* 1999;9(3):751–71.
3. N. J. Ashbolt. Microbial Contamination of Drinking Water and Disease Outcomes in Developing Regions. *Toxicology.* 2004;198(1):229–38.
4. M. W. LeChevallier, W. D. Norton, R. G. Lee. Occurrence of *Giardia* and *Cryptosporidium* spp. in Surface Water Supplies. *Applied and Environmental Microbiology.* 1991;57(9):2610–16.
5. D. K. Nordstrom. Worldwide Occurrences of Arsenic in Ground Water. *Science.* 2002;296(5576): 2143–5.
6. R. H. De Bruin, R. M. Lyman. Coalbed Methane in Wyoming. In *Coalbed Methane and the Tertiary Geology of the Powder River Basin, Wyoming and Montana, 50th Annual Field Conference Guidebook.* (American Association of Petroleum Geologists, 1999): pp. 61–72.
7. J. McArthur, P. Ravenscroft, S. Safiulla, M. Thirlwall. Arsenic in Groundwater: Testing Pollution Mechanisms for Sedimentary Aquifers in Bangladesh. *Water Resources Research.* 2001;37(1): 109–17.
8. R. Nickson, J. McArthur, W. Burgess, et al. Arsenic Poisoning of Bangladesh Groundwater. *Nature.* 1998;395(6700):338.

9. H. F. Clark, G. Benoit. Legacy Sources of Mercury in an Urbanised Watershed. *Environmental Chemistry.* 2009;6(3):235–44.
10. J. L. Riera, J. E. Schindler, T. K. Kratz. Seasonal Dynamics of Carbon Dioxide and Methane in Two Clear-Water Lakes and Two Bog Lakes in Northern Wisconsin, USA. *Canadian Journal of Fisheries and Aquatic Sciences.* 1999;56(2):265–74.
11. J. McPhee. *The Control of Nature.* (Farrar, Strauss, and Giroux; 1989).
12. J. R. Craig, D. J. Vaughan, B. J. Skinner. *Resources of the Earth: Origin, Use, and Environmental Impacts,* fourth edition. (Prentice Hall, 2011).
13. W. K. Hamblin, E. H. Christiansen. *The Earth's Dynamic Systems,* tenth edition. (Prentice Hall, 2001).
14. D. Huddart, T. Stott. *Earth Environments: Present, Past, and Future.* (Wiley-Blackwell, 2012).
15. D. Meadows. Places to Intervene in a System. *Whole Earth.* 1997;91:78–84.
16. G. M. Hornberger, P. L. Wiberg, J. P. Raffensperger, P. D'Odorico. *Elements of Physical Hydrology.* (JHU Press, 2014).
17. M. Sturm, A. Matter. *Turbidites and Varves in Lake Brienz (Switzerland): Deposition of Clastic Detritus by Density Currents.* (Wiley Online Library, 1978).
18. J. S. Monroe, R. Wicander. *Physical Geology.* (Brooks/Cole, 2001).
19. C. C. Plummer, D. McGeary, D. H. Carlson. *Physical Geology,* ninth edition. (McGraw-Hill, 2003).
20. M. M. Brinson. Changes in the Functioning of Wetlands Along Environmental Gradients. *Wetlands.* 1993;13(2):65–74.
21. US Environmental Protection Agency. What Is a Wetland? 2016 (Accessed November 22, 2016). Available from: www.epa.gov/wetlands/what-wetland.
22. P. Engstrom, K. Brauman. *Global Aquifers Map.* (Ensia, 2013 [Accessed July 6, 2016]). Available from: www.ensia.com/features/groundwater-wake-up.
23. W. Struckmeier, A. Richts, U. Philipp, A. Richts, cartographers. *Groundwater Resources of the World.* (UNESCO, BGR, IAEA, IAH, 2008).
24. A. Richts, W. F. Struckmeier, M. Zaepke. Whymap and the Groundwater Resources Map of the World 1: 25,000,000. In *Sustaining Groundwater Resources.* (Springer, 2011): pp. 159–73.
25. UNESCO-IHP. *Atlas of Transboundary Aquifers: Global Maps, Regional Cooperation and Local Inventories.* (UNESCO, ISARM Programme, 2009).
26. T. Gleeson, Y. Wada, M. F. Bierkens, L. P. van Beek. Water Balance of Global Aquifers Revealed by Groundwater Footprint. *Nature.* 2012;488(7410):197–200.
27. US Department of Energy. *Groundwater Contamination Illustration, Hanford.* (US DOE Pacific Northwest National Laboratory, 2004).
28. R. E. Peterson, M. P. Connelly. Water Movement in the Zone of Interaction between Groundwater and the Columbia River, Hanford Site, Washington. *Journal of Hydraulic Research.* 2004;42(S1):53–8.
29. R. P. Osborn. *Hanford Nuclear Reservation: Black Rock Groundwater Could Affect Movement of Radioactive Contamination under Hanford.* (Columbia Institute for Water Policy, 2007 [Accessed March 25, 2016]). Available from: columbia-institute.org/blackrock/Issues/Hanford .html.
30. P. K. Barten, J. A. Jones, G. L. Achterman, et al. *Hydrologic Effects of a Changing Forest Landscape.* (National Research Council of the National Academies of Science, USA, 2008).
31. E. Custodio. Aquifer Overexploitation: What Does It Mean? *Hydrogeology Journal.* 2002;10(2): 254–77.
32. Global Runoff Data Center, cartographer. *Major River Basins of the World.* (Federal Institute of Hydrology [BfG], 2007).
33. BGR, UNESCO. *Global Groundwater Maps*: (BGR, 2016 [Accessed August 4, 2016]). Available from: www.whymap.org/whymap/EN/Downloads/Global_maps/globalmaps_nod e_en.html.
34. F. van Weert, J. van der Gun, J. Reckman. *Global Overview of Saline Groundwater Occurrence and Genesis.* (International Groundwater Resource Assessment Centre [IGRAC], 2009).

35. UN FAO. *AQUASTAT Main Database*. (United Nations Food and Agriculture Organization; 2016 [Accessed April 11, 2016]). Available from: www.fao.org/nr/water/aquastat/data/query/index.html?lang=en.

36. M. Waterloo, F. Beekman, L. Bruijnzeel, K. Frumau. The Impact of Converting Grassland to Pine Forest on Water Yield in *International Association of Hydrological Sciences Special Publication*, 1993:149–.

37. G. P. Marsh. *Man and Nature, Physical Geography as Modified by Human Action*. (Charles Scribner, 1865).

38. C. R. Kutzleb. *Rain Follows the Plow: The History of an Idea*. (University of Colorado, 1968).

39. R. Betts, P. Cox, M. Collins, et al. The Role of Ecosystem-Atmosphere Interactions in Simulated Amazonian Precipitation Decrease and Forest Dieback under Global Climate Warming. *Theoretical and Applied Climatology*. 2004;78(1–3):157–75.

40. D. Ellison, M. N Futter, K. Bishop. On the Forest Cover–Water Yield Debate: From Demand-to Supply-Side Thinking. *Global Change Biology*. 2012;18(3):806–20.

41. L. Morello. *Cutting Down Rainforests Also Cuts Down on Rainfall*. (*Scientific American*, 2012 (Accessed May 21, 2015). Available from: www.scientificamerican.com/article/cutting-down-rainforests/.

42. D. Spracklen, S. Arnold, C. Taylor. Corrigendum: Observations of Increased Tropical Rainfall Preceded by Air Passage over Forests. *Nature*. 2013;494(7437):390–.

43. L. E. Aragao. The Rainforest's Water Pump: An Investigation of Naturally Occurring Water Recycling in Rainforests Finally Marries the Results of Global Climate Models with Observations. *Nature*. 2012;489(7415):217–9.

44. M. J. Ferrill. *Rainfall Follows the Plow*. (University of Nebraska, 2011 [Accessed July 28, 2017]). Available from: http://plainshumanities.unl.edu/encyclopedia/doc/egp.ii.049.

45. X. Lee, M. L. Goulden, D. Y. Hollinger, et al. Observed Increase in Local Cooling Effect of Deforestation at Higher Latitudes. *Nature*. 2011;479(7373):384–7.

46. M. D. Abrams. Where Has All the White Oak Gone? *BioScience*. 2003;53(10):927–39.

47. J. A. Jones, D. A. Post. Seasonal and Successional Streamflow Response to Forest Cutting and Regrowth in the Northwest and Eastern United States. *Water Resources Research*. 2004;40(5).

48. A. R. Hibbert. *Forest Treatment Effects on Water Yield*. (US Forest Service, 1965).

49. E. S. Veny. *Forest Harvesting and Water: The Lake States Experience 1*. (Wiley Online Library, 1986).

50. W. J. Liu, Y. P. Zhang, H. M. Li, Y. H. Liu. Fog Drip and Its Relation to Groundwater in the Tropical Seasonal Rain Forest of Xishuangbanna, Southwest China: A Preliminary Study. *Water Research*. 2005;39(5):787–94.

51. J. Cavelier, G. Goldstein. Mist and Fog Interception in Elfin Cloud Forests in Colombia and Venezuela. *Journal of Tropical Ecology*. 1989;5(03):309–22.

52. H. Vogelmann. Fog Precipitation in the Cloud Forests of Eastern Mexico. *BioScience*. 1973;23(2):96–100.

53. T. E. Dawson. Fog in the California Redwood Forest: Ecosystem Inputs and Use by Plants. *Oecologia*. 1998;117(4):476–85.

54. J. P. Hardy, P. M. Groffman, R. D. Fitzhugh, et al. Snow Depth Manipulation and Its Influence on Soil Frost and Water Dynamics in a Northern Hardwood Forest. *Biogeochemistry*. 2001;56(2):151–74.

55. J. Kittredge. *Influences of Forests on Snow in the Ponderosa, Sugar Pine, Fir Zone of the Central Sierra Nevada*. (University of California, 1953).

56. A. Gelfan, J. Pomeroy, L. Kuchment. Modeling Forest Cover Influences on Snow Accumulation, Sublimation, and Melt. *Journal of Hydrometeorology*. 2004;5(5):785–803.

18

Annual Hydrographs and Water Use

18.1 Patterns of Water Flow

The availability of most fresh water varies dramatically by seasons and years as stream and river flows, water table depths, and soil moisture change. Only aquifers and large freshwater lakes have water available evenly throughout the year.

At times, water is too scarce; and other times it is too plentiful, with monsoons and other seasonal rains creating the greatest variation. Areas affected by El Niño have large between-year variations in water (Chapter 7).

Nearly all country groups theoretically have sufficient fresh water based on their calculated "renewable water" (Figure 18.1). Renewable water is an optimistic calculation of net annual incoming water. All renewable water may not be completely usable. Water's spatial distribution and its allocation leave shortages. More important, the seasonal variation creates times of water shortages. A challenge is to develop ways to distribute water throughout the year – to save the excess of wet seasons to provide sufficient water during dry seasons.

The temporal patterns of water flow vary within rivers and even parts of rivers (Figure 18.2–3). The pattern of water flow at any place on a river or stream is the result of water draining from the area above it. For convenience, water flows are generally measured at key changes in topography so the area that water drains from can be readily identified. This area is known as the "drainage basin," "catchment area," or "watershed" (technically, watershed is semantically the opposite of a drainage basin). Drainage basins can be of different shapes depending on the geomorphology. Smaller drainage basins ("subcatchment basins") can be nested within larger catchments (Figure 18.2). In some places, however, there is no river removing the water from an area because the area is too dry or too frozen, the water drains underground (for example, in karstic topography or glacial outwash), or the waters close to the oceans drain in small rivers and streams (Figure 17.4).

Drainage basins are also conveniently bounded areas for other resource and political considerations since they often demark areas of common water interests, resources, cultures, and commerce.

Each basin has a unique seasonal water flow pattern [3, 4] measured at the downstream river or stream exit. Although annual flows can vary, the patterns are regular enough that

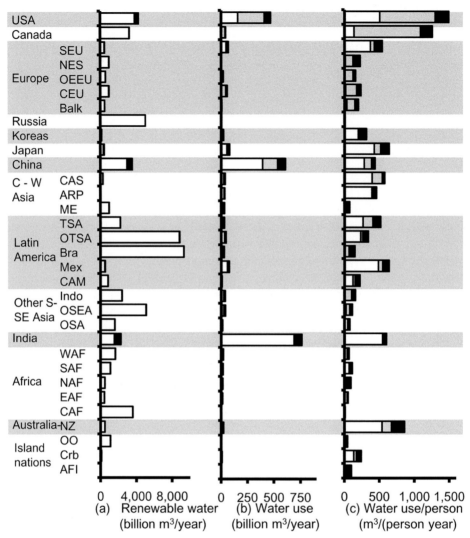

Figure 18.1 (a) Renewable water by country groups (white = unused, black = used); (b) water use; and (c) water use per capita for country regions. (For (b) and (c): white = agriculture, gray = industrial, black = household and municipal.) [1].

they can be anticipated in the long term, predicted months ahead, and somewhat managed in various ways. The graph of this flow over the course of a year is known as an "annual hydrograph" (Figure 18.3).

Hydrographs for different areas vary for many reasons, as described for Figures 18.3a–j:

- The Kansas River is in a dry grassland and receives most precipitation in the summer as convective storms (Figure 7.4a).

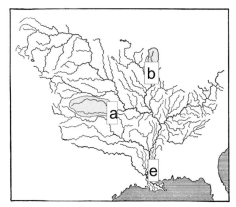

Figure 18.2 Mississippi River, USA, system and subcatchment areas (gray). Letters refer to Figure 18.3 [2].

- The Wisconsin River has highest water flow with spring snowmelt.
- The Willamette River is dominated by advective weather systems coming from the North Pacific Ocean Kuroshio; in winter, the Willamette River is in the rainy convergence of the Hadley and Polar cells (Figures 7.5, 7.9). In summer, it lies within the dry Hadley cell.
- The Verde River flow comes from both spring snow melt and late-summer convective storms (monsoons).
- The Mississippi River near its mouth is the combined influence of many smaller drainages (Figure 18.2) dominated by snowmelt, advective, and convective weather patterns.
- The Nile water flow is greatest when the rainy equatorial convergence zone moves northward during the northern hemisphere's summer (Figure 7.9a), raining in the Nile River's drainage basin.
- The Amazon water flow is similarly greater when the convergence zone moves northward.
- The Congo River is on the equator. Its northern, Ubangi drainage receives much rain when the convergence zone shifts northward in the northern hemisphere's summer (Figure 7.9a); its southern, Kasai drainage receives rain when the convergence zone shifts southward (Figure 7.9b).
- Rhine River: The low elevation, Chochem drainage of the Rhine River is at the same latitude as the Willamette River in Oregon (Figure 18.3c); it shows the same advective system, but coming from the Atlantic Gulf Stream. The Rhine River's Ilanz subbasin is higher in the Alps; and its water flow is largely determined by the snowmelt, similar to the Wisconsin River (Figure 18.3b).
- The Ganges and Brahmaputra seasonal rains (monsoons) are caused by both the convergence zone moving northward in summer and the convective rising of air from dry lands and deserts to the north. The high flows of these two rivers occur at the same time and can create extreme floods where they converge and flow through Bangladesh.

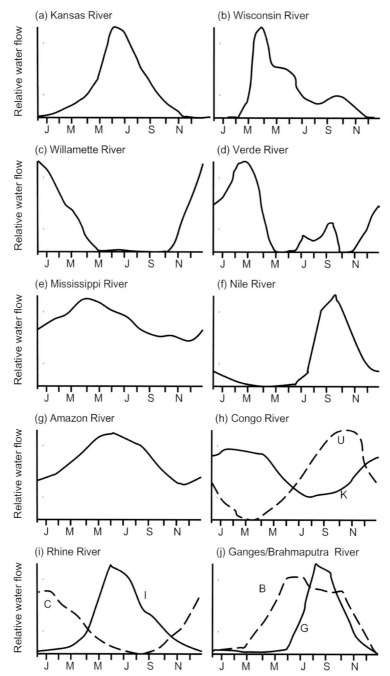

Figure 18.3 Annual hydrographs of several rivers. (a) Kansas River, Junction City, Kansas, USA; (b) Wisconsin River, Merrill, Wisconsin, USA; (c) Willamette River, Albany, Oregon, USA; (d) Verde River, McDowell, Arizona, USA (a–d from [5] using analysis of [6]); (e) Mississippi River, Vicksburg, Mississippi, USA [7]; (f) Nile River at Aswan Dam, Egypt [8]; (g) Amazon River at mouth, Brazil [9]; (h) Congo River (U = Ubangi drainage; K = Kasai drainage) Democratic Republic of Congo and Republic of Congo [10]; (i) Rhine River, Europe (C = low elevation Cochem drainage; I = mountainous Ilanz drainage) [11]; (j) Ganges (G) and Brahmaputra (B) Rivers, India and Bangladesh [12].

Water flows can be affected by dry weather, common at the middle latitudes on the interiors and western sides of continents and occasionally found elsewhere. Rivers and streams can dry during some seasons. Sometimes either no water flows or the water may recede below the bed's surface and flow as subsurface water.

Variations in local weather can change the water's yearly flow pattern. Extra precipitation as snow, rain, sleet, freezing rain, or a combination; droughts of various durations; unusually high or low temperatures; and especially these events occurring out of their usual season can greatly affect the amount and pattern of water flow. A rain-on-snow event can create unseasonal floods, described earlier.

Disturbances can affect water flow patterns. Fires, insects, or timber harvest in forests or shrublands can allow more water to flow into the soil and increase streamflow rates. Similarly, cessation of periodic fires can allow the vegetation to regrow and reduce stream-flow and aquifer recharge [13]. Sometimes fires create a hydrophobic layer in the soil that reduces water infiltration and so increases surface runoff and erosion.

18.2 Transboundary Water Use and Conflict

Trying to sustain unpolluted terrestrial water illustrates the issues and opportunities to sustaining ecosystem services. Disputes for many centuries have been over such issues as:

- Who owns the water – the first person to use it, the person who owns the land it originated on, or downstream beneficiaries?
- Who bears the cost of maintaining it (keeping it clean, flowing, and not flooding) – the first person to use it, the person who owns the land it originated on, or all downstream beneficiaries?

Like the issue of ensuring habitats for species (Chapter 16), another ecosystem service, the sharing of benefits and costs of water has been contentious.

Water issues have been solved in different ways within political jurisdictions, with some places giving priority water rights to the first user and others giving rights to the landowner where the water originated or to upstream users.

Transboundary water rights is a large international issue [14]. The 263 international river basins account for nearly half of the world's terrestrial surface area and involve 145 countries. Up to 17 countries can share a single river. Despite the potential for violent conflict, most international water sharing has been done peacefully, possibly because principles for international freshwater management have been established and frequently modified for over one hundred years. The relatively few conflicts have been over water supply and infrastructure. Agreement seems to be more forthcoming on water quantity, quality, economic development, and hydropower [15].

Issues with transboundary water are being exacerbated by the changing climate, increasing populations, and increasing economic development [14, 16]. At the same time, new agreements are being forged to address them [17].

18.3 Water Use, Land Use, and Water Flow

The full potential to provide water can be realized by managing a catchment basin in a coordinated way that sustains water flows throughout the year better. Such management can entail avoiding rapid runoffs shortly after rains or snowmelt and prolonging moderately high river flows and infiltrations to aquifers. This management can be done in several ways. Judicious placement of different land uses can include areas that promote water percolating into the soil and moving slowly as groundwater to rivers and aquifers instead of rapidly to rivers as overland flow. Using appropriate farming practices, avoiding overgrazing, and otherwise avoiding compacting soils can help, as can buffer areas along streams and rivers that promote water infiltration and underground flow instead of overland flow to streams.

Other ways besides land use changes can also provide more water at needed times: reservoirs, treating and recycling municipal wastewater, and desalinization (Figure 18.4). Treating wastewater and desalinization are discussed in Chapter 19. Reservoirs of various sizes and numbers are the dominant way that extra water is provided [18]. These reservoirs catch water during high water times and release it later during drought periods. Farmland displaced by reservoirs can be balanced by opportunities to grow second, irrigated crops during the dry season on the remaining lands. Reservoirs have environmental concerns (Chapters 18, 19). Newly designed reservoirs and modifications to older ones can help avoid issues with older ones that prevent the flow of fish and bed load sediment [18–24].

18.3.1 Industry and Transportation

Water use varies by industries and their geographical locations. Much water can be expended for cooling during thermal (fossil fuel) and nuclear electric power production [25] and for generating hydroelectric power. Water and energy can be conserved by generating heat and steam while condensing, cooling, and reusing the water – a process known as Combined Heat and Power (CHP, Chapter 24) – instead of releasing it as steam. Alternatively, photovoltaic or wind energy can generate electricity without water loss (Chapter 21). Water is also used in the production of foods, beverages, textiles, metals, paper, and commodities such as wood pulp, paper, rubber, and plastic. Except for the beverage industry, much of the water is used in processing and cooling, and released after use.

Recently, water is being used for extracting fossil fuels from oil shale [26] through hydraulic fracturing of the shales ("fracking," Chapter 27). Fracking has led to water contamination concerns related to gas migration, contaminant transport through fractures, wastewater discharges, earthquakes, and accidental spills [27].

Two industrial issues are the cleanliness of water upon release and the relation between where the water is taken in where it is discharged. Industries are increasingly being required to clean their wastewater; and industrial facilities are finding it convenient to recycle the water internally since it has already been cleaned. New facilities being installed in countries

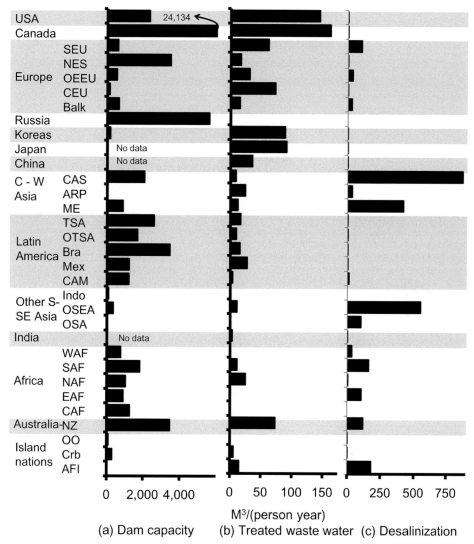

Figure 18.4 Water use extension methods and amounts by country groups: (a) dam capacity; (b) treated municipal water; (c) desalinization [1].

without currently effective water pollution laws often include wastewater treatment and internal recycling, anticipating that laws will become stricter and clean water, scarcer. Consequently, industrial polluters are sometimes older facilities in developed countries that were installed before treatment and recycling were needed.

Where not recycled, industrial or municipal water can deplete rivers and aquifers by discharging cooling water as vapor into the atmosphere or by discharging water from aquifers into rivers or lakes. Alternatively, where industrial areas, or cities, remove and return water from rivers or lakes (reservoirs) that are parts of rivers, the use creates much less change to the hydrologic cycle.

Industries often locate – and use water – near energy and/or water centers. Future industrial centers may be in deserts if solar power becomes efficient there (Chapter 23). Some desert areas could shift their water use and economic base from agriculture to industry. Limiting factors will then be water, raw materials, and shipping facilities. Industrial use of water in deserts could be much more efficient than agricultural use; and solar energy facilities in deserts may promote desalinization either while cooling the steam generated by Concentrated Solar Power (CSP) technologies (Chapter 27) [28, 29] or using some electricity generated by photovoltaic (PV) technologies. Desert areas close to oceans or seas for easy shipping of bulk industrial goods, such as Morocco, are already being considered for such enterprises.

River transport using barges is a very cost- and energy-efficient for shipping industrial goods over long distances but relies on reliable water flows in rivers (Figures 4.4a, 24.12b).

18.3.2 Household and Municipal Water

Household and municipal water is generally used for drinking, food preparation, cleaning and sanitary purposes, and other household chores. Many urban and some rural households receive water through a municipal distribution system, usually from regional freshwater sources. Water distribution and treatment are often absent in rural and developing areas; and the water issues are quite different. There, household water is obtained from rooftops, public taps, wells, or individual surface water withdrawals. Some less developed urban areas deliver water by tanker trucks to a central source for collection by individual households. Privatization of water distribution has led to problems of equitable access to water by poor communities [30]. Waste water treatment is often limited in rural and developing areas.

The water available per person varies dramatically by geographic region (Figure 18.5). Fifty liters per person-day (lpd; 18.3 m^3 per person year) is an estimated minimum global water target for human hydration, hygiene and sanitation, and food preparation – of which 5 lpd is needed for hydration alone [31]. US urban areas consume up to 400 lpd, while countries such as Gambia, Cambodia, and Somalia use 10 lpd or less [32].

Some urban and rural people do not have access to clean drinking water – primarily in less developed areas. Urban water can come from contaminated sources and containers, while rural area streams can be polluted by upstream villages and farms. Developed countries can use water more efficiently by recycling through water treatment facilities. North America, Western Europe, Japan/Korea, and Australia/New Zealand are the primary recyclers of wastewater at present (Figure 18.4).

Cities and industrial areas commonly cover much of their soil with pavement or roofs and drain rainwater and wastewater rapidly to streams and rivers as overland flow ("runoff"). Cities and industrial areas also destroy the soil's structure and so further exacerbate the soil's ability to infiltrate water. These actions limit the recharge of aquifers. Where trees and

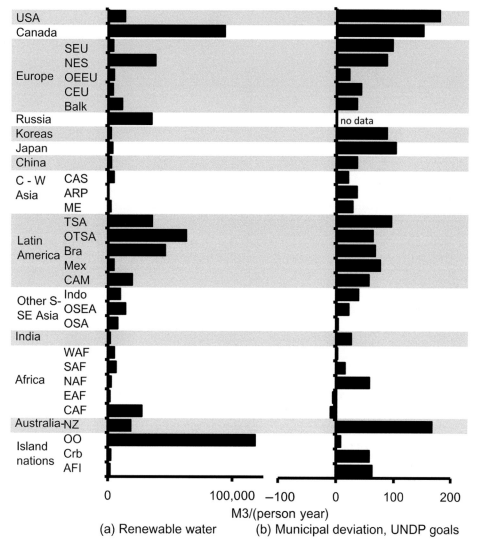

Figure 18.5 (a) Renewable (theoretically available) water per person and (b) municipal water deviation from UNDP goal of 18.3 m³/(person year) by country groups [1].

other vegetation are removed, less water evaporates to the atmosphere; but the absence of roots often means soil infiltration rates are poor.

18.3.3 Agriculture, Grazing, and Forestry

Agriculture is the greatest human use of water, primarily for irrigation but also for animal husbandry and crop handling (Chapter 21).

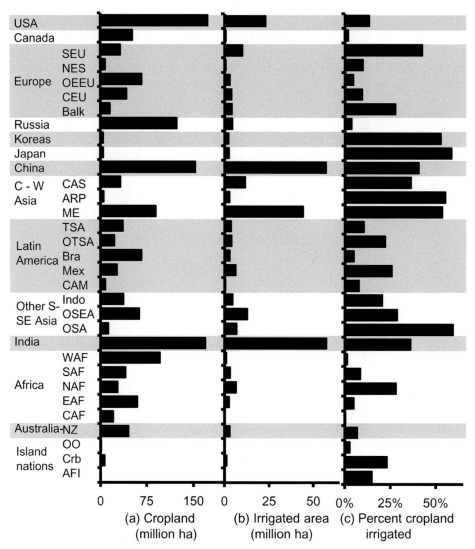

Figure 18.6 (a) Total cropland (farmland for annual and perennial crops), (b) area irrigated, and (c) percentage of cropland irrigated by country groups [33].

Irrigation, the artificial watering of plant crops, has been undertaken for about 10 thousand years. Currently, it is used in most of the world, with many countries irrigating a large proportion of their agriculture land (Figure 18.6). Four regions – China, the Middle East, India, and the United States – irrigate the greatest area because these areas contain much arable land. Irrigation can be used to grow crops in arid areas; to change the water distribution in areas of seasonal rainfall; and to provide water during occasional droughts.

Irrigation is probably a large contributor to the global sufficiency of food and its efficiency of land area use of the past 50 years (Chapter 5). Irrigation water flushes excess salts from the soil, accumulates in plant material, and evapotranspires. Extra irrigation water is used to flush chemical fertilizers and pesticides from the soil.

Non-arid crops grown in desert regions require much more water than in other regions. Consequently, desert agriculture may not be efficient compared to growing dry-land crops in deserts and/or growing more crops in areas of more rainfall where irrigation systems are much more effective. Irrigating deserts may also divert water from other uses, including protecting biodiversity (Chapter 21).

Agriculture land uses also affect the hydrologic cycle, with different measures sometimes counteracting each other. Ditches can force water to flow rapidly into streams and rivers before it can infiltrate the soil. Plowing can sometimes compact the soil below the plow layer, reducing the ability of water to infiltrate, although a good structure and infiltration capacity is maintained within the plow layer.

Removing trees and other vegetation except for crops reduces the evapotranspiration of moisture from deep within the soil and so allows this moisture to replenish groundwater and aquifers. Farmers in arid areas often conserve soil moisture when not growing crops by plowing ("fluffing" the soil) or otherwise inhibiting vegetation and evaporation.

Irrigation water can deplete the source water from rivers, lakes, or aquifers as it evaporates during transport and application, infiltrates the soil, and evapotranspires as crops grow. Only a portion of the water is returned to groundwater. The relationship of water and agriculture will be further discussed in Chapter 21.

Improper grazing can reduce the soil's structure by compaction. Grazing can be enhanced by irrigating pastures; more commonly, fodder is raised, harvested, stored, and fed to cattle during winter if the area is to be irrigated. Water in drinking troughs distributed throughout a grassland can control the movement of grazing animals. A single drinking trough in a large grassland will concentrate the animals around the water, where they can overeat the vegetation and turn the grassland to desert in an increasingly widening radius around the water.

Much of the relationship between forests and water was discussed in Chapter 17. Forests are occasionally irrigated in agroforestry or intensive plantation systems (Chapter 30).

18.3.4 Recreational Uses

Water for recreation has recently become recognized and often conflicts with more traditional uses. Swimming, fishing, and boating advocates can demand that lakes be maintained full of water that were formerly drawn down in summer for agriculture irrigation. Sightseeing, rafting, kayaking, fishing, and similar river activities need sufficient water flows in summer and often mean less water is diverted to irrigation and industrial uses. Water recreation is also demanding greater cleanup of industrial water before being discharged.

Golf courses and desert swimming pools can be direct competitors with agriculture and biodiversity for irrigation and other water, as can artificial snowmaking for skiing.

18.3.5 Ecological Uses

Environmental awareness of the past few decades has led people to value water for less anthropocentric uses [34]. Terrestrial and aquatic ecosystems depend on water, and there is increasing emphasis on leaving water in rivers and lakes during droughts for these uses. Similar concerns of river dams blocking migration of anadromous fish have led to both retrofitting dams and designing new ones differently [18–24].

The Aral Sea [35], central Asia, and Lake Chad [36], Western Africa, represent important biodiversity issues. Both Lakes are drying from both human use of water and the changing climate; and species endemic to these lakes are threatened. (See "Aral Sea Basin" case study, Chapter 3.) Since these lakes and their unique species survived the dry period of 18 thousand years ago (Figure 9.5a), it is probable that stopping human use would prevent these lakes from drying.

Other concerns are that dams stop the previous downstream movement of sand, silt, and clay. These sediments accumulate behind dams; and the sediment-depleted outflowing water then picks up more sediment below the dam, scouring the river bed and sometimes eroding alluvial flood plains. Fertilizers in runoff water from agricultural lands can also create excess nutrients, eutrophication, and other problems to downstream aquatic ecosystems. The curtailing of forest and grassland fires may lead to shortages of nutrients that leach from these fires into downstream rivers and help the fish survive. Discharge of treated waste water into lakes and rivers requires that water already exists to dilute the treated effluent, reducing the overall concentration of any residual contaminants. Fish populations also require certain other environmental conditions such as water of a particular quality and temperature and flood events of a certain size and periodicity. The above concerns require that waters be managed in a sensitive manner [37].

18.4 Protection from Water

Water management is often done to protect people and property [38]. Farms are often established in floodplains because the costs of floods are worth the benefits of growing crops on fertile ground (Chapter 12). Cities are often established where people did not initially consider the costs and dangers of periodic floods.

Dams and levees (or dikes) are often constructed to keep flooding rivers away from low-lying areas (Figure 12.8). Flood channels and diversion canals are also used to divert excess water away from sensitive areas, often by shifting water to another channel (Figure 18.7a). And, meandering rivers are often straightened to allow the water to flow off the floodplain quicker, to make navigation easier, and to reduce the length of levees (Figure 12.6). The rapid runoff means less time for water to percolate to aquifers, so more flows to oceans.

Figure 18.7 (a) Diversion channel 6 meters deep created by Emperors Vespasian and Titus (AD 69–81) to divert silty storm water from Sileucia, the harbor of Antioch. (b) Avalanche control fences in the European Alps. (A black and white version of this figure will appear in some formats. For the color version, please refer to the plate section. Color plate 8 (Figure 18.7a only).)

Artificial structures are also created to stop erosion or siltation caused by water. Terraces, tree planting, and concrete drains are sometimes created in highly erosive soils to prevent water from moving at high speeds and eroding the soils (Figure 12.4). Contour plowing of fields and control of grazing, logging, and farming near streams are also used to minimize erosion and subsequent siltation.

Siltation of reservoirs, harbors, and ship canals has been a concern for thousands of years (Figure 10.4). Land use control to avoid siltation of water reservoirs and diversion canals to keep floodwaters from flowing into harbors have been used (Figure 18.7a).

Avalanches commonly occur in predictable locations as snow accumulates. Avalanches can be quite destructive, but can often be mitigated either by identifying avalanche hazard areas and triggering avalanches before they become destructive or by constructing barriers that prevent snow from sliding (Figure 18.7b).

Floods, GLOFs, and other water hazards have been discussed earlier (Chapter 12).

References

1. UN FAO. *AQUASTAT Main Database*. (United Nations Food and Agriculture Organization, 2016 [Accessed April 11, 2016]). Available from: www.fao.org/nr/water/aquastat/data/query/index .html?lang=en.
2. J. W. Redway, cartographer. *Mississippi River System*. (American Book Company, 1901).
3. J. C. Dooge. A General Theory of the Unit Hydrograph. *Journal of Geophysical Research*. 1959;64(2):241–56.
4. V. K. Gupta, E. Waymire, C. Wang. A Representation of an Instantaneous Unit Hydrograph from Geomorphology. *Water Resources Research*. 1980;16(5):855–62.
5. C. S. Jarvis. Floods in the United States: Magnitude and Frequency. *Report*. 1936 771.
6. C. O. Wisler, E. F. Brater. *Hydrology*. (John Wiley and Sons, Inc., 1959).
7. H. L. Schramm Jr, editor. Status and Management of Mississippi River Fisheries. In *Proceedings of the Second International Symposium on the Management of Large Rivers for Fisheries*. (Citeseer, 2004).

8. J. V. Sutcliffe, Y. P. Parks. The Hydrology of the Nile. *International Association of Hydrological Sciences Special Publication no* 5. 1999:179.

9. J. Mortatti, J. M. Moraes, J. C. Rodrigues Jr., R. L. Victoria, L. A. Martinelli. Hydrograph Separation of the Amazon River Using [18]O as an Isotopic Tracer. *Scientific Agriculture.* 1997;54 (3 Piracicaba Sep/Dec.).

10. M. Becker, J. S. da Silva, S. Calmant, et al. Water Level Fluctuations in the Congo Basin Derived from Envisat Satellite Altimetry. *Remote Sensing.* 2014;6:9340–58.

11. J. U. Belz. *The Runoff Regime of the River Rhine and Its Tributaries in the 20th Century Analysis, Changes, Trends.* 2010 (Accessed July 8, 2017). Available from: www.chr-khr.org/en/project/discharge-regime-rhine-and-its-tributaries-20th-century.

12. P. G. Whitehead, E. Barbour, M. N. Futter, et al. Impacts of Climate Change and Socio-Economic Scenarios on Flow and Water Quality of the Ganges, Brahmaputra and Meghna (Gbm) River Systems: Low Flow and Flood Statistics. *Environmental Science: Processes and Impacts.* 2015;17(6):1057–69.

13. P. K. Barten, J. A. Jones, G. L. Achterman, et al. *Hydrologic Effects of a Changing Forest Landscape.* (National Research Council of the National Academies of Science, USA, 2008).

14. P. H. Gleick. Water and Conflict: Fresh Water Resources and International Security. *International Security.* 1993;18(1):79–112.

15. M. A. Giordano, A. T. Wolf. *Atlas of International Freshwater Agreements.* (United Nations Environment Programme, 2002).

16. D. Michel. Troubled Waters, in Pandya A., editor, *Climate Change, Hydropolitics, and Transboundary Resources.* (Stimson Institute, 2009).

17. A. Litke, A. Fieu-Clarke. The UN Watercourse Convention: A Milestone in the History of International Water Law 2015 (Accessed January 16, 2017). Available from: www.globalwaterforum.org/2015/02/02/the-un-watercourses-convention-a-milestone-in-the-history-of-international-water-law.

18. D. M. Rosenberg, P. McCully, C. M. Pringle. Global-Scale Environmental Effects of Hydrological Alterations: Introduction. *BioScience.* 2000;50(9):746–51.

19. N. L. Poff, D. D. Hart. How Dams Vary and Why It Matters for the Emerging Science of Dam Removal. *BioScience.* 2002;52(8):659–68.

20. B. D. Richter, G. A. Thomas. Restoring Environmental Flows by Modifying Dam Operations. *Ecology and Society.* 2007;12(1):12.

21. C. Katopodis. Developing a Toolkit for Fish Passage, Ecological Flow Management and Fish Habitat Works. *Journal of Hydraulic Research.* 2005;43(5):451–67.

22. G. Čada, J. Loar, L. Garrison, R. Fisher Jr, D. Neitzel. Efforts to Reduce Mortality to Hydroelectric Turbine-Passed Fish: Locating and Quantifying Damaging Shear Stresses. *Environmental Management.* 2006;37(6):898–906.

23. G. F. Čada. The Development of Advanced Hydroelectric Turbines to Improve Fish Passage Survival. *Fisheries.* 2001;26(9):14–23.

24. P. McCully. *Silenced Rivers: The Ecology and Politics of Large Dams.* (Zed Books, 2001).

25. I. A. Shiklomanov, J. C. Rodda. *World Water Resources at the Beginning of the Twenty-First Century.* (Cambridge University Press, 2004).

26. G. A. Burton, N. Basu, B. R. Ellis, et al. Hydraulic "Fracking": Are Surface Water Impacts an Ecological Concern? *Environmental Toxicology and Chemistry.* 2014;33(8):1679–89.

27. R. D. Vidic, S. L. Brantley, J. M. Vandenbossche, D. Yoxtheimer, J. D. Abad. Impact of Shale Gas Development on Regional Water Quality. *Science.* 2013;340(6134):1235009.

28. H. H. Rogner, R. F. Aguilera, C. L. Archer, et al. Energy Resources and Potentials. In *Global Energy Assessment – Toward a Sustainable Future.* (Cambridge University Press, 2012): pp. 425–512.

29. T. M. Pavlović, I. S. Radonjić, D. D. Milosavljević, L. S. Pantić. A Review of Concentrating Solar Power Plants in the World and Their Potential Use in Serbia. *Renewable and Sustainable Energy Reviews.* 2012;16(6):3891–902.

30. S. Galiani, P. Gertler, E. Schargrodsky. Water for Life: The Impact of the Privatization of Water Services on Child Mortality. *Journal of Political Economy.* 2005;113(1):83–120.

31. United Nations. *The Millennium Development Goals Report 2009*. (United Nations Publications, 2009). 9210542967.
32. P. H. Gleick, N. Ajami. *The World's Water, Volume 8*. (Island Press, 2014).
33. UN FAO. *FAOSTAT: Download Data* 2009 (Accessed September 16, 2012). Available from: http://faostat3.fao.org/download/.
34. S. L. Postel. Aquatic Ecosystem Protection and Drinking Water Utilities. *American Water Works Association Journal*. 2007;99(2):52.
35. P. Micklin. The Aral Sea Disaster. *Annual Review of Earth and Planetary Sciences*. 2007;35:47–72.
36. The World Bank. *Restoring a Disappearing Giant: Lake Chad*. (The World Bank Group, 2014 [Accessed May 21, 2016]). Available from: www.worldbank.org/en/news/feature/2014/03/27/restoring-a-disappearing-giant-lake-chad.
37. S. Postel, B. Richter. *Rivers for Life: Managing Water for People and Nature*. (Island Press, 2012).
38. K. Smith, R. Ward. *Floods: Physical Processes and Human Impacts*. (Wiley, 1998).

19

Managing and Mitigating Hydrologic Systems

19.1 Changing the Hydrologic Systems

Water is a resource whose availability and cleanliness is largely affected by the management of other resources – agriculture, energy, forests, and minerals. People manage water for many purposes and in different ways in different seasons and locations. In parts of Canada and the Amazon, water is plentiful, while it is scarce in inland western North American, central Asia, and parts of Africa. People try to get rid of excess water during flood seasons, but conserve it in dry seasons.

Water is difficult to distribute – to store and move in large quantities, prevent or divert its flow, or ship for long distances. Further complicating water management is the unequal distribution of people relative to available water (Figure 18.1c). For example, few people live in Canada, where water is abundant, but many live in dry parts of Africa. People also unintentionally change water flows when they change land uses.

Water management can be done through dismantling an aquifer system or lake and exploiting water stocks, altering but retaining the hydrologic system, or accepting only the water delivered by an unchanged system. Dismantling a hydrologic system can be done by "drawing down" an aquifer or lake – reducing the water it contains. Such drawdowns can lead to drying of lakes [1] or land subsidence in the case of aquifers [2]. Drawdowns will obtain much water quickly but will eventually leave little for future decades or generations, often permanently destroying the ecosystem service of providing sustainable fresh water. Such destruction is justified by people who presume new technologies will emerge to create and/or exploit water more efficiently [3] – or the climate/geology will have changed and new or replenished stocks will become available.

Alternatively, thousands of years of experience have shown that less seasonal and annual fluctuations of available water can be achieved by retaining the hydrologic system but adjusting it with management and infrastructures. Unlike the atmospheric system (Chapter 9), the hydrologic system probably can provide more ecosystem services and fewer negative consequences with careful adjustments than with preservation. There are trade-offs to changing a hydrologic system – and not changing it – that need to be considered carefully.

An area's hydrologic cycle will not necessarily remain stable over time. Some changes can be observed and/or predicted with sociological changes at large and small scales. For

298

example, the drying of shallow farm wells and drawdowns of aquifers in the western United States beginning in the 1930s coincided with the prevention of forest fires and the increases in forest density [4]. The expansion of urban areas and concomitant runoff directed to the ocean instead of to the aquifer in New Jersey has led to declining water tables and changes to the aquifer [5]. The spring floods on the Yangtze and Yellow rivers in China are probably exacerbated by the very heavy yak and sheep grazing on the Tibetan Plateau, where both rivers begin (Figure 14.8b). The abandonment of excessive yak herding – a sociological change – could moderate these spring floods.

The rate of human modification of hydrologic systems has increased recently. Agricultural ditches, storm drainage systems, straightened river channels, and other systems that rapidly move fresh water back to oceans have reduced the world's total stock of freshwater. This reduction can probably be reversed in most cases.

Dams, levees, and other hydrologic infrastructures do not last forever; and some will need to be dismantled or rebuilt. Others may remain as historic relics (Figure 18.7a). New river systems may provide values with alternative designs [6]. Important trade-offs will need to be analyzed and decisions made on a case-by-case basis both on whether and how to build new hydrologic infrastructures and whether and how to dismantle or rebuild existing, aging ones. Some dams have been built poorly or on unsuitable landforms and are becoming increasingly in danger of bursting and flooding downstream cities [7]. Issues not anticipated when the dam was built need to be included – recreation and ecological issues, for example – in addition to the traditional values of irrigation and industrial water, flood control, and water power. Social issues created by the dam and the impacts and cost of each alternative will also need to be considered.

19.2 Aquifers

Aquifers have the advantage of not occupying land surface area. Water infiltrating to aquifers is often filtered by the soil and so free of many chemicals. On the other hand, sometimes aquifers can also be naturally polluted and harmful to people [8]. They can be rendered useless or harmful by misuse – overly drawing down, allowing salt water intrusion, or polluting them.

19.2.1 Aquifer Drawdown

Aquifers are drawn down if water is pumped out through wells more rapidly than it flows back into the aquifer [9]. Drawdown is accelerated when potential recharge water is prevented from percolating into the soil, such being diverted as overland flow directly to rivers and the sea. Measuring and managing the aquifer level is difficult because of such things as seasonal changes, lag times between rain events and water reaching the aquifer, and the aquifer surface readjusting itself after a well draws down one place [10, 11].

Under many circumstances, an aquifer can be recharged by allowing a lot of water to infiltrate through the soil to the aquifer. Parts of some aquifers cannot accept new water once previous water has been drawn off; however, there do not seem to be major areas unable to recharge [2].

People have been depleting some of the world's key aquifers (Figure 17.2) [12]; consequently, lifestyles that depend on these aquifers cannot be sustained without different behaviors or a change in climates. Highest drawdowns (0.3–1 meters/year) are occurring in northwest India, with slightly lower drawdowns (0.1–0.3 meters/year) in eastern China, central and western India, Iraq and Iran, and the United States central and Pacific areas [13]. Nationally, five countries are (net) drawing down their aquifers: Saudi Arabia, Libya, Egypt, Pakistan, and Iran [14]. Changes of actions can probably recharge some of them. For example, the Sacramento aquifer may begin recharging again as fires or forest management remove the excess forest trees from its Sierra Nevada mountain drainage basin [4].

On the other hand, the Ogallalla aquifer of the western United States may be a fossil aquifer; and human use has recently been drawing it down – some areas are now devoid of the aquifer's water.

19.2.2 Salt Water Intrusion and Aquifer Pollution

Once the "halo" of fresh water in coastal aquifers (Figure 17.1) is depleted and salt water moves into the aquifer [9], reversing the process is difficult or impossible. Some places may mandate that permeable surfaces such as forests and other open spaces be maintained on a large proportion of the landscape to allow water infiltration to the aquifer. Such mandates could limit suburban sprawl and promote concentrated living (Figure 4.7).

Some aquifers are naturally polluted (Chapter 17) and others can be polluted by human action. Agricultural chemicals can be percolated through aquifers, as can industrial (Figure 17.3), commercial, and household chemicals. These can contaminate wells as they move through the aquifer, destroying the aquifer's use for people in distant places and times.

Three major mitigation approaches are: avoiding chemicals being put where they can infiltrate the aquifer, cleaning up past chemical infiltration areas, and breeding and using microbes that can decompose harmful chemical put into the environment – "bioremediation." Measures for mitigating arsenic in aquifers are being developed [15].

19.3 Surface Waters

Surface waters offer more benefits than aquifers; but dams and irrigation systems can create problems that were not anticipated when the dam was built.

19.3.1 Dams and Reservoirs

Dams have existed in the Middle East since 3000 BC [16]. Ponds and lakes in agriculture areas can have multiple benefits (Figure 19.1). If allowed to fill during seasons of high water flow, they can decrease the peak flow and flooding downstream. Also, the water can be

Figure 19.1 Small reservoir in Haryana, India, that protects the downstream city from flooding and provides irrigation water for village agriculture.

released later during a dry season and irrigate crops, provide drinking and industrial water, and provide enough water in rivers for ecological, recreation, and transport purposes. They can also provide hydroelectric power.

With increasing demands among resources and with developing technologies, dams are having other positive and negative impacts beyond preventing floods and increasing water supplies [17]. Large dams in particular pose several controversial management issues as trade-offs to their benefits [18–20]. They often disrupt the previous hydrologic cycle of rivers, affecting fish populations and riparian plant communities. They can impact downstream agriculture that has been adapted to seasonal flooding, such as in the lower Nile River basin [21]. Dams can also alter stream temperatures and nutrient loads and prevent sediment from traveling downstream. Socially, construction of large dams can displace people whose land becomes flooded for the reservoir. After this flooding, the standing water can increase disease vectors such as mosquitoes that carry malaria and snails that carry schistosomiasis. Where large dams encourage people to settle downstream in the former floodplain, they subject the settlers to rare floods too large to be contained by the dam or catastrophic floods if the dam bursts either from poor construction or an earthquake.

Dams also have benefits. They provide very much renewable energy (Figure 25.6); prevent floods; provide aquatic habitats, water for irrigation and water for drinking; and present recreation opportunities. Existing and new reservoirs present different opportunities and challenges. Existing reservoirs have already changed land uses above and below the dam; altered regimes of the water, power, navigation, and recreation; and adjusted ecosystems and species interactions. Installation of new reservoirs will eliminate some habitats but create others.

Approximately 800 thousand dams existed worldwide, with 40 thousand categorized as "large dams" [20]; a "large dam" is taller than 15 meters from its foundation with storage capacity over 3 million cubic meters. Seventy-five percent of these large dams were in five countries as of the year 2000: China, the United States, India, Japan, and Spain [20]. All but 5,000 of them have been constructed in the past half-century [16].

Such modifications to rivers undoubtedly threatened or destroyed some species and favored others both by changing habitats and reducing the area needed for food production [22]; however, the awareness of potential harms and ways to avoid them have also improved (Chapter 18) [16, 22–29].

Even while large dams are still being constructed, a global movement is expressing concern about the ecological and social costs of dams when designing new projects. Some groups are successfully promoting the removal of dams [26]. Removing dams will create new ecosystems and some that may resemble their former conditions. In some cases, dam removal may restore key species such as anadromous fish that could not migrate above the dams as well as sediment to downstream floodplains, but it will affect other values [30].

The large infrastructures are increasingly being complemented with "soft path" solutions such as lower-cost community-scale systems, equitable pricing, and efficient technologies [31] similar to innovations in energy (Figure 24.4).

19.3.2 Managing Watersheds

Watersheds and places that feed aquifers can be managed in ways that provide water and other values (Chapter 28). A "regulated" forest or shrub land in which each year a portion of the area is harvested, burned in a controlled manner, or otherwise reduced in vegetation density and allowed to regrow provides more water flow and reduction in siltation through a reduction in evapotranspiration, safety from wildfires, and maintenance of soil structure (Chapter 29). A management approach that integrates vegetation management with other land uses is probably most helpful [32].

Eliminating roots and other organic matter in soil – such as through over grazing, forest harvesting without regeneration, poor plowing, or hot fires can cause erosion in watersheds that fill reservoirs with silt [33]. Silted reservoirs become useless for flood control, irrigation, or energy production (Figure 19.2) [34]. Siltation by surrounding farms is also a concern in the Panama Canal [35].

Changing to tree species that are "drought avoiders" (Chapter 17), to deciduous species that do not evapotranspire in winter, and to grasses in some areas can help avoid water loss but maintain the infiltration capacity of the soil.

19.3.3 River Diversions and Irrigation

Water has been diverted from rivers for irrigation in many parts of the world for millennia. Recently, irrigation canals have extended many kilometers. Rivers have also been diverted for transportation canals and water supplies, at times sending water hundreds of miles through pipes and aqueducts.

Figure 19.2 Completely silted-in reservoir caused by inappropriate land use in the drainage basin (Armenia).

As values change, features are being designed into new systems and retrofitted into old ones that try to accommodate new values while maintaining existing ones:

• Mechanical devices allow fish to cross irrigation dams but prevent them from being swept into the ditches and fields [36]:

- River diversions are being adjusted to leave water in the river for ecological purposes during dry seasons [37];
- Irrigation techniques are becoming more water-efficient [38];
- Canals are being made impermeable to decrease seepage loss and fit with covers to reduce evaporation loss [39].

19.4 Conserving Existing Water

Places with seasonally or annually limited water flows have several opportunities for extending their water by either storing, reusing (Figure 18.4), or using less. Dam storage has been discussed. Other ways will be discussed below. Some of these methods require sophisticated technologies.

19.4.1 Using Less Water

Some urban areas implement strategies that reduce the individual users' demands for water. These include progressive water pricing policies [40], mandating the use of efficient plumbing fixtures and appliances, and updating the urban water infrastructure to minimize leaks and other waste.

Irrigation technologies have progressed dramatically [41]; and alternate sources of irrigation water are being investigated, including reused or recycled waste water. Spot applications of fertilizers and pesticides could necessitate less irrigation water to flush them (Chapter 21).

Home lifestyles can affect water conservation. Landscaping that requires little watering, appropriate home cooling systems, water-conserving faucets and toilets, and use of appropriate cooking and eating utensils (hands, chopsticks, or forks and spoons) can affect water use in washing.

Thermoelectrical generators can avoid water loss if the heated water is cooled and reused after it is used for steam generation ("combined heat and power" [CHP]) [42].

19.4.2 Recycling, Treating, and Reusing

The same water can be used and returned to the river and used and recycled many times by a series of users as it moves downstream (Figure 18.5). The water can be reused more frequently if the wastewater is purified before being returned to the river. Treating waste water requires technologies and infrastructure for "primary," "secondary," and "tertiary" treatment and is most common in urban, developed areas. Primary treatment removes large debris and initially separates water from sludge and grease. Secondary treatment is a biological treatment that further reduces contaminants in the water. Disinfection, often by chlorination, occurs before the treated water is discharged. Tertiary treatment reduces the water's nutrient content.

Reuse of waste water has occurred indirectly for many years as downstream users withdrew water from the same river where upstream users discharged their waste water. However, direct reuse of treated waste water within the same community has been increasing [43]. Waste water is also being used for agriculture irrigation [44] and is increasingly being viewed as a commodity [45]. Waste water contains contaminants, and treatments must match the needed quality for the desired end use. Drinkable water, for example, requires more treatment than water used for landscape irrigation [45].

19.5 Obtaining More Water Sustainably

Other technologies have developed from the need to secure water in regions with scarce supply and increasing demand. While these issues can intertwine, they will be discussed individually here.

19.5.1 Rainwater Rooftop Harvesting

Direct use of precipitation can be done by collecting water from rooftops and storing it in cisterns instead of sending it directly to the ground or, worse yet, to drains that lead to rivers [46]. The water can be used for washing, garden irrigation, and sewage flushing. With appropriate roof and storage systems, the water can also be used safely for drinking. An added efficiency is to shift the water sequentially to several pools of progressive dirtiness as it is used.

19.5.2 Artificial Glaciers and Tapping Glaciers

Glaciers have been used to generate water in droughty summers in some areas by spreading ashes on them; the dark ashes absorb heat and melt glacial ice [47]. In arid regions with high mountain glaciers, droughts occur in late spring after low elevation water is depleted and before higher elevation snows melt. Innovative, inexpensive "artificial glaciers" – snow packed in shady places – can be created at low elevations from winter snow that melts after low elevations are dry but before high elevation glaciers melt [48].

Rapid melting of glaciers with changing climates is increasing the danger of GLOFs (Chapter 12) – in the Himalayas, Andes, and elsewhere [49]. Recently, glacial lakes within and adjacent to the ice are being fit with artificial water drains [50] to reduce the danger of floods. In some cases, these drains are used for irrigation and hydroelectric power generation [51].

19.5.3 Water Canals and Pipelines

Canals and pipes have long been used to ship water moderate distances – to southern California from the Colorado River and northern California and from upstate to New York City [9]. A recently installed water pipe under the Mediterranean Seas ships water from a reservoir in Turkey to the island of Cyprus 80 kilometers away [52]. With the drying climate of the Middle East, people will probably either need to migrate elsewhere or obtain water to sustain them in place. Shipping water to them in pipelines may be a viable option. Pipeline routes and rights-of-way have already been established for oil and gas which could be used for water pipes. With a reduction in fossil fuel use, reconstructing the pipelines for water transport may be a new use for this long-term infrastructure (Chapter 10). Shipping water in tankers, in large plastic bags, or as icebergs to needed ports have also been considered [9, 53].

19.5.4 Virtual Water

An alternative to shipping water is to trade "virtual water" [54, 55]. That is, since agriculture crops require much water – especially for irrigation – countries and regions with water shortages would forgo producing foods, concentrate on less water-consuming enterprises, and purchase their food from places without such water shortages. In this way, food becomes "virtual water."

19.5.5 Desalinization

Desalination is possible for communities with saline and brackish waters and affordable energy. Curacao in the Netherlands Antilles has been operating desalination facilities since the late 1920s and Saudi Arabia built its first facility in the late 1930s [56]. Over one hundred countries use some desalination (Figure 18.5).

The two major desalinization technologies are:

- Multistage flash distillation (MSF) heats saline water and distills fresh water in stages of different temperatures and pressures that efficiently distill the increasingly saline brine. It requires much energy [56].
- Reverse osmosis (RO) separates water from brine using a semipermeable membrane through which water passes, but not the salts. RO desalinization plants are becoming increasingly larger and cheaper [57].

Demand management and waste water reclamation have generally been more cost effective than desalinization [56]. However, if solar-thermal energy production in deserts becomes efficient, desert areas near oceans could become industrial areas and access fresh water by using some energy for desalinating salt water – especially using the energy released in cooling the turbine water [58].

References

1. P. Micklin. The Aral Sea Disaster. *Annual Review of Earth and Planetary Sciences.* 2007;35:47–72.

2. J. A. Nunn. Seasonal Groundwater Withdrawal in Southwestern Louisiana: Implications for Land Subsidence and Resource Management. *Gulf Coast Association of Geological Societies Transactions*. 2010;60:515–24.

3. J. L. Simon. *The Ultimate Resource 2*. (Princeton University Press, 1998).

4. P. K. Barten, J. A. Jones, G. L. Achterman, et al. *Hydrologic Effects of a Changing Forest Landscape*. (National Research Council of the National Academies of Science, USA, 2008).

5. H. Sun, D. Grandstaff, R. Shagam. Land Subsidence Due to Groundwater Withdrawal: Potential Damage of Subsidence and Sea Level Rise in Southern New Jersey, USA. *Environmental Geology*. 1999;37(4):290–6.

6. S. Postel, B. Richter. *Rivers for Life: Managing Water for People and Nature*. (Island Press, 2012).

7. N. Adamo, N. Al-Ansari. Mosul Dam Full Story: Safety Evaluations of Mosul Dam. *Journal of Earth Sciences and Geotechnical Engineering*. 2016;6(3):185–212.

8. R. Nickson, J. McArthur, W. Burgess, et al. Arsenic Poisoning of Bangladesh Groundwater. *Nature*. 1998;395(6700):338–.

9. J. R. Craig, D. J. Vaughan, B. J. Skinner. *Resources of the Earth: Origin, Use, and Environmental Impacts*, fourth edition. (Prentice Hall, 2011).

10. V. Batu. *Aquifer Hydraulics: A Comprehensive Guide to Hydrogeologic Data Analysis*. (John Wiley & Sons, 1998).

11. N. Brozović, D. Sunding, D. Zilberman. Optimal Management of Groundwater over Space and Time. *Frontiers in Water Resource Economics*. (Springer, 2006), pp. 109–35.

12. L. F. Konikow, E. Kendy. Groundwater Depletion: A Global Problem. *Hydrogeology Journal*. 2005;13(1):317–20.

13. Y. Wada, L. P. van Beek, C. M. van Kempen, et al. Global Depletion of Groundwater Resources. *Geophysical Research Letters*. 2010;37(20).

14. J. Rockström, M. Falkenmark, T. Allan, et al. The Unfolding Water Drama in the Anthropocene: Towards a Resilience-Based Perspective on Water for Global Sustainability. *Ecohydrology*. 2014;7(5):1249–61.

15. R. R. Shrestha, M. P. Shrestha, N. P. Upadhyay, et al. Groundwater Arsenic Contamination, Its Health Impact and Mitigation Program in Nepal. *Journal of Environmental Science and Health, Part A*. 2003;38(1):185–200.

16. P. McCully. *Silenced Rivers: The Ecology and Politics of Large Dams*. (Zed Books, 2001).

17. A. Brismar. River Systems as Providers of Goods and Services: A Basis for Comparing Desired and Undesired Effects of Large Dam Projects. *Environmental Management*. 2002;29(5):598–609.

18. D. Trussell, editor. *The Social and Environmental Effects of Large Dams*. (Pergamon, 1992).

19. E. Goldsmith, N. Hildyard. *The Social and Environmental Effects of Large Dams. Volume 2: Case Studies*. (Wadebridge Ecological Centre, 1986).

20. World Commission on Dams. *Dams and Development: A New Framework for Decision-Making: The Report of the World Commission on Dams*. (Earthscan, 2000).

21. G. F. White. The Environmental Effects of the High Dam at Aswan. *Environment: Science and Policy for Sustainable Development*. 1988;30(7):4–40.

22. D. M. Rosenberg, P. McCully, C. M. Pringle. Global-Scale Environmental Effects of Hydrological Alterations: Introduction. *BioScience*. 2000;50(9):746–51.

23. G. Čada, J. Loar, L. Garrison, R. Fisher Jr., D. Neitzel. Efforts to Reduce Mortality to Hydroelectric Turbine-Passed Fish: Locating and Quantifying Damaging Shear Stresses. *Environmental Management*. 2006;37(6):898–906.

24. G. F. Čada. The Development of Advanced Hydroelectric Turbines to Improve Fish Passage Survival. *Fisheries*. 2001;26(9):14–23.

25. C. Katopodis. Developing a Toolkit for Fish Passage, Ecological Flow Management and Fish Habitat Works. *Journal of Hydraulic Research*. 2005;43(5):451–67.

26. N. L. Poff, D. D. Hart. How Dams Vary and Why It Matters for the Emerging Science of Dam Removal. *BioScience*. 2002;52(8):659–68.

27. B. D. Richter, G. A. Thomas. Restoring Environmental Flows by Modifying Dam Operations. *Ecology and Society.* 2007;12(1):12.

28. B. D. Richter, S. Postel, C. Revenga, et al. Lost in Development's Shadow: The Downstream Human Consequences of Dams. *Water Alternatives.* 2010;3(2):14.

29. S. J. Clarke, L. Bruce-Burgess, G. Wharton. Linking Form and Function: Towards an Eco-Hydromorphic Approach to Sustainable River Restoration. *Aquatic Conservation: Marine and Freshwater Ecosystems.* 2003;13(5):439–50.

30. E. H. Stanley, M. W. Doyle. Trading Off: The Ecological Effects of Dam Removal. *Frontiers in Ecology and the Environment.* 2003;1(1):15–22.

31. P. H. Gleick. Global Freshwater Resources: Soft-Path Solutions for the 21st Century. *Science.* 2003;302(5650):1524–8.

32. N. R. Council. *New Strategies for America's Watersheds.* (National Academies Press, 1999).

33. G. G. Ice, D. G. Neary, P. W. Adams. Effects of Wildfire on Soils and Watershed Processes. *Journal of Forestry.* 2004;102(6):16–20.

34. B. Bowonder, K. Ramana, T. H. Rao. Management of Watersheds and Water Resources Planning. *Water International.* 1985;10(3):121–31.

35. R. S. Harmon. An Introduction to the Panama Canal Watershed. in *The Río Chagres, Panama.* (Springer, 2005): pp. 19–28.

36. S. B. Gale, A. V. Zale, C. G. Clancy. Effectiveness of Fish Screens to Prevent Entrainment of Westslope Cutthroat Trout into Irrigation Canals. *North American Journal of Fisheries Management.* 2008;28(5):1541–53.

37. B. D. Richter, R. Mathews, D. L. Harrison, R. Wigington. Ecologically Sustainable Water Management: Managing River Flows for Ecological Integrity. *Ecological Applications.* 2003;13(1):206–24.

38. T. A. Howell. Enhancing Water Use Efficiency in Irrigated Agriculture. *Agronomy Journal.* 2001;93(2):281–9.

39. P. K. Swamee, G. C. Mishra, B. R. Chahar. Design of Minimum Water-Loss Canal Sections. *Journal of Hydraulic Research.* 2002;40(2):215–20.

40. A. Ruijs. Welfare and Distribution Effects of Water Pricing Policies. *Environmental and Resource Economics.* 2009;43(2):161–82.

41. J. Wallace. Increasing Agricultural Water Use Efficiency to Meet Future Food Production. *Agriculture, Ecosystems & Environment.* 2000;82(1):105–19.

42. E. Thorin, J. Sandberg, J. Yan. Combined Heat and Power. *Handbook of Clean Energy Systems.* 2015.

43. J. Haarhoff, B. Van der Merwe. Twenty-Five Years of Wastewater Reclamation in Windhoek, Namibia. *Water Science and Technology.* 1996;33(10):25–35.

44. F. Pedrero, I. Kalavrouziotis, J. J. Alarcón, P. Koukoulakis, T. Asano. Use of Treated Municipal Wastewater in Irrigated Agriculture – Review of Some Practices in Spain and Greece. *Agricultural Water Management.* 2010;97(9):1233–41.

45. S. W. Hermanowicz, T. Asano. Abel Wolman's "the Metabolism of Cities" Revisited: A Case for Water Recycling and Reuse. *Water Science and Technology.* 1999;40(4):29–36.

46. V. Meera, M. M. Ahammed. Water Quality of Rooftop Rainwater Harvesting Systems: A Review. *Journal of Water Supply: Research and Technology-AQUA.* 2006;55(4):257–68.

47. M. Freeberne. Glacial Meltwater Resources in China. *The Geographical Journal.* 1965;131(1):57–60.

48. C. Norphel, editor. Artificial Glacier: A High Altitude Cold Desert Water Conservation Technique. In *Defense of Liberty Conference Proceedings.* (New Delhi: Defense of Liberty Conference, 2012).

49. M. Carey, C. Huggel, J. Bury, C. Portocarrero, W. Haeberli. An Integrated Socio-Environmental Framework for Glacier Hazard Management and Climate Change Adaptation: Lessons from Lake 513, Cordillera Blanca, Peru. *Climatic Change.* 2012;112(3–4):733–67.

50. J. M. Reynolds, A. Dolecki, C. Portocarrero. The Construction of a Drainage Tunnel as Part of Glacial Lake Hazard Mitigation at Hualcán, Cordillera Blanca, Peru. *Geological Society, London, Engineering Geology Special Publications.* 1998;15(1):41–8.

51. A. Byers, J. Recharte. *New Security Beat [Internet]*. (Wilson Center Environmental Change and Security Program, 2015 [Accessed 2016]). Available from: www.newsecuritybeat.org/2015/04/ glacial-floods-threaten-mountain-communities-global-exchange-fostering-adaptation/?q=1.

52. P. Tremblay. *Turkey's Peace Pipe to Cyprus*. (Al Monitor, 2015 [Accessed December 1, 2016]). Available from: www.al-monitor.com/pulse/originals/2015/10/turkey-cyprus-water-pipeline-delivers-fears.html.

53. V. Smakhtin, P. Ashton, A. Batchelor, et al. Unconventional Water Supply Options in South Africa: A Review of Possible Solutions. *Water International*. 2001;26(3):314–34.

54. J. A. Allan. Virtual Water-the Water, Food, and Trade Nexus. Useful Concept or Misleading Metaphor? *Water International*. 2003;28(1):106–13.

55. J. N. Galloway, M. Burke, G. E. Bradford, et al. International Trade in Meat: The Tip of the Pork Chop. *AMBIO: A Journal of the Human Environment*. 2007;36(8):622–9.

56. P. H. Gleick. *The World's Water 2000–2001*. (Island Press, 2000).

57. D. Talbot. *Megascale Desalination*. (Massachusetts Institute of Technology, 2015 [Accessed March 26, 2016]). Available from: www.technologyreview.com/s/534996/megascale-desalination.

58. S. Licht. Efficient Solar-Driven Synthesis, Carbon Capture, and Desalinization, Step: Solar Thermal Electrochemical Production of Fuels, Metals, Bleach. *Advanced Materials*. 2011;23 (47):5592–612.

Part VII

Agriculture

20

Food Groups and Nutrition

20.1 Agriculture, Food, and Nutrients

Agriculture is the deliberate breeding, growing, and utilization of plants and animals under conditions largely modified by people. For this book, agriculture will include cultivation of annual and perennial plants; orchard trees and other shrubs; and terrestrial and aquatic plants in farms, gardens, ponds, and greenhouses. It will also include raising animals in pens, rangelands, and aquatic farms. It will not include collecting wild plants; street tree management, landcare, or forestry, bushmeat harvesting, zoo or wildlife management, or hunting and fishing.

The amount of food can change by production methods. A certain amount is necessary for people to survive and prosper; however, amounts beyond an upper threshold do not enable people to prosper more. Rather, it can create problems of health and disposal.

This chapter will first focus on food, which is over 99% of the world's agriculture production (Table 20.2). Nonfood crops are discussed at the end of this chapter.

Agricultural terminology differs by user. Some confusing terms are:

- "Arable" commonly refers to land capable of being farmed, even if it not currently farmed; however, the United Nations Food and Agriculture Organization ("UNFAO") uses "arable" to refer to land currently in annual agricultural crops. The term "arable" will be avoided.
- "Permanent crop" is used by the UNFAO to refer to land in perennial crops, such as orchards. The term "permanent crop" will be avoided.
- "Calorie" used in food is actually 1,000 calories (a kilocalorie) in other energy analyses. Sometimes "food calorie," "Calorie" (where "C" is capitalized), or kilocalorie is used in food discussions. Calorie (capital "C"; abbreviated "kcal") will refer to food calories in this book.
- "White meat" refers only to poultry by some authorities; to poultry and white-colored fish by others; and yet others also include pork.
- "Mutton" refers to meat from sheep or lambs and may or may not include goat meat, depending on the authority.
- "Ruminants" may or may not include camels, but does not include horses.

313

20.2 Food Groups, Nutrients, and Consumption

Plants need seventeen elements; all are taken up in nonorganic forms (Table 10.2). Animals need the same elements and more; however, animals ingest most elements as parts of organic compounds.

Foods and other animal and plant parts are generally classified in groups:

- *Carbohydrates:* sugars and starches contain carbon, hydrogen, and oxygen. They are an energy source. (Alcohol acts like a carbohydrate, but has side effects of inebriation.)
- *Proteins* are very complex combinations of amino acids made largely of carbon, hydrogen, oxygen, and nitrogen.
- *Fats and oils* are *hydrocarbon chains* (contain carbon and hydrogen). Oils are liquid: fats contain longer chains and so are solid at room temperature. They are energy sources.
- *Aromatic chemicals* can be useful in protecting against predation and attracting pollinators or fruit disseminators.
- *Mineral nutrients* are elements required in small amounts and are usually present in food or water (Chapter 10).
- *Fiber:* cellulose, hemicellulose, and lignin are components of wood and other plant structures; once formed, are nearly impossible to break down by living creatures other than certain microorganisms.
- *Bones, hides, hair, wood*, and other nonedible animal parts can have various uses.

People need to take in water, oxygen, energy, specific proteins, and minerals to survive. The greatest volumes of food are in carbohydrates, proteins, and fats. Carbohydrates, proteins, fats, mineral nutrients, and fibers are referred to as "nutrients" in this book.

Energy from food comes largely from breaking the chemical bonds in organic compounds. These are primarily the carbon and hydrogen bonds in fats and oils and the carbon, hydrogen, and oxygen bonds in sugars, starches, proteins, celluloses, lignins, and alcohol. Different compounds yield different amounts of energy (Table 20.1); carbohydrates yield about of 3.5 Calories per gram (kcal/g); proteins, 3.7 kcal/g; fats, 8.6 kcal/g. Some animals can also obtain energy by hosting microbes in their gut (gastrointestinal tract) that digest cellulosic materials and lignin.

Proteins are eaten and then separated into amino acids through digestion and then resynthesized into many thousands of new proteins. There are 22 amino acids that people and other animal and plant species need for survival. Amino acids are the building blocks of DNA, RNA, and other complex cellular molecules. The human body can produce some of these amino acids, can produce some others to limited extents, but cannot produce others. Consequently, people need foods containing the amino acids they are lacking in order to be healthy. These amino acids must to be consumed in specific proportions because excesses of some do not compensate for shortages of others.

Many minerals and vitamins are also required in small amounts, can be present in foods or water, and are often ingested when eating. Some geographic locations and foods lack

Table 20.1 *Food energy and proteins per 100 grams of various foods at moisture content generally purchased [8]*

	Food energy (calories)	Protein (grams)	Moisture (%)
Oils and fats	842	0	4%
Nuts and seeds	408	12	26%
Cereals and grain products	336	10	15%
Grain legumes and products	299	20	24%
Sugars and syrups	277	0	20%
Milk and products	216	14	58%
Meat, poultry, and game	181	19	72%
Starchy roots, tubers	116	2	70%
Beverages	113	4	83%
Fish and shellfish	103	18	77%
Fruits	72	1	79%
Vegetables and products	57	3	82%

some minerals and vitamins; and their absences generally cause weaknesses, diseases, or poor growth unless supplements are eaten that compensate for the deficiencies.

20.3 Physiology of Food Synthesis

Food initially comes from photosynthesis by plants, which first capture part of the sun's energy by bonding carbon, hydrogen, and oxygen into sugar (generally glucose) molecules (Figure 20.1). The plants then convert this sugar to starches, celluloses and lignins, proteins, fats and oils, aromatic compounds, and other chemicals (Figure 20.1).

The chemical conversions from sugars require energy, and part of the solar energy absorbed through photosynthesis is used by the plant to make these chemical changes. This energy comes through the process of respiration, which is the opposite of the photosynthesis reaction (Figure 8.4) and is done by both plants and animals. Photosynthesis and respiration are not automatically reversible, "mass balance" chemical reactions that create an equilibrium. A plant's highest priority for use of photosynthate (glucose and compounds derived from it) is energy to stay alive. Consequently, when photosynthesis is slow, the plant satisfies its use of photosynthates for energy before it grows by producing celluloses, lignins, starches, proteins, and aeromatics [1]. Photosynthesis can slow because of insufficient sunlight (shade), water, soil oxygen, or certain mineral nutrients. Excess heat can also accelerate respiration (luxury respiration) and leave little photosynthate for other uses.

Plants have slightly different photosynthesis mechanisms that make some ("C3 plants") efficient when water is plentiful; but others ("C4 plants") can continue to photosynthesize when water is scarce [2]. A complete plant canopy of C3 plants can convert up to about

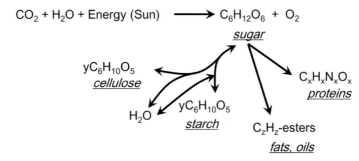

Figure 20.1 Food products from green plant photosynthesis. (Solid lines = chemical reactions.) Element names in Appendix II.

4.5% of the incoming solar energy to energy in organic compounds; and C4 plants, about 6% [3]. Most of the remaining energy is reflected, passed through to the soil, reradiated, exited by conduction and convection, or absorbed by evapotranspiring water (Figure 23.2).

Organic matter can be decomposed through respiration by free-living microbes that convert some of it directly to carbon dioxide, water, and energy and some to other organic compounds – primarily carbohydrates and proteins, as in the case of mushrooms. Some non-free-living microbes live in the guts of all animals and humans and help decompose organic compounds to molecules that can be absorbed by the host. Non-free-living microbes also live symbiotically with plant roots (mycorrhizae and others). Heat energy produced by some decomposing microbes can generate fires inside well-insulated places such as hay lofts.

Herbivorous animals can readily extract sugars, starches, proteins, and aromatic compounds from plants either directly or through the microbes in their guts (Figure 20.2). Animals such as ruminants (cattle, sheep, giraffes, water buffaloes, deer) and horses and rabbits have guts with cellulose-digesting microbes; they can digest fiber to a much greater extent than other animals. Even for them, digesting cellulosic fibers is a slow process (Chapter 14).

Some microorganisms – the fermenting yeasts and some bacteria to a lesser extent – can convert sugars and fibers to alcohol. This fermentation can occur by microorganisms in overripe fruits and some other foods and artificially in human-made devices. Some alcohol (ethanol: CH_3OH) can be a source of energy for people and other animals; however, methanol (C_2H_7OH) can be poisonous. Ethanol, methanol, propanol, and butanol can be used as biofuels.

20.4 Plant Food Groups and Production

The United Nations and other organizations analyze food nutrition by dividing foods into groups that reflect nutrient sources and concentrations [4–8]. Plant food groups vary, but generally consist of the following:

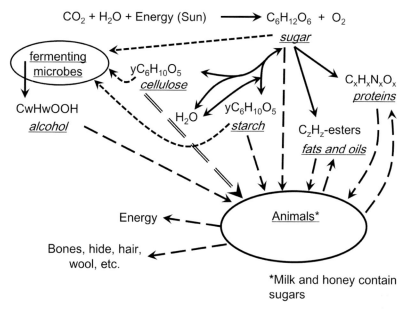

Figure 20.2 Herbivorous animal food ingestion and product production. Cellulose is ingested only by ruminants and other digesters of cellulosic materials (double line arrow). Element names in Appendix II.

- Cereals and grains
- Fruits, excluding melons
- Oil crops
- Pulses
- Roots and tubers
- Tree nuts
- Vegetables, melons, cane

These groupings are not taxonomically precise. The food groups generally come from different parts of the plant and so have different concentrations of energy and proteins (Figure 20.3), with evolutionary causes for the differences. For this book, sugarcane, sugar beets, and melons are included with vegetables unless stated differently.

20.4.1 Carbohydrates, Proteins, Oils, Celluloses, and Aromatics

Cereals, pulses, treenuts, and most oil crops are primarily seeds. Seeds are generally quite dry when mature to avoid premature germination and to be light so they can be disseminated far. On the other hand, fruits, roots and tubers, and vegetables are generally full of moisture when harvested. These natural moisture differences are commonly reflected in storage and trade conditions (Figure 20.4).

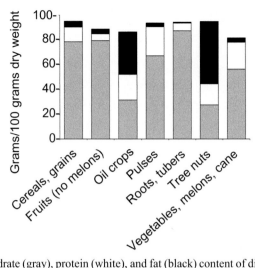

Figure 20.3 Carbohydrate (gray), protein (white), and fat (black) content of different plant foods, dry weight basis [4, 5].

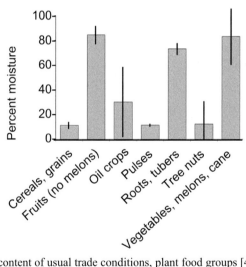

Figure 20.4 Moisture content of usual trade conditions, plant food groups [4, 5].

Sugars are found throughout the living parts of plants, and especially growing parts that need ready access to the sugar's energy for respiration. They are also found in fruits and flowers where their energy, combined with aromatics, entice birds and insects to eat the fruit or nectar, in the process distributing seeds or pollinating flowers.

Plants often convert sugars to starches by removing water, thus making them less heavy, less water soluble, and less attractive to predators. Starches are converted back to sugars that dissolve in the plants' evapotranspiration stream and protoplasm to transport the carbohydrate within the tree and to protect cells from freezing [9, 10]. Starches can be found throughout the plant in place of sugar during some seasons. Starches are also stored with seeds as fleshy cotyledons or endosperm that provides energy for the seed before it can become self-sufficient in photosynthesis. Starches are often concentrated in the inner bark (phloem) of woody stems and twigs, in and around the dormant buds of perennial twigs, and in specialized roots known as tubers and bulbs of some plants. The energy is stored and used for spring growth. On a dry weight basis, cereal and pulse seed crops, fruits, and tubers contain high concentrations of carbohydrates as either sugars or starches (Figure 20.3).

Plants develop fibrous cell walls to various extents. At one extreme are woody trees, whose xylem (woody stems and branches inside the bark and cambium; Chapter 28) is so hard to digest that primarily microbes and insects consume them. Other plants and plant parts such as leaves contain fewer unpalatable fibers.

Oils and fats – hydrocarbons – take more energy for the plant to produce than sugars and starches (carbohydrates); however, once formed, they provide more energy per weight (Table 20.1). High oil contents are especially common in tree nuts and oil crops (Figure 20.3). Under wild conditions, energy-efficient nuts can travel in light air currents; or they can use their energy reserves and grow quite large or survive a long time before they need to receive energy from sunlight.

Proteins are needed in growing parts of plants. They contain proportionately large amounts of nitrogen, which is commonly the most limiting element for plant growth. Plants often concentrate proteins in seeds; so grains and nuts have relatively high concentrations (Figure 20.3). Food groups with high concentrations of carbohydrates – fruits, roots, and tubers – generally have low concentrations of proteins because the carbohydrates dilute the proteins. To some extent, pulses and grains are exceptions. Pulses are legumes and so have less nitrogen limitation because they symbiotically fix their own nitrogen with microbes that take it from the atmosphere to their roots. Consequently, pulses have evolved to be less frugal with their proteins and have greater concentrations in their seeds than in other carbohydrate-laden seeds.

Aromatic plants contain molecules that change the taste, smell, toxicity, and effects of the plants on humans. These exist in different concentrations in different parts of the plant and are responsible for the toxicity of wild almonds, the stimulating effects of coffee, tea, and tobacco, the acrid taste of lemon pulp, the skin reaction from touching poison ivy, the sweet taste of lemon zest, and the poisonous effect of walnut trees on other plants.

20.4.2 *Patterns of Food Availability*

Plants and animals evolved in synchrony with climatic cycles of their local area. Plants usually become dormant and use little energy in winter in temperate, boreal, and polar climates. In late summer and autumn, they store some energy in tree stems, buds, and inner bark (usually sugars and starches); in roots and tubers (usually starches); and in seeds (carbohydrates or fats and oils). Plants use this energy when weather warms in spring to produce new leaves and flowers and so begin photosynthesizing again. The plants then grow and produce new fibers and some carbohydrates as stems, twigs, and roots. They also produce nectar (sugars) in their flowers and carbohydrates and hydrocarbons in fruits, nuts, and other seeds throughout the growing season. As summer progresses, more energy and proteins are stored in perennial stems, twigs, buds, and roots as well as in fruits and seeds.

The annual cycle of temperature and rainfall – the "climatic rhythm" [11] – differs among ecological zones (Chapter 7). Plants have evolved their timing of germination, growth, flowering, fruiting, seed dispersal, and dormancy to be congruent with these climatic rhythms. The actual triggers for these physiological mechanisms are varied but include day length and cumulative temperature (heating degree days).

Plants sometimes evolve new genetic variations that enable them to live beyond the boundaries of their previous climate. They can evolve or be bred so they are stimulated by greater or lesser day lengths and other stimuli as climates change or people move the plants.

Some aspects of climatic rhythms are not as easily changed. Wheat, for example, evolved in a Mediterranean climate of cool, wet winters and dry summers. It has been grown in central China for over 3 thousand years, but it still creates problems because its growth is not in synchrony with central China's heavy summer rains [11].

Food crops vary in their ability to survive in different temperature and soil moisture regimes. For example, sugar cane, oranges, and taro do not survive freezing temperatures well, while potatoes and sugar beets do. Others such as alfalfa, cassava, and peanuts can survive dry conditions; but rice, soybeans, and sunflowers grow under moist conditions. Rice is often raised in water because it can tolerate flooding better than many of its competitive plants and predatory animals.

Animals respond to seasons in their food consumption, reproduction cycles, and migrations – often as plant food availabilities seasonally change (Chapter 14). People have modified these seasonal responses somewhat in domestic animals.

20.5 Animal Food Groups and Production

Terrestrial animal food groupings also vary [4–8]. Groups used in this book are:

- Eggs
- Milk
- Poultry
- Beef
- Pork
- Mutton
- Other red meat

The aquatic food groups are primarily animal groups. They include:

- Demersal fish (generally "bottom feeders")
- Pelagic fish (generally "surface fish")

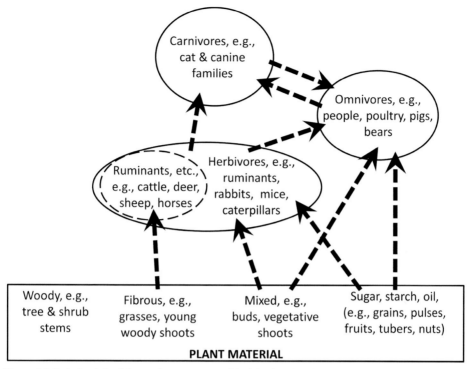

Figure 20.5 Animal food flows shown on a modified food web (Figure 13.3).

- Freshwater fish
- Seaweeds
- Mollusks and similar animals
- Cephalopods (squids and others)
- Crustacean

Animals, including humans, get their food energy from eating plants and/or other animals (Figure 20.5).

Except for carbohydrates in milk, honey, and a small amount in eggs, crustaceans, and mollusks, animals primarily produce proteins, fats, and oils (Figures 20.6a–b). Animal food groups vary in proportions of proteins and fats, with all aquatic animals being high in proteins while pork, mutton, and to a lesser extent beef are high in fats. The proportion of water in animal products is less varied than in plant products.

As currently produced, meat is a very inefficient use of land and fossil fuels. About 17% of the Calories and 39% of the proteins people consume are from animal products – primarily meat and dairy products; however, at least 60% of the plant food calories produced by agriculture are used to provide this meat (Figure 22.11). Animals consume much of the energy they take in for metabolism, movement, and warmth. They return only 25 to 50 of the energy in proteins and fats that people can then consume, depending on the animal (Chapter 22).

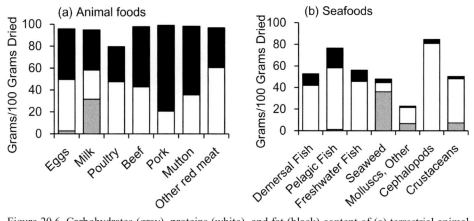

Figure 20.6 Carbohydrates (gray), proteins (white), and fat (black) content of (a) terrestrial animal products and (b) seafood groups [4, 5].

On the other hand, eating meat is a convenient way to eat balanced meals. Each piece of meat generally contains high concentrations of nearly all amino acids people need, whereas each plant generally contains only a few. It is possible to obtain the needed balance of amino acids with a vegetarian diet, but it requires more attention to ensure the proper mixture of plants – and balance of amino acids – are consumed [12].

Raising ruminants and other animals that digest fibers can indirectly expand available food sources if they feed in grasslands (Figure 20.5). Animals also provide food at times and in forms unavailable directly from plants, such as in winter or dry seasons when plant foods are not freshly available; however, modern storage, preservation, and shipping technologies are minimizing food seasonality concerns.

20.6 Human Food Requirements

Healthy, large people can consume about 2.5- to 3 thousand Calories of energy per day in a moderately active lifestyle and 9 thousand Calories or more if extremely active (Chapter 22) [13]. Children and smaller people consume fewer. Eating more Calories than the body consumes tends to add weight – usually fat; and consuming fewer causes weight loss and then sicknesses associated with malnutrition. Different food groups provide different concentrations of Calories at the moisture contents generally purchased (Table 20.1).

People should consume about 55 to 65 grams of proteins per day to remain healthy [13].

A challenge for human nutrition is to balance Calorie and protein intakes so that sufficient proteins are consumed without excess Calories. A diet of just foods such as fruits, roots and tubers, or pork (Figure 20.7) would have been suitable for physically active subsistence farmers who probably consumed 6,000 Calories per day. They would have eaten enough potatoes or pork to gain the needed proteins but would have burned the excess Calories. On the other hand, eating just these foods would saturate a modern, sedentary

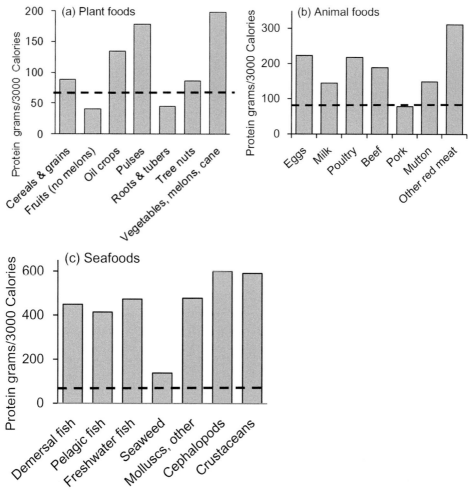

Figure 20.7 Proteins provided from targeted human Calorie intake (3,000 Calories) by (a) plants food groups, (b) terrestrial animal food groups, and (c) aquatic food groups. Dashed line indicates recommended minimum daily protein intake (65 grams) [4, 5].

person's Calorie needs before he/she consumed enough proteins to remain healthy; the sedentary person on this diet would be either protein deficient or fat. Food agriculture, therefore, strives to provide a mixture of foods that enable people to eat a balanced diet of sufficient Calories and proteins that agrees with their lifestyle.

20.7 Nonfood Agriculture Commodities

Nonfood agriculture commodities comprise less than 1% of the total agriculture production on a trade–weight basis (Table 20.2 and Figure 20.8) [14]. Despite their small

Table 20.2 *Weight of nonfood agricultural crops compared to plant and animal food crops (weights in normally traded basis) [14]*

		(millions of tons)	(%)
Food crops			
Plant foods		8,240	97%
Animal foods		298	3%
	Total	*8,538*	*99.28%*
Tobacco		3	0.04%
Coffee and tea		12	0.14%
Fibers			
Cotton lint		19	71%
Jute		3	10%
Other		5	19%
	Total	*28*	*0.32%*
Wool and hides			
Wool		2	21%
Hides		7	79%
	Total	*9*	*0.10%*
Rubber		10	0.12%
All crops		8,600	100%

proportion of total agriculture production, many important and/or luxury nonfood items come from crops.

Nonfood agricultural products include consumable, non-consumable, and work and hobby products. Consumable products include fibers, rubber, wool and hides, tobacco, and coffee and tea. Non-consumables include ornamental plants and animals (fish, birds, cats, dogs). Work and hobby products include horses, mules, camels, llamas, buffalos, and others as well as plants for flower gardens.

Fibers for cloth and ropes are the largest group of nonfood agricultural products, with cotton and jute being the primary crops. Many other fiber crops are locally important as well, such as abaca (manila hemp, Philippines) and agave fibers (southwestern United States); and many plants can be made into bast fibers. Flax (linen fibers), hemp (Cannabis fibers), kapok (fiber filling [mattresses] from Sieba tree), ramie (fabric fiber), sisal (a cord and rope fiber, from agave), and silk (from silkworm cocoons, on mulberry trees) are also made. Except for silk, the cellulose fibers in plants are the basis of fibers' usefulness. The tension strength of this cellulose makes fibers useful for ropes and cloth.

Leather and wool are generally byproducts of raising animals for meat and/or dairy products.

Coffee, tea, honey, and chocolate could be classified as "food"; however, their use is more for stimulation than nutrition. They each provide less than one-tenth of 1% of the agriculture crops globally. Coffee and chocolate come from the fruits of coffee and cacao

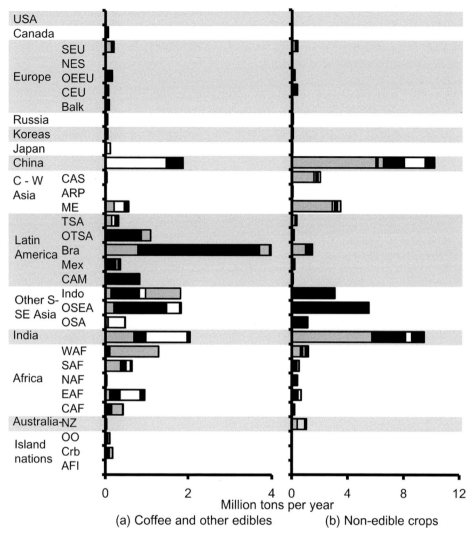

Figure 20.8 Miscellaneous (a) "edible" and (b) nonedible agriculture products by country groups [14]. (a) From left: left gray = tobacco; left black = coffee; white = tea; right gray = cocoa bean; right black = honey. (b) From left: first gray = cotton lint; left black = other fibers; left white = raw silk; gray = greasy wool; black = goat and sheep skins; right white = animal hides; right gray = bees wax; right black = rubber. Some countries did not report.

plants. Tea comes from the leaves of tea plants. Tobacco is a smaller proportion of agriculture production; it comes from tobacco plant leaves.

Latex, used to make rubber products, comes from the stems of rubber trees, commonly grown in plantations. Rubber can be made synthetically from fossil fuels, but the high quality needed in such products as airplane tires requires latex from rubber trees.

Many products made from fossil fuels, such as polyester ropes and cloth, plastics, and synthetic rubber, can also be made from the cellulose or other chemicals in agricultural products or wood. The environmental trade-offs of different products requiring farmland, forests, or fossil fuels could be assessed.

References

1. C. D. Oliver, B. C. Larson. *Forest Stand Dynamics*, updated edition. (Wiley, 1996).
2. R. Pearcy, J. Ehleringer. Comparative Ecophysiology of C3 and C4 Plants. *Plant, Cell and Environment*. 1984;7(1):1–13.
3. X.-G. Zhu, S. P. Long, D. R. Ort. What Is the Maximum Efficiency with Which Photosynthesis Can Convert Solar Energy into Biomass? *Current Opinion in Biotechnology*. 2008;19(2):153–9.
4. UN FAO. *Food Composition Table for Use in East Asia 1972*. (United Nations Food and Agriculture Organization, 1972 [Accessed July 13, 2016]). Available from: www.fao.org/doc rep/003/X6878E/X6878E02.htm#ch4.1.
5. UN FAO. *A Handbook on Food Balance Sheet*. (United Nations Food and Agriculture Organization, 2001).
6. UN FAO. *Production, Crops*. (United Nations Food and Agriculture Organization, 2014 [Accessed November 17, 2015]). Available from: http://faostat.fao.org/site/567/DesktopDefault .aspx?PageID=567#ancor.
7. A. L. Merrill, B. K. Watt. *Energy Value of Foods: Basis and Derivation*. (US Department of Agriculture, Agricultural Research Service, 1955).
8. B. MacKenzie. *Food Composition* 1999. (Accessed July 14, 2016). Available from: www .brianmac.co.uk/food.htm.
9. J. J. Sauter, B. van Cleve. Biochemical and Ultrastructural Results During Starch-Sugar-Conversion in Ray Parenchyma Cells of Populus During Cold Adaptation. *Journal of Plant Physiology*. 1991;139(1):19–26.
10. D. Siminovitch, F. Gfeller, B. Rheaume. The Multiple Character of the Biochemical Mechanism of Freezing Resistance of Plant Cells. *Cellular Injury and Resistance in Freezing Organisms*. 1967;2:93–117.
11. T. Wang, J. Wu, X. Kou, et al. Ecologically Asynchronous Agricultural Practice Erodes Sustainability of the Loess Plateau of China. *Ecological Applications*. 2010;20(4):1126–35.
12. P. Foster, H. D. Leathers. *The World Food Problem*. (Lynne Rienner Publishers Inc., 1999).
13. UN World Health Organization. *Protein and Amino Acid Requirements in Human Nutrition*. (United Nations World Health Organization, 2007).
14. UN FAO. *FAOSTAT: Download Data*. (United Nations Food and Agriculture Organization, 2009 [Accessed September 16, 2012]). Available from: http://faostat3.fao.org/download/.

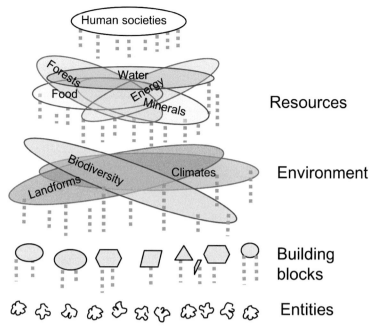

Plate 1 General model discussed in this book. [Also Figure 1.1.]

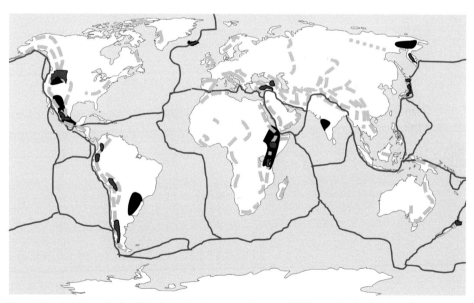

Plate 2 Predominant faults (thin dark lines), mountain areas (thick gray dashes), ash soils (gray with black borders), basalt flows (black) [3, 5, 20–22; Chapter 10]. [Also Figure 10.3.]

Global Ecological Zones

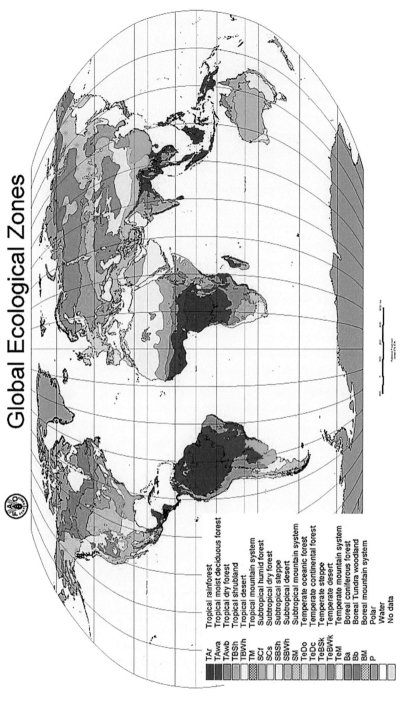

TAr	Tropical rainforest
TAwa	Tropical moist deciduous forest
TAwb	Tropical dry forest
TBSh	Tropical shrubland
TBWh	Tropical desert
TM	Tropical mountain system
SCf	Subtropical humid forest
SCs	Subtropical dry forest
SBSh	Subtropical steppe
SBWh	Subtropical desert
SM	Subtropical mountain system
TeDo	Temperate oceanic forest
TeDc	Temperate continental forest
TeBSk	Temperate steppe
TeBWk	Temperate desert
TeM	Temperate mountain system
Ba	Boreal coniferous forest
Bb	Boreal Tundra woodland
BM	Boreal mountain system
P	Polar
	Water
	No data

Plate 3 Locations of "Global Ecological Zones." [Also Figure 7.1.] Reproduced with permission, United Nations, Food and Agriculture Organization.

Plate 4 (a) Deeply weathered, red laterite soils in a shield landform (excavation for a house), Ghana, Africa. (b) Fence of weathered, red rocks taken from beneath deeply laterized soils, southwestern India. [Also Figure 11.1a-b.] Figure 11.1a courtesy of Ms. Dora Cudjoe.

Plate 5 (a) Decomposing granite bedrocks creating several meters of red clay soil; granite bedrock in center. (b) Two large, parallel quartz seams running diagonally from lower left. Other rock has decomposed to soil. Both from western South Carolina, USA. [Also Figure 11.3a-b.]

Plate 6 Glacier on higher elevation in Iceland, with tongues extending to low elevations. Farmhouses and road at bottom center show scale. [Also Figure 9.3.] Used with permission of Professor Roger Mesznik.

Plate 7 Open structure in Ukraine, created by clear-cutting that could grow into forest, shrubland, or grassland. [Also Figure 14.9c.]

Plate 8 Diversion channel 6 meters deep created by Emperors Vespasian and Titus (AD 69–81) to divert silty storm water from Sileucia, the harbor of Antioch. [Also Figure 18.7a.]

Plate 9 (a) Pine savanna kept open by grazing, Montana, USA. (b) Oak savanna kept open by fire, Russian Far East. [Also Figure 14.11a-b.]

Plate 10 (a) Continuous closed forest that supports little tiger prey (and tigers) in Northeast China [56]. (b) Savanna/open/dense mosaic in Jim Corbett National Park (Uttarakhand, India) maintained by managed burning that supports tiger prey and tigers; note deer in mid-ground. [Also Figure 14.12a-b.]

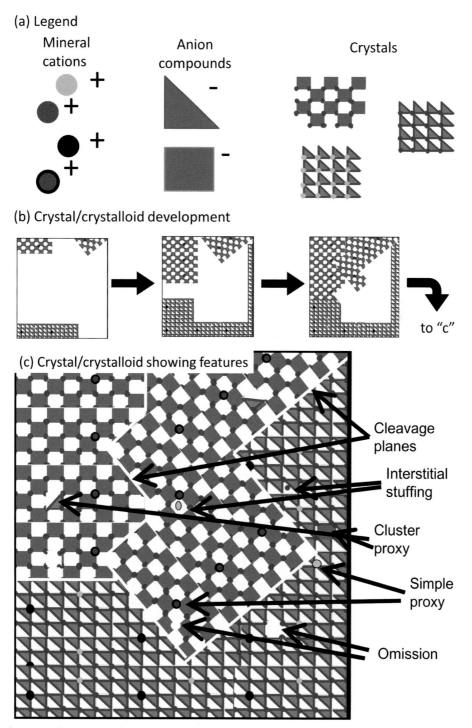

(a) Legend

Mineral cations

Anion compounds

Crystals

(b) Crystal/crystalloid development

to "c"

(c) Crystal/crystalloid showing features

Cleavage planes

Interstitial stuffing

Cluster proxy

Simple proxy

Omission

Plate 11 Schematic development of a saturated solution into a mixture of crystals, mixed crystals, crystalloids, and solid solutions. [Also Figure 26.3a-c.]

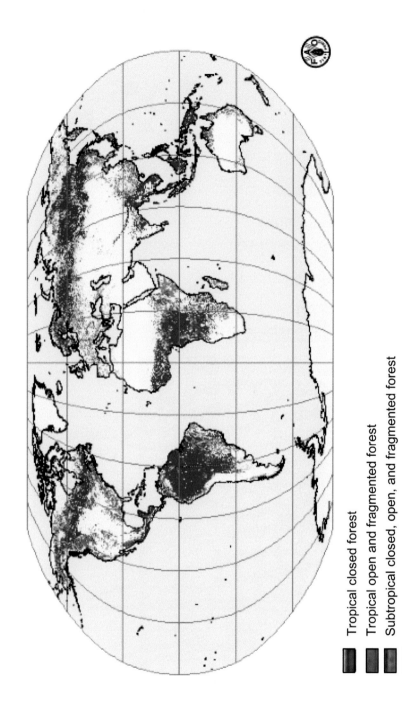

Plate 12 World forests by major ecological domains (Figure 7.1). [1, 2; Chapter 29]. [Also Figure 29.1, with legend.] Reproduced with permission, United Nations, Food and Agriculture Organization.

Tropical closed forest

Tropical open and fragmented forest

Subtropical closed, open, and fragmented forest

Temperate closed, open, and fragmented forest

Boreal closed forest

Boreal open and fragmented forest

21

Agriculture and the Green Revolution

21.1 Development of Agriculture and Associated Technologies

Some people apparently began to cultivate crops and shift from hunter-gatherers to more stationary, agricultural lifestyles about 10 thousand years ago, shortly after the glaciers retreated (Chapter 5). Causes of this shift are uncertain, but many climate and sea level changes were occurring then.

Plant domestication seems to have been a diffuse and gradual process [1]. In the Fertile Crescent of southwest Asia, several plants suitable for cultivation, now known as "founder crops," were cultivated [2]. They included flax (*Linum usitatissimum*); the pulses, lentils (*Lens culinaris*), pea (*Pisum sativum*), chickpea (*Cicer arietinum*), and bitter vetch (*Vicia ervilia*); and the cereals emmer wheat (*Triticum dicoccum*), einkorn wheat (*Triticum monococcum*), and barley (*Hordeum vulgare/sativum*).

Another approximately contemporary agriculture site is China, which apparently began cultivating millet species (family *Poaceae*, or *Gramineae*) [3]. India gradually changed from hunter-gatherer to agrarian societies beginning at the same time [4]; and crops were being cultivated in the Papua New Guinea highlands shortly afterward [5]. Cattle were apparently domesticated about 10,000 years ago in the African Sahel (sub-Sahara region) and sedentary plant-crop lifestyles developed thousands of years later [6]. Cultivation of corn (maize; *Zea mays*) in Central America apparently began about 6,000 years ago and sunflowers (*Helianthus*) and cotton (*Gossypium*) were cultivated by 2,000 years later. Potatoes (*Solanum*) were grown in the Andes over 10,000 years ago [7]; and beans (*Phaseolus*), tomatoes (also *Solanum genus*), and other crops were domesticated much later.

Human foods and agriculture have been constantly changing – what species are eaten, the genetic composition of the species, how they are grown and processed, and in what combination with other foods [8]. Not all foods were suitable for cultivation; the Brazil nut (*Bertholletia excela*) is still collected in the wild. Some plants that appeared to be foods were not healthy – wild almonds (*Sterculia foetida*), horse chestnuts (*Aesculus hippocastanum*), some mushrooms, and honey from rhododendron (*Rhododendron*). Other foods do not taste very good – bracken fern tubers (*Pteridium aquilinum*) and "poverty bread" from tree bark. Some foods are easier and more productive to grow or process – sugar cane

(*Saccharum officinarum*) compared to sugar beets (*Beta vulgaris* cultivar). Still other plants and animals had, or still have, myths or cultural taboos. The tomato was once considered poisonous in Europe. Even today, various cultures prefer not to eat pig, cow, or dog meat.

As agriculture spread, crops were moved to new locations, new crops were cultivated, some old crops were abandoned, and crops abandoned in some places were re-cultivated elsewhere. Rice (*Oryza*), cotton, millet, and walnuts (*Juglans*) seem to have been independently domesticated in several places. The Columbian Revolution (colonization of the Americas beginning about 500 years ago) generated exchanges of many animal and plant crops among the Americas, Eurasia, and Africa. The later Australian colonization introduced many crops there, but few were taken away.

Domesticated animals were sometimes part of the stationary agriculture lifestyle. Milk cows, draft animals, and poultry were kept close to dwellings. Cattle, sheep, goats, and similar meat animals needed to be rotated among distant pasturelands, and kept away from gardens and croplands – difficult tasks before use of barbed wire. Pastoral lifestyles developed when people moved with animals to seasonally suitable pastures. Domestication of the horse and development of wheels for carts enabled a blend of pastoral and stationary agriculture lifestyles [9]. People with carts and horses could move with the herds and carry many possessions. Or, herders could travel great distances with the herds and return to stationary settlements.

21.1.1 Agrarian Culture

Incremental modifications to plant and animal systems continued, especially after the abundant use of iron began about 2,500 years ago [10, 11]; but most people still lived rural, agrarian and pastoral lifestyles. Rapid dissemination of information, seeds, and animals increased production where large areas were administered by a knowledgeable group such as the Roman, Chinese, Indian Mogul, Islamic, Incan, Mesoamerican, Mongolian, Ottoman, and various European empires and colonies. Books were written in ancient China and Rome on agriculture. Efficiency improved with plows, scythes, draft animals, and yoking systems. Irrigation systems were used many thousand years ago, and water power lifted irrigation water and thrashed grains. People studied botanical and mechanical bases of agriculture and machinery. Specializations in metallurgy led to improved plows. Irrigation, planting, frost-avoiding, and (more recently) greenhouse techniques enabled people to grow plants in climates that they had not evolved to grow in.

People modified lifestyles and animal and plant behaviors to make food available longer during the year. They domesticated, imported, and/or cultivated new plants and animals, modified the growth environment, and developed storage and preservation mechanisms. Meat animals were kept alive until needed so the meat would not spoil. Meats were also preserved through various smoking, salting, spicing, ice storage, or drying techniques. Domestic chickens were bred to produce eggs frequently; and domestic cattle were bred to produce offspring at any season.

The informal and formal interactions among farmers – sharing ideas, seeds, and animal sires – contributed very much to agriculture development [8]. Informal farmers' markets in much of the world, more organized fairs, farmers' societies [12], reputations of local experts, and more recent government "extension" agents and agriculture research centers have all led to a synergistic increase in information.

Plant breeding was done empirically for thousands of years before being scientifically understood [8]. It was done for greater yields, easier processing, and superior pest resistance as farmers sought better-growing individual plants and animals. Mendelian genetics confirmed many former breeding practices and allowed genetic improvements to proceed systematically, especially in agriculture research and education facilities. Genetic breeding increased as the impacts of inbreeding and outbreeding, hybrid vigor, and recessive genes became understood. Cloning technologies allowed propagation of many "clones" with many advantages but almost no ability to evolve to changing climates or competition.

An impending problem with plant breeding is that both the costs and successes of improving a limited number of species and genotypes may lead to a narrowing of the range of species being widely cultivated to those that have been improved (Figure 22.10).

21.1.2 The "Green Revolution" and Beyond

The many technological and educational factors had caused global food production to outpace the rising population by the 1960s. The "Green Revolution" was pronounced by W. S. Gaud, USA Department of State in 1968 [13].

These and new factors are still advancing (Chapters 5, 22). Data collection and information sharing is accelerating with satellites, the internets, webcasts, and other modern means of measuring and networking. Irradiating seeds is rapidly creating mutant seeds, some of which prove resistant to crop diseases and are reproduced in bulk. "Genetically Modified Organism" (GMO) methods can alter targeted plants and animals very rapidly for specific, inheritable characteristics. One method is transgenics, in which a desired gene – a specific DNA sequence – from one organism is inserted into the DNA sequence of the target plant or animal. Another method is to remove a gene from an organism.

Except for GMO technologies, plant and animal breeding is essentially forcing evolution to fit an environment chosen by people. Since life began, transgenics and gene deletions have apparently occurred only rarely without recent human inducement. There are hopes and concerns with both the rate and targeted direction of human-induced GMO technologies [14, 15]. The gene technologies can create pesticide-resistant plants and plants that turn color when in need of irrigation. They can also increase the protein contents of foods [16]. On the other hand, the modified organism may breed with related wild plants and give the wild plants characteristics that may be desirable in crop production but not in the wild.

Much of human existence has been a quest to ensure there is enough to eat. Even recently, hundreds of thousands of people have died from starvation [17] – more related to food distribution than to a global inability to produce enough. The freedom from worry about food is remarkable considering the uncertainties of weather; insects and diseases; erosion; urban encroachment on farmland; introductions of new crops, breeds, and cultivation techniques; changing farm and food regulations; changes in food preferences; and often overproduction and subsequent low prices.

The complexity of foods and their preparations available to people in developed countries has reduced concerns of malnutrition, famine, and spoiled foods; but it has brought about issues of chemical pollutants, depletion of soil productivity, monocultures and genetic improvements, and others. Food shortages, poverty, and famines still occur in parts of the world; but these issues will not be solved by returning to past agriculture practices. The issues with intensive agriculture can be solved with new, innovative solutions involving technology, policy, and lifestyle changes. To craft effective solutions, it is important to understand present agriculture practices.

21.2 Growth and Use Efficiency of Crops

Plant crops differ dramatically in their efficiencies of growing Calories and proteins per area (Figure 21.1).

Nearly all crop plants have increased in output per hectare because of many human influences during the past century [8].

21.2.1 Corn as an Example

Corn production for the past one hundred and fifty years in the United States will be used as an example of some efficiencies (Figure 21.2).

Corn was already grown in eastern North America when European- and African-Americans settled there in the seventeenth century. Native Americans and the recent immigrants used corn for animal feed, corn meal, and other foods. The westward migration of people in the eighteenth and nineteenth centuries allowed local corn varieties to be shared and informally bred [8].

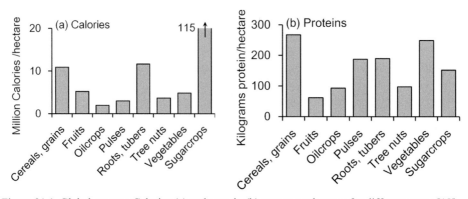

Figure 21.1 Global average Calories (a) and protein (b) grown per hectare for different crops [18].

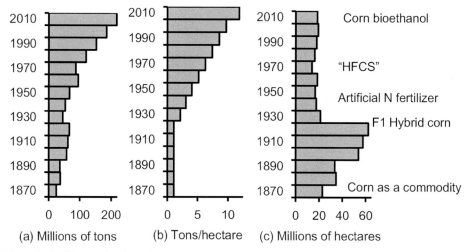

Figure 21.2 Changes in (a) total corn production, (b) tons per hectare, and (c) hectares of corn grown in the United States from 1870 until 2010 [19]. Right shows some breakthroughs affecting production efficiency.

Railroads of the nineteenth century enabled corn to be grown on productive lands and shipped far distances to cities. Distribution was awkward because corn was shipped by individual farmers to specific buyers until systematic grading was introduced in 1856. Then, corn was traded efficiently in bulk. Barbed wire was introduced about 20 years later. It assured farmers that livestock could be kept out of fields, and so they could profit from the added effort of growing corn more intensively. Professional agriculture advisors, colleges, and research centers increased the efficiency of farming in the late nineteenth century. Large irrigation projects were undertaken in the United States beginning about 1900 and continuing through the 1980s. These increased the growth per hectare of corn and many other crops.

The automobile, pickup truck, and tractor began replacing horses, mules, oxen, and other draft animals in the early 1900s [20, 21]. This replacement freed many hectares of pasture and hayfields and much corn previously used as supplemental livestock feed.

By the 1920s the United States was producing so much corn and many other plant and animal crops that prices fell very low, leading to unique problems (Chapter 5). Farmers and ranchers with poor soil and pastures and no irrigation could not profitably sell their produce. They commonly could not gain enough cash by selling crops, livestock, or the land to enable them to move elsewhere and find non-agriculture jobs. Finally, the government intervened, purchased the excess harvest, and paid farmers not to grow crops – often paying them to grow trees, instead, to rebuild the depleted soil structures [22]. These measures reduced surplus corn and other crops, raised prices, and so encouraged renewed agriculture investments on the remaining productive farmland. Corn production had been improving since the mid-1800s, but the

Figure 21.3 (a) F1 hybrid corn growing evenly when planted close together (Central China). (b) Non-F1 hybrid corn growing unevenly and so planted widely (Central China).

increasing average yields per hectare became apparent only as people abandoned poorer farmlands in the 1930s [8, 19].

Meanwhile, genetic research produced F1 hybrid corn with two important characteristics (Figure 21.3) [19]. First, F1 hybrid corn plants physically grow very uniformly, rather than some individuals outgrowing and shading others. Consequently, many more individuals could be grown per hectare; and all of them would survive and produce grain (ears of corn). Second, seeds of F1 hybrid corn do not have the same growth uniformity. Consequently, new F1 corn seeds needed to be purchased each year; and seed companies had an annual market for F1 seeds that enabled them to invest in producing and improving corn seeds.

After World War II, ammonium nitrate, previously used in military explosives, was diluted and converted into an efficient fertilizer. In the late 1970s, a biochemical process development enabled corn to be converted into fructose (HFCS; High Fructose Corn Sugar) and used in place of other sugars [19]. GMO corn is now being grown. The results of over 100 years of innovation (Figure 21.2) have been a steady increase in corn production per area, creating an increase in total food without a concomitant increase in land devoted to agriculture. And again, the corn became a surplus. This surplus corn was commonly shipped abroad until other countries objected to its interfering with their agriculture sector. Instead of the area of corn growth being reduced again, much is now being used for biofuel (Chapters 9, 24).

Another genetic improvement has been breeding shorter corn, cotton, and wheat plants that blow over less commonly in windy rainstorms and also divert less energy and carbohydrates into stems and so more into fruits.

21.2.2 Recent Changes in Other Plant Crops

Other plant crops have also undergone changes (Figure 21.4), with different factors affecting their production [8]. Of the thousands of plants and many possible animals available for food, the world primarily relies on about 200 plants, 20 terrestrial animals, and quite a few aquatic animals.

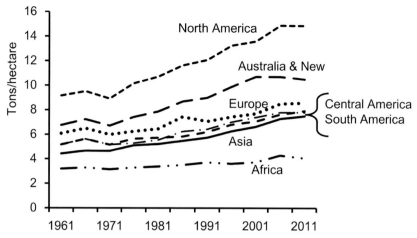

Figure 21.4 Changes in average production of twelve plant food groups in different geographic areas [18]. Groups: total cereals, citrus fruits, coarse grains, fiber crops, fruit (excluding melons), oilcakes, primary oilcrops, pulses, roots and tubers, treenuts, vegetables, and melons.

Different crops are produced in great abundance either because their production efficiency (yield: weight per hectare) is high, because large areas are planted in this crop, because people are accustomed to its taste, because farmers are familiar with how to grow it, or for a mixture of the above and other reasons. In the case of wheat, greatest yield efficiency is not always in places where much wheat is produced. In North America, wheat is grown on level land where large mechanized operations can offset the high labor costs; but it is relatively unproductive land that is not used for other things deemed more valuable. By contrast, wheat can be produced in the loess soils of Ukraine that are highly productive and the large, level expanses enable efficiencies of mechanization. Wheat can be produced in China in areas too small for mechanization because of cheap labor and/or the culture of harvesting at a time of extended family gatherings. It is possible that Africa's average plant food production efficiency is the lowest because they only recently acquired crop scientists and technologies, they have many subsistence farmers working less productive soils (which would bring their average productivity down), or Africa may not have many areas of productive land-forms (Chapter 10).

Production efficiency, area planted, and thus abundance of the different crops also changes with time (Figure 21.5). Wheat increased primarily because of production effi-ciencies; maize, paddy rice, and soybeans increased because of both more area used and greater production efficiencies. Barley declined in area planted, but remained relatively stable in total production because its efficiency of production increased.

Many productive farms became unproductive and subsistence by poor soil management. Improper plowing or tilling can ruin a soil by eroding or compacting it, leaching its minerals, or making it saline (Chapters 17–19).

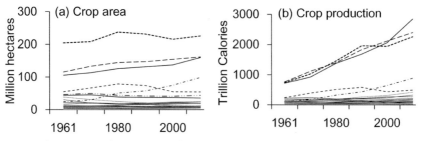

Figure 21.5 Changes in world (a) area and (b) production of major plant crop species [18]. From top: wheat = short dash; paddy rice = long dash; maize = solid; soybeans = dash/dot; barley = dash; other lines = sugar cane, cassava, potatoes, sorghum, coconuts, sunflower seeds sweet potatoes, millet, groundnuts, bananas, oats.

On the other hand, many unproductive soils can be countered with remedial action. Compaction can be countered by deep plowing with a subsoil plow and plowing in organic matter, as well as by avoiding grazing, plowing, or tilling when wet. Erosion can be countered by contour plowing; and mineral leaching can be countered by rotating legume crops, fertilizing, and plowing in organic matter including "biochar" [23] to restore the cation exchange capacity. Saline soils have been discussed in Chapters 18–19).

Many technologies are also increasing agriculture efficiencies:

- Moveable electric wire fences allow efficient animal control.
- Communication – radios, telephones, television, computers – enable rapid dissemination of weather, technology, price, pest, and other information.
- Plastics enable light, inexpensive greenhouses and irrigation systems.
- Portable machines – pumps and sprayers – enable quick responses to irrigation, fertilizer, and pest control needs.
- Plowing and otherwise "working" the soil has become more efficient as both technologies and techniques have improved. For example, modern tractors are equipped with GPS systems which, when linked with satellites, allow farmers to identify and treat precise locations as needed.
- Airplanes facilitate chemical and seed applications.
- Fixed-wing drones are also being used for chemical applications. They save both the energy and soil compaction of using piloted aircraft or heavy ground machinery [24].

Farm workers – employees and contractors – are increasingly educated, knowledgeable, have interesting jobs, and justify reasonable wages.

21.3 Energy Consumption in Agriculture

An estimate of the world's available energy spent on agriculture "to the farm gate" varies from 2% [25] to 9% [26], depending on what items are included. The energy expended per farmed hectare and per ton of food produced increases as countries develop [25]. The greatest farm energy input in the developed world is nitrogen fertilizer, consuming as much as one-third of the total energy. Fuel, machinery, irrigation, and seeds also consume significant amounts [27]. Farm energy efficiency in developed countries seems inversely related to the country's area in farmland and directly related to farm size and energy costs [28].

21.4 Plowing, Irrigation, and Chemicals

Technologies have increased the efficiency of food production in three ways: increased crop production, reduced possibility of crop failure, and reduced food waste. Both farming more areas and making each area more productive can increase crop production. Food production technologies can be synergistic, such as when fertilization enhances irrigation's effectiveness. They can be interdependent, such as where genetic improvement is dependent on weed control. Many technologies are specific to particular soils and climates – sufficient irrigation in a loess soil may be too much in a clay soil. Or, technologies can complicate each other, such as when extra irrigation is needed to flush fertilizers and pesticides applied to soils.

Reducing the possibility of crop failure means the farmer can invest more in gaining greater productivity per area while confident that the risk of a destructive drought, freeze, or insect/disease outbreak will not destroy the investment. Consequently, investments in frost protection and irrigation systems can increase productivity in moist and warm climates even if the systems are never turned on. The public also benefits because crop failures or surpluses and resulting food price fluctuations are less likely. Also, less excess crop growth is needed to hedge against crop failures.

Farming without chemical fertilizers or pesticides is most desirable and is increasingly being done as "organic farming" [29]. Some studies suggest that organic farming can be as productive as other farming and provide other benefits as well [30–33]. Other studies show organic farming to be less productive [34–36]. Food production could move away from chemical pesticides and fertilizers and toward organic farming, perhaps using chemicals in spot applications and emergencies similar to Integrated Pest Management, discussed in Section 21.4.3. Organic farming and its integration will be discussed in more detail in Chapter 22.

Reducing food waste will be discussed in Chapter 22.

21.4.1 Equipment for Plowing, Tilling, and Other Uses

"Plowing" generally means mixing the first few decimeters of soil to improve its structure and mix in organic matter and other soil amendments. "Tilling" generally means mechanically smoothing, breaking clods, and otherwise preparing the soil surface for placing seeds, reducing competing plants, incorporating amendments, and improving the soil's structure (Chapter 10). "Fluffing" fills the soil surface layer with large pores so water will not

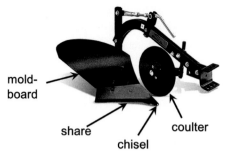

mold-
board

share   coulter

chisel

Figure 21.6 The moldboard plow. Courtesy of Brinly-Hardy Company, Jefferson, IN, USA.

evaporate and weed seeds will not germinate. Plowing and tilling also shape the soil into rows and ditches as needed for specific crops.

Long ago, plowing/tilling generally used sticks to create holes by hand. Plowing was later done by pulling a heavy stick using human or animal labor. Over time, sticks were displaced by metal plows; and highly productive soils with thicker roots could be tilled. Romans first introduced the "moldboard" on the plow about 2,000 years ago (Figure 21.6). Refined and still used, the moldboard cuts beneath the vegetation's roots and soil and turns it over in a smooth, flowing motion. The overturned sod kills the unwanted vegetation by drying the roots and burying the leaves.

Tractors began replacing draft animals in about 1900 as countries developed [20, 21, 37–39]. A convenient three-point hitch now allows diverse farm implements to be attached to tractors and raised, lowered, turned, powered, and removed easily.

Tractors allow a variety of specialized plowing and tilling equipment to be attached:

- Discs that are modifications of moldboard plows;
- Groups, or "gangs," of discs;
- Subsoilers ("rippers"), long chisel plows to break up compact, deep soil layers;
- Smaller gangs ("harrow") of chisels or small discs;
- Tilling gangs, similar to disc gangs, used for tilling and "fluffing."

Various other tractor attachments are used for seed planting, spreading of liquid and dry fertilizers and pest chemicals, weed control, and harvesting. They commonly replaced many "hand labor" farm workers with a few machine operators.

Some farms use herbicides instead of tilling to control weeds. "No-till" agriculture reduces both erosion and the energy needed to drag the plows [40, 41], but increases herbicide use [42].

21.4.2 Irrigation

Irrigation is currently a dominant water use in most countries (Figure 18.1c), with two-thirds of the world regions irrigating over 20% of their farmland (Chapter 18).

Irrigation timing is critical based on the crop, soil condition, and natural rainfall pattern. The key is to balance the needs of plant roots for water and air (Chapter 7).

Plants vary in their ability to tolerate water-saturated and/or water-deficit soils. All crops benefit from sufficient water at the right times; but, the efficiency of water use varies by species and C3–C4 plant groups (Chapter 20) [43].

The climatic rhythm in which each plant evolved helps determine its water needs during the different stages. Sugarcane, for example, utilizes large amounts of water during all stages, while the peanut is moderately tolerant to droughts during the food (kernel) maturation, but not during the flowering period. High irrigation reduces crude protein concentrations in some grasses and so reduces its usefulness for animal fodder, especially if irrigation occurs late in the growing season.

The first irrigation sent water through ditches to fields, where they either flooded the fields ("flood irrigation") or sent water between crop rows ("ditch irrigation"). These techniques are still used in places. Sprinklers that distribute water from overhead pipes are now sometimes used. Large gangs of coordinated sprinklers travel in circles or straight lines and can cover fifty hectares or more with a single installation. "Drip irrigation" releases water slowly from small hoses at each plant's base. A refinement is a plastic envelope with perforations on the lower side that is linked to a water system and placed between crop rows. It reduces evaporation from the soil and irrigates the soil beneath.

"Dryland farming" – farming arid areas without irrigation – allows water to accumulate in soil for 2 years. Crops are only planted every second year, and the land is kept "fallow" – free of vegetation that depletes the soil water – during the other year.

Improper irrigation can lead to or exacerbate soil salinity and thus kill crops. Heavy irrigation or heavy rains can help avoid and recover saline soils unless the groundwater or underlying aquifer is saline. (Chapter 17; See "Goulburn–Broken Catchment" case study, Chapter 3.)

Water use efficiency during irrigation can be improved by lining and covering irrigation ditches to prevent water seepage and evaporation; satellite and computer monitoring to determine optimum timing and locations for irrigation; and planting crops at different times or with different climatic rhythms so not all fields need irrigation at the same time.

Irrigation has enabled many crops to be grown in deserts, but the large amount of scarce water needed for irrigation is creating biodiversity problems around the Aral Sea [44, 45] and desertification around Lake Chad [46]. Crop production is less harmful to the environment – and less water is needed – where it is shifted to moist climates where water is more plentiful, less irrigation water is needed, and irrigation is used for occasional watering when an unexpected rainless period occurs.

Applying fertilizers and pesticides also necessitates more irrigation to dilute or wash away these chemicals' effects.

21.4.3 Weeds, Pests, Fertilizers, and Chemicals

Weeds (unwanted plants) and animal "pests" eat, compete with, or otherwise interfere with crops. These can be native species or, increasingly, nonnative species inadvertently transported through international shipping of organic raw materials (Chapter 16). It is not always necessary to eliminate a weed or pest completely, but to reduce its impact. Pest management commonly attacks one stage of the pest population's lifecycle so that it cannot perpetuate or expand. Plant pests can be prevented from germinating, growing, pollinating, flowering, or setting seeds through appropriate interventions. Insect pests are often not allowed to lay eggs, mature to adults, reproduce, or overwinter.

Plant, animal, and microbe pests are often controlled through quarantine of the pest, planting a field with multiple crops (intercropping), using a highly resistant crop, rotating or changing crops, destroying crop residues after harvest, staggering planting times, and/or using pesticides. Such activities necessitate cooperation among farmers [8] and are facilitated with modern satellite, GPS, and other technologies.

Microbial insecticides and parasitic insects have reduced pest populations [47]; however, introducing a predator to control a pest can exacerbate problems if the new predator becomes a new pest, such as the cane toad when introduced into Australia [48].

Chemical ("pesticides") are being used to combat these pests, but these chemicals are creating issues (Chapter 7) [19, 49]. Pests and weeds evolve resistances, so chemicals that were effective at one time may not be at another [50]; and herbicides and insecticides often prove to have harmful side effects. Consequently, new chemicals are continually needed. An infrastructure of new research and development is constantly creating new weed- and pest-avoiding techniques.

Pesticide chemicals can commonly be classified as: Arsenicals, Chlorinated hydrocarbons, Triamines, and glysophates. Early chemicals such as the insecticide DDT and the herbicides 2,4,5-T and paraquat often created enough harmful side effects that they have been replaced (Chapter 8).

"Systemic pesticides" commonly interrupt the normal growth processes of target pests. "Contact killers" generally kill plant or animal tissue upon contact. Pesticides are often selective for specific genetic groups. 2,4-D, for example is more lethal to angiosperms (broadleaf plants and grasses) than gymnosperms (conifers), while triazines are harmful to broadleaf weeds but less so to grasses.

Pesticides have commonly been added to entire fields in a broadcast manner to areas both needing them and not. Precise applications of pesticides facilitated by satellite monitoring and then using drones with GPS systems for delivery and recording are now feasible [24]. These innovations can save fuel, time, and irrigation water (for flushing the pesticide).

"Integrated Pest Management" (IPM) is a pragmatic approach that combines all techniques [51]. IPM first monitors pest populations and tries to stop threatening ones before they get large with mild techniques, but defaults to narrowly targeted, increasingly severe techniques as pest buildups and trade-offs warrant.

A strong infrastructure of technologies and knowledgeable people could develop that uses IPM to displace large machinery and broad-scale applications. This infrastructure could be a social driver to increase rural employment (Chapters 5–6).

21.4.4 Fertilization

Management of soil fertility developed slowly. At first, people abandoned agricultural land periodically to allow the fertility to recover ("swidden agriculture"). They also added fertility by planting legumes and adding animal manure, human manure ("night soil"), bird manure ("guano"), ashes, fish, and lime. Artificial fertilization is now done in many parts of the world (Figure 21.7) [52].

Modern, inorganic fertilizers are generally listed by the proportions of Nitrogen ("N"), Phosphorus ("P"), and Potassium ("K") elemental weights. For example, a label of "20–10–10" would have 20% N, 10% P, and 10% K.

21.4.4.1 Nitrogen

Microbes associated with legumes and some other plant roots can convert diatomic atmospheric nitrogen (N_2) into molecular forms that go into the plant – and into the soil when the plant or its leaves die. Lightning can also "fix" small amounts of atmospheric nitrogen into the soil. Soil nitrogen can be converted back to diatomic atmospheric nitrogen (N_2) by denitrifying bacteria.

Nitrates (ammonium, sodium, and potassium nitrates) were found useful for both nitrogen fertilization and gunpowder. Nitrate deposits in South America were first mined [53]; then, the Haber–Bosch method [54] was developed to produce ammonium nitrate from atmospheric nitrogen. The deposits have declined, and refinements of the Haber–Bosch method are now the major sources of nitrogen fertilizer. The process consumes much natural gas, with concomitant carbon dioxide release to the atmosphere.

Global nitrogen fertilizer use is stabilizing as people learn to use it more efficiently, as energy costs and water pollution from excess nitrogen rise, and as people plant more legumes. Other methods of producing nitrates include using solar energy and ash or lime, recovering nitrogen from sewage water and animal feedlots, and flooding croplands in alluvial floodplains. Just as crops were bred to respond to high nitrogen additions, they could be bred to thrive with less nitrogen.

21.4.4.2 Phosphorus and Potassium

Phosphate rock minerals are the only large sources of phosphorus. Potassium fertilizer uses primarily "potash," a variety of mined and manufactured salts containing the element potassium in water-soluble forms. Future supplies of phosphate and potash are becoming of concern (Chapter 27). Methods are needed that apply these important minerals to agriculture land more sparingly by identifying precise areas of need, that avoid leaching

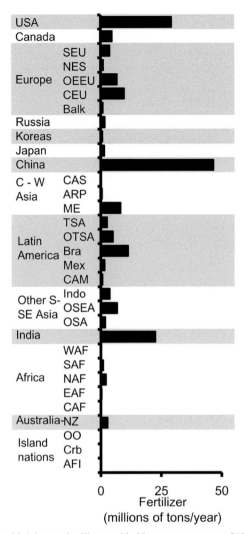

Figure 21.7 Weight of artificial crop fertilizers added by country groups [18].

them out through excess irrigation, and that retain them in soils by increasing the cation exchange capacity (Chapter 7).

21.4.4.3 Soil pH

Acid soils – soils with low pH – generally have few available cations, relatively high populations of fungi, and low populations of bacteria. Acid soils occur naturally where the landform's bedrock contains few cations, where the cations have been leached because the landform is very old or highly leached, or where abundant organic matter creates organic

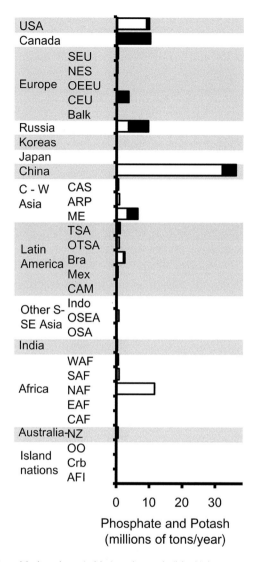

Figure 21.8 Origin of world phosphate (white) and potash (black) by country groups [55].

acids. Most crops do not grow well on very acid soils. Applications of "lime" (usually Calcium or magnesium carbonate) can increase both the pH and the calcium or magnesium availabilities.

A very high pH can also ruin crops in karst landforms (Chapter 9) or with over-fertilization or under-irrigation ("fertilizer burn"). Sulfur compounds are sometimes added to increase acidity (reduce the pH).

21.4.4.4 Other Minerals and Soil Organic Matter

Plants use other needed elements in such low concentrations that they are generally sufficiently available in the soil except in extremely old landforms (shields) or highly leached, sandy soils.

Organic matter has long been recognized as an appropriate soil addition for improving its structure and fertility. Biochar (charcoal) is being added to some soils to increase the soil structure, cation exchange capacity, and carbon sequestration [23]. Pulse grazing, discussed below, can add organic matter to grazing lands.

21.5 Other Activities

Other farming activities include general maintenance, marketing produce, and keeping updated on technologies and crops in demand. Various crops require a field to be "laid out" in different ways, with ditches, furrows, and irrigation equipment arranged according to the crop's needs. This "laying out" is usually not done to every field each year. Consequently, a farmer makes a commitment to one or several crops of similar needs for several years when organizing a field. Other activities are discussed below.

21.5.1 Frost Management

Late-spring freezing temperatures for a few minutes can kill vulnerable flowers and so eliminate a crop's food production for the entire year. Such freezes occur somewhat unexpectedly during cool nights with heat radiating outward through a cloudless sky and the cold, heavy air settling in low places ("frost pockets") or concave slopes ("frost drains").

Several techniques can combat these "killer frosts:" layers of thick, dark smoke from burning oil in "smudge pots"; creating turbulent air using strategically placed stationary fans, human fans, or helicopters; irrigation that freezes water on the flowers and releases heat that keeps the flowers from freezing. Identifying when and where frosts will occur is becoming more precise with modern sensing and predicting technologies.

21.5.2 Greenhouses

Greenhouses have transparent roofs that transmit sunlight in but prevent longwave heat from radiating out (Figure 23.2). Consequently, greenhouses enable plants to be grown in colder climates and seasons than field cultivation. They extend both the range of many foods and the times of year they are available. The cost and weight of glass for transparent roofs limited the widespread use of greenhouses until recently, when transparent plastics are making greenhouses lighter, easier to assemble, and less expensive.

21.5.3 Animal Husbandry

"Animal husbandry" is the raising and care of farm animals for work and food and nonfood products. Except for dogs, domestication probably began with herding of wild females for milk, wool, meat, and possibly dung for fuel and fertilizer [9]. Male animals are hard to manage. Over time, gentle males were found and domesticated as well. In many domestic species, a few males are kept for breeding, with the rest castrated or killed. Aggressive characteristics are only desired in race horses, some hunting dogs, and falcons – and at one time in European warhorses.

Managing the carrying capacity of the habitat has been a major issue in grazing animals (Table 13.2), and overgrazing has been common when people shared the grassland and grazing was not well understood (Figure 14.10) [56]. Rangelands were at one time grazed lightly by dispersing cattle or moving sheep and goats quickly. A recent innovation – "pulse grazing" – has been to graze an area intensively, even destroying some sod (grass and the soil held by its roots), and then move the animals elsewhere [57–59]. Pulse grazing increases the plant diversity by allowing annual and biennial plants to become established where sod is destroyed (Chapter 14). Over time, the sod thickens through growth, fires, and lighter grazing until intensive grazing again destroys it (Figure 14.5). This technique takes advantage of the herding instinct of grazers. It is also gaining popularity where predators necessitate keeping the animals close together for safety.

Keeping animals in pens or stalls ("feedlots"), as opposed to open conditions ("free range") has gained popularity by increasing the efficiency of feeding, collecting eggs or milk, and slaughtering the animals [19]. There is concern about the animals' health and the amount of hormones and antibiotics needed to keep them alive and growing rapidly when confined. Feedlots cause urine and feces to seep into the soil in a concentrated area [19].

Attempts to accentuate certain characteristics through inbreeding led to "breeds" – groups with similar genotypes and predictable behaviors and appearances within a species. Different animal breeds are still being developed for cattle with different qualities of meat or milk; dogs for hunting, tracking, herding, retrieving, and protecting; horses for racing and beauty; and cats for beauty.

A useful old breeding practice has been to produce mules by crossing a male donkey with a female horse. The mule is sterile but strong, hardworking, and relatively intelligent.

References

1. D. R. Harris, G. C. Hillman. *Foraging and Farming: The Evolution of Plant Exploitation.* (Routledge, 2014).
2. T. A. Brown, M. K. Jones, W. Powell, R. G. Allaby. The Complex Origins of Domesticated Crops in the Fertile Crescent. *Trends in Ecology & Evolution.* 2009;24(2):103–9.
3. L. Barton, S. D. Newsome, F.-H. Chen, et al. Agricultural Origins and the Isotopic Identity of Domestication in Northern China. *Proceedings of the National Academy of Sciences.* 2009;106 (14):5523–8.
4. D. Q. Fuller. Agricultural Origins and Frontiers in South Asia: A Working Synthesis. *Journal of World Prehistory.* 2006;20(1):1–86.

5. T. P. Denham, S. G. Haberle, C. Lentfer, et al. Origins of Agriculture at Kuk Swamp in the Highlands of New Guinea. *Science*. 2003;301(5630):189–93.
6. F. Marshall, E. Hildebrand. Cattle before Crops: The Beginnings of Food Production in Africa. *Journal of World Prehistory*. 2002;16(2):99–143.
7. D. Ugent, T. Dillehay, C. Ramirez. Potato Remains from a Late Pleistocene Settlement in Southcentral Chile. *Economic Botany*. 1987;41(1):17–27.
8. A. L. Olmstead, P. W. Rhode. *Creating Abundance*. (Cambridge University Press, 2008).
9. D. W. Anthony. *The Horse, the Wheel, and Language: How Bronze-Age Riders from the Eurasian Steppes Shaped the Modern World*. (Princeton University Press, 2010).
10. D. R. Haidar. *Assyrian Iron Working Technology and Civilization*. (The University of Wisconsin, Madison, 2011).
11. D. B. Wagner. The Earliest Use of Iron in China. *BAR International Series*. 1999;792:1–9.
12. R. L. Tontz. Memberships of General Farmers' Organizations, United States, 1874–1960. *Agricultural History*. 1964;38(3):143–56.
13. W. S. Gaud. *The Green Revolution: Accomplishments and Apprehensions. AgBioWorld*, 1968 (Accessed November 11, 2015). Available from: www.agbioworld.org/biotech-info/topics/bor laug/borlaug-green.html.
14. M. Schermer, J. Hoppichler. GMO and Sustainable Development in Less Favoured Regions – The Need for Alternative Paths of Development. *Journal of Cleaner Production*. 2004;12(5): 479–89.
15. T. Twardowski. Chances, Perspectives and Dangers of GMO in Agriculture. *Journal of Fruit Ornamental Plant Research*. 2010;18(2):63–9.
16. I. Potrykus. "Golden Rice", a GMO-Product for Public Good, and the Consequences of Ge-Regulation. *Journal of Plant Biochemistry and Biotechnology*. 2012;21(1):68–75.
17. P. Foster, H. D. Leathers. *The World Food Problem*. (Lynne Rienner Publishers Inc., 1999).
18. UN FAO. *FAOSTAT: Download Data*. (United Nations Food and Agriculture Organization, 2009 [Accessed September 16, 2012]). Available from: http://faostat3.fao.org/download/.
19. M. Pollan. *The Omnivore's Dilemma: A Natural History of Four Meals*. (Penguin, 2006).
20. C. D. Oliver, L. L. Irwin, W. H. Knapp. *Eastside Forest Management Practices: Historical Overview, Extent of Their Application, and Their Effects on Sustainability of Ecosystems*. General Technical Report PNW-GTR–324. (Pacific Northwest Research Station, US Forest Service, 1994).
21. A. L. Olmstead, P. W. Rhode. Reshaping the Landscape: The Impact and Diffusion of the Tractor in American Agriculture, 1910–1960. *The Journal of Economic History*. 2001;61(03): 663–98.
22. US Economic Research Service. *History of Agricultural Price-Support and Adjustment Programs, 1933–84*. (US Department of Agriculture, Economic Research Service, 1984).
23. J. Lehmann, S. Joseph. *Biochar for Environmental Management: Science, Technology and Implementation*. (Routledge, 2015).
24. Technology Quarterly. Civilian Drones. *The Economist*. June 10, 2017: 3–13.
25. UN FAO. Energy for Agriculture. *The Energy and Agriculture Nexus*. Environment and natural resource working paper no. 4. (United Nations Food and Agriculture Organization; 2000).
26. UN FAO. *Energy-Smart: Food at FAO*, 53. (UN FAO Climate, Energy, and Tenure Division, 2012).
27. D. Pimentel, M. H. Pimentel. *Food, Energy, and Society*. (CRC Press, 2007).
28. C. J. Cleveland. The Direct and Indirect Use of Fossil Fuels and Electricity in USA Agriculture, 1910–1990. *Agriculture, Ecosystems and Environment*. 1995;55(2):111–21.
29. J. Guthman. *Agrarian Dreams: The Paradox of Organic Farming in California*. (University of California Press, 2014).
30. E. Holt-Gimenez, A. Shattuck, M. Altieri, H. Herren, S. Gliessman. We Already Grow Enough Food for 10 Billion People . . . And Still Can't End Hunger. *Journal of Sustainable Agriculture*. 2012;36(6):595–8.

31. L. C. Ponisio, L. K. M'Gonigle, K. C. Mace, et al., editors. Diversification Practices Reduce Organic to Conventional Yield Gap. *Proceedings of the Royal Society B;* 2015.
32. P. Ramesh, N. Panwar, A. Singh, et al. Status of Organic Farming in India. *Current Science.* 2010: 1190–4.
33. V. Seufert, N. Ramankutty, J. A. Foley. Comparing the Yields of Organic and Conventional Agriculture. *Nature.* 2012;485(10 May):229–32.
34. D. J. Connor. Organically Grown Crops Do Not a Cropping System Make and nor Can Organic Agriculture Nearly Feed the World. *Field Crops Research.* 2013;144:145–7.
35. V. Smil. *Enriching the Earth: Fritz Haber, Carl Bosch and the Transformation of World Food Production.* (The MIT Press, 2001).
36. D. Melnychuk. Opportunities for Organic Agriculture in Ukraine. *Invited Seminar by Rector DO Melnychuk, National University of Life and Environmental Sciences of Ukraine.* (Yale University School of Forestry and Environmental Studies, 2012).
37. K. C. Abercrombie. Agricultural Mechanisation and Employment in Latin America. *International Labour Review.* 1972;106:11.
38. S. R. Bose, E. H. Clark. Some Basic Considerations on Agricultural Mechanization in West Pakistan. *The Pakistan Development Review.* 1969;9(3):273–308.
39. F. Krausmann, K.-H. Erb, S. Gingrich, et al. Global Human Appropriation of Net Primary Production Doubled in the 20th Century. *Proceedings of the National Academy of Sciences.* 2013;110(25):10324–9.
40. R. Lal, D. Reicosky, J. Hanson. Evolution of the Plow over 10,000 Years and the Rationale for No-Till Farming. *Soil and Tillage Research.* 2007;93(1):1–12.
41. D. R. Huggins, J. P. Reganold. No-Till: The Quiet Revolution. *Scientific American.* 2008;299(1): 70–7.
42. R. Brown, J.-P. Steckler. Prescription Maps for Spatially Variable Herbicide Application in No-Till Corn. *Transactions of the ASAE.* 1995;38(6):1659–66.
43. R. Pearcy, J. Ehleringer. Comparative Ecophysiology of C3 and C4 Plants. *Plant, Cell and Environment.* 1984;7(1):1–13
44. P. Micklin. The Aral Sea Disaster. *Annual Review of Earth and Planetary Sciences.* 2007;35:47–72.
45. P. P. Micklin, W. D. Williams. *The Aral Sea Basin.* (Springer; 1996).
46. The World Bank. *Restoring a Disappearing Giant: Lake Chad.* (The World Bank Group, 2014 [Accessed May 21, 2016]). Available from: www.worldbank.org/en/news/feature/2014/03/27/restoring-a-disappearing-giant-lake-chad.
47. C. F. Chilcutt, B. E. Tabashnik. Effects of *Bacillus thuringiensis* on Adults of *Cotesia plutellae* (Hymenoptera: Braconidae), a Parasitoid of the Diamondback Moth, *Plutella xylostella* (Lepidoptera: Plutellidae). *Biocontrol Science and Technology.* 1999;9(3):435–40.
48. B. L. Phillips, G. P. Brown, J. K. Webb, R. Shine. Invasion and the Evolution of Speed in Toads. *Nature.* 2006;439(7078):803–.
49. R. Carson. *Silent Spring.* (Houghton Mifflin, 1962).
50. S. R. Palumbi. Humans as the World's Greatest Evolutionary Force. *Science.* 2001;293(5536): 1786–90.
51. M. L. Flint, R. van den Bosch. The Philosophy of Integrated Pest Management. In *Introduction to Integrated Pest Management.* (Springer, 1981): pp. 107–19.
52. US Department of Agriculture. *Fertilizer Use and Price.* 2013 (Accessed July 15, 2016). Available from: www.ers.usda.gov/data-products/fertilizer-use-and-price.aspx.
53. G. J. Leigh. *The World's Greatest Fix: A History of Nitrogen and Agriculture.* (Oxford University Press, 2004).
54. J. R. Craig, D. J. Vaughan, B. J. Skinner. *Resources of the Earth: Origin, Use, and Environmental Impacts,* fourth edition. (Prentice Hall, 2011).
55. US Geological Survey. *International Minerals Statistics and Information.* (US Department of the Interior, 2014 [Accessed 2016 October 14, 2016]). Available from: http://minerals.usgs.gov/minerals/pubs/country/.

56. G. Hardin. The Tragedy of the Commons. *Science*. 1968;162(3859):1243–8.
57. E. J. Jacobo, A. M. Rodríguez, N. Bartoloni, V. A. Deregibus. Rotational Grazing Effects on Rangeland Vegetation at a Farm Scale. *Rangeland Ecology and Management*. 2006;59(3): 249–57.
58. C. Jones. Building New Topsoil. *Stipa Native Grasses Forum*; 3 May 2002.
59. C. White. *Revolution on the Range: The Rise of a New Ranch in the American West*. (Island Press, 2012).

22

Future Agriculture Production and Distribution

22.1 Global Distribution of Agriculture Land

Having overcome tremendous challenges of providing sufficient food efficiently, future agriculture challenges will include distributing food equitably to all regions, keeping agriculture areas small to avoid displacing biodiversity, identifying and addressing current practices that need changing, and navigating between the risks and opportunities that existing and new food production technologies offer.

Agriculture science and technology have kept nutrition ahead of the rising world population without dramatically expanding the agriculture area needed, so that forests and biodiversity have not yet been severely depleted. The continued global population increase and currently expanding meat demand may change this and lead to more cropland and fewer forests for biodiversity. This chapter discusses global and regional agriculture production and distribution of food. It ends with a brief discussion of how agriculture might change in the future. Figure 22.1 shows the global distribution of agricultural land.

22.2 Hunger and Food Shortages

People do not eat the same things or amounts in all parts of the world. Some people in all regions receive too few Calories, and some regions have a greater proportion of people receiving too few Calories (Figure 22.2–3).

People in some regions also receive too few proteins and fats based on world standards (Figure 22.3). These regions are somewhat related to areas with too few total Calories. In most places with insufficient food, people generally raise fewer animals and so eat less meat. They eat starches with few proteins or do not balance their vegetarian diets to provide sufficient proteins. In all regions, some people choose not to eat meat for religious, cultural, or personal reasons and may ingest insufficient proteins.

Fat ingestion seems to be regionally specific but only partly related to protein intake. Some wealthy countries (the United States and Europe) consume much fat. People in cold climates or who exercise a lot sometimes develop a culture of high fat consumption to obtain energy. They can become fat when the culture becomes sedentary but the eating habit continues.

347

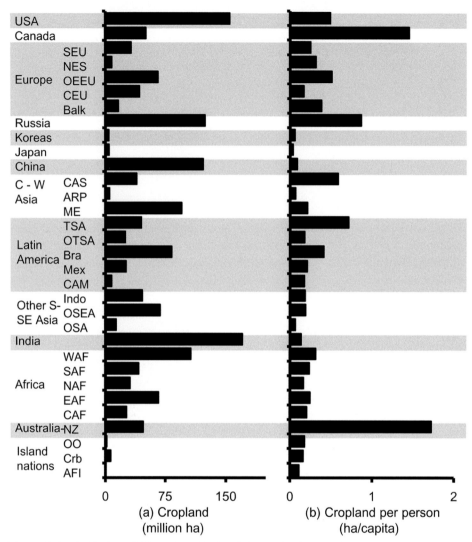

Figure 22.1 Distribution of (a) total cropland and (b) cropland per capita by country groups (both annual and perennial crops) [1].

22.3 Food Distribution

The total food produced in the world has generally been consistently rising, but can be impacted by economic and climatic fluctuations. A "dip" in wheat production in the 1990s (Figures 22.4, 22.10a) was probably caused by the collapse of the Soviet Union. It created a dip in total food produced and a slight decline in milk production but a rise in chicken production (Figure 22.12). It did not markedly affect the world's Calorie or protein consumption (Figure 22.10b).

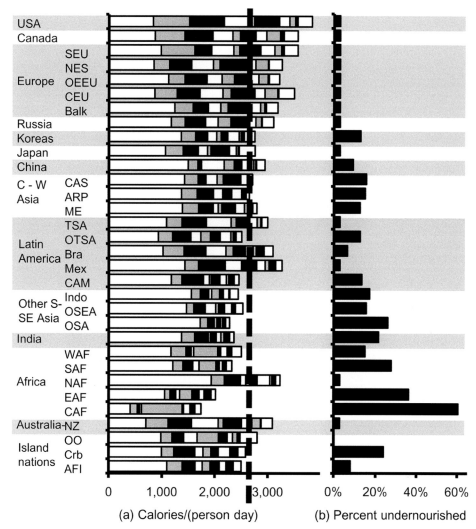

Figure 22.2 (a) Average food energy consumption per capita by country groups and food types. Vertical dashed bar = target Calories (2,700/person-day). (b) Proportion of people receiving insufficient food Calories by country groups [2–4]. (a) From left: first white = cereals; first gray = vegetable oils; first black = sugars and sweeteners; second white = meat and offal; second gray = roots and tubers; second black = milk, eggs, and fish; third white = fruits and vegetables; right gray = animal fats; right black = pulses; right white = others (including beer and wine)

Annual production of a particular crop for a country can fluctuate as much as 25% (Figure 22.4) because of weather, insects or other pests, and other factors; however, the world's food supply is well buffered by food storages, excess production, international trading, and abilities to shift food from one use to another – such as from animal feed to

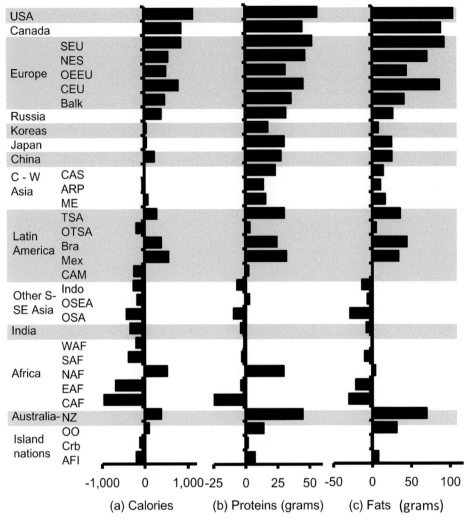

Figure 22.3 Average deviation of people by country groups in daily consumption from target ["T"] (a) Calories [T = 2,700], (b) protein [T = 65], and (c) fat [T = 60] [1].

human consumption. The storage and ability to ship foods has reduced the global need for a large excess annual production to guard against crop failures.

Most of the world's food is eaten in the region where it is grown (Figure 22.5a); but a few regions such as Japan, Korea, and North Africa import much more food than they produce (Figure 22.5b). Although not self-sufficient in food and energy, Japan and South Korea are quite prosperous. By contrast, Uzbekistan and Nigeria should be quite prosperous based on their potential to produce food and energy, but are not (Chapter 6). Australia and New Zealand, the United States, parts of Europe, and Brazil have exported comparatively large

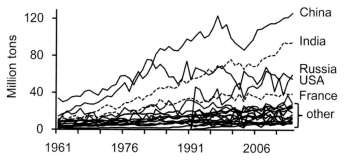

Figure 22.4 Annual wheat production fluctuations by the twenty leading wheat-producing countries [1].

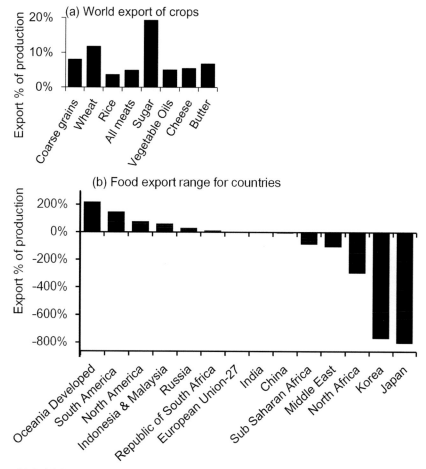

Figure 22.5 (a) Percentage of world production of major foods that are shipped abroad. (b) Most extreme importers (−) and exporters (+) of food, and selected intermediate country groups [1].

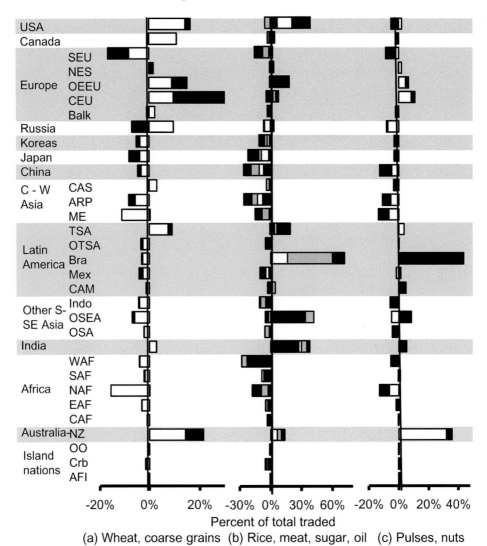

Figure 22.6 Exports (+) and imports (−) of major food groups by country groups as percentage of total world trade [1]. (a) 160 million tons traded annually each (white = wheat, black = coarse grains; (b) 38–70 million tons traded annually each (inner black = rice, white = meat, gray = sugar, outer black = vegetable oil; (c) 1.5–5.5 million tons traded annually each (white = butter, black = cheese)

food quantities (Figure 22.6), often to the displeasure of other countries who felt the low prices of these incoming foods harmed their domestic agricultural economy; international pressure may have forced recent changes [5]. Countries and regions can lack food because they are not culturally, socially, and economically organized to trade, import, and/or distribute food appropriately for internal and external reasons (Chapter 6).

In addition to wheat and coarse grains (including maize and potatoes), a large proportion of the world's sugar is traded (Figures 22.5a, 22.6b). Sugar can be produced from crops grown in many parts of the world; however, most sugar comes from sugar cane grown only in warm climates because it is very productive (in Calories/ha) and labor there is inexpensive.

22.4 Food Production

The predominant plant foods grown and consumed in each region are largely the result of its climate and soils; most root crops (potatoes), cereals, and coarse grains are grown nearly everywhere. Sugar cane is grown primarily in tropical countries (Figure 22.7). Animal production seems to be based more on culture than are plant foods. India produces very little meat – and especially little cattle meat for religious reasons. China and Europe produce much more pig meat and less cow meat than North and South America (Figure 22.8). The Moslem and Jewish faiths do not promote pig meat for religious reasons.

22.4.1 Food Production over the Past 50 Years

The global average consumption of food has been increasing at a greater rate than the population for the past 50 years (Figure 22.9). The average world consumption of foods is about 2700 Calories and 80 grams of protein per person per day (Figure 22.9) – very close to the recommended amounts (Chapter 20) if the food were equally distributed to all people. Cereals comprise about 50% of the Calories consumed and 40% of the proteins. Animal products provide about 17% of the Calories and 39% of the proteins. The Calorie and protein intakes, the amount of non-cereal plant foods, and the proportion of protein-rich meats have been increasing for the past 50 years.

Currently, the world produces about 12 quadrillion Calories of plant food each year and 1.2 quadrillion Calories of terrestrial and aquatic animal food (Figures 22.11a, 22.12). About 60% of the plant food (Calories) is not directly eaten; most is used to produce animal food, with some used for biofuels (Figure 22.11b). The animals are then generally eaten by people.

Most plant food consumption is grains – rice, wheat, and maize – with roots and tubers, oil crops, and vegetables contributing as well (Figures 22.9–10). For most crops, production has increased much more than the area planted (Figure 21.5).

Much more dramatic shifts have occurred in meat production (Figure 22.12) compared to plants in the past 50 years. The largest shares of animal foods are now from pigs, milk, poultry, and cattle, with eggs also contributing. Both pork and chicken meat increased and surpassed milk products and cattle meat in global production. Cow and buffalo milk products, cattle meat, and hen eggs increased in production, but not as dramatically as pork and chicken meat. Global production of other meats remained nearly static at a much smaller proportion of total production.

354 Future Agriculture Production and Distribution

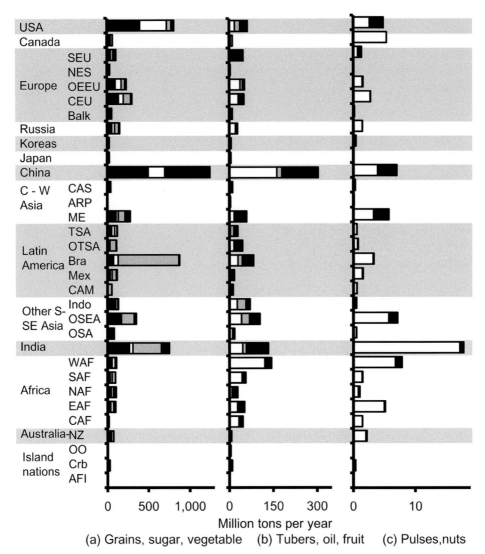

Figure 22.7 Major plant food groups grown by country groups [1, 2–4]. (a) From left: black = cereals, white = coarse grains, gray = sugar, right black = vegetables; (b) white = roots and tubers, gray = oil crops, black = fruits; (c) white = pulses, black = nuts.

Aquatic food production changed dramatically, although its impact was not great since it is a small part of the world's food production (Figure 22.13). The total amount of fish increased because of dramatic increases by inland water fish farms (aquaculture). There has been a recent stabilization and possible decline of marine water fish capture. Crustacean and mollusk productions increased, in part because of marine and inland water aquaculture. Aquatic plant production for food increased dramatically because of marine aquaculture.

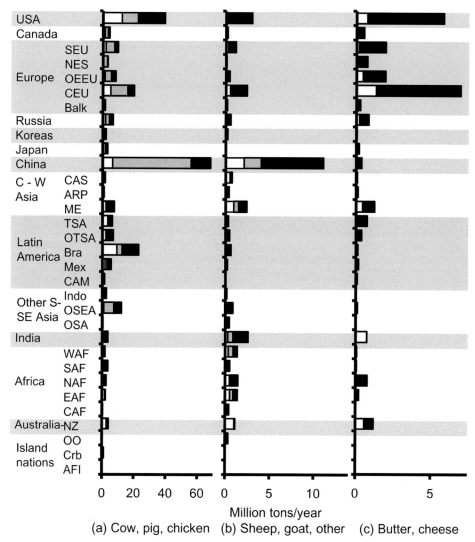

Figure 22.8 Major meats, cheese, and milk produced by country groups [1, 2–4]. (a) white = Cow, gray = pig, black = chicken; (b) white = sheep, gray = goat, black = other meat; (c) white = butter, black = cheese.

22.4.2 Increasing Demand for Meat

A recent increase in meat demand has been occurring as developing countries become wealthy enough to want more meat [7, 8]. This increase could reduce the forest area available for biodiversity and could continue for decades as more people mature and more countries develop (Figure 5.12). Extra meat consumption is exacerbated when

Table 22.1 *Legend for Figure 22.9. Numbers begin at bottom and are the same for (a) and (b) (for example, #1 at bottom = cereals)*

1	cereals (gray)	9	alcohol (black)
2	vegetable oils (white)	10	pulses (gray)
3	meat (black)	11	animal fats (white)
4	sugar & sweeteners (gray)	12	oil crops (black)
5	starchy roots (white)	13	eggs (gray)
6	milk (black)	14	fish, seafood (white)
7	vegetables (gray)		Top: tree nuts, spices, offal,
8	fruits (white)		stimulants, sugar crops (black)

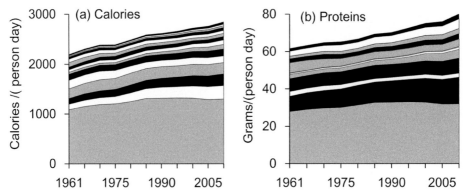

Figure 22.9 Global average (a) Calorie and (b) protein consumption per person over time [1]. Legend in Table 22.1.

animals are raised in feedlots from feed produced from croplands. Chicken meat yields only one-half of the Calories needed to feed the chicken, and beef yields only yields one-fourth of the Calories. Consequently, when using croplands to produce chicken food and feeding chickens in feedlots, twice as much area is needed to produce the same Calories in chicken meat that can be provided in a vegetarian diet – and four times as much area is needed to produce the same Calories in beef as in a vegetarian diet.

This meat demand is causing an increased clearing of forests and grasslands to raise animal feed. Coupled with the increased food demand is a concern about shrinking water availability as aquifers are drawn down to grow these extra crops just as the changing climate is reducing rainfall in some areas.

22.5 Options: Intensive, Organic, Range, Pen

The concerns over agriculture chemicals, narrowing of the food crop gene pool, drawdowns of aquifers, and rising meat demand could be met in different ways [7, 9, 10]. Some are:

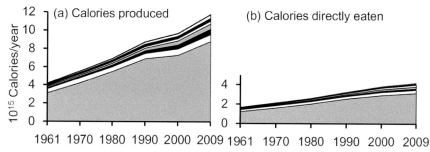

Figure 22.10 Changes in (a) plant foods produced and (b) plant foods directly eaten globally by people [1]. (From bottom: gray = cereals and grains; white = roots and tubers; black = oil crops; middle gray = vegetables; middle white = fruits; middle black = pulses; upper gray = tree nuts; upper white = sugar crops.)

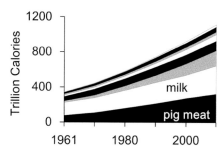

Figure 22.11 Changes in world food consumption from terrestrial and aquatic animals [1]. (From bottom: pig meat, milk, lowest gray = poultry, center black = cattle meat, center white = eggs, next gray = sheep and goats, upper black = marine fish, upper white = freshwater fish.)

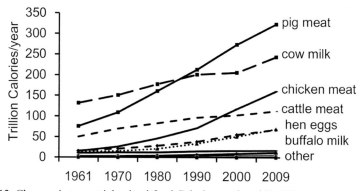

Figure 22.12 Changes in terrestrial animal food Calories produced [5, 14].

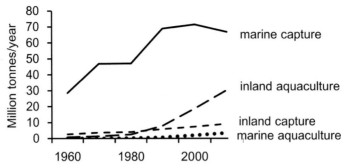

Figure 22.13 Changes in aquatic food crop production [1].

- clearing more forests and grasslands to grow the extra food needed;
- increasing agriculture efficiencies that may require more technologies, chemicals, energy, and water but produce more food per hectare;
- utilizing the world's grasslands to raise animals for meat rather than feeding them from cropland;
- changing to more vegetarian diets;
- finding other alternatives, such as many "urban gardens" or other innovative ideas;
- moving toward information-based farming that uses technologies and educated people to minimize chemicals, energy, and water.

Clearing more forests and grasslands and increasing agricultural efficiencies would continue present trends. The other alternatives will be discussed below.

22.5.1 Feedlots, Grasslands, and Vegetarian Diets

Table 22.2 compares the global area in farmland, forests, and grassland under various scenarios of raising meat and using vegetarian diets. All scenarios assume the current world population and amount of protein consumed; and, feed from agriculture waste is not considered. Grassland grazing is assumed to be done in ways that protect or enhance the biodiversity [11–13, 10].

Compared to the current baseline, feeding no animals on grasslands (pastures) would reduce the forest area by 25%, while obtaining 70% of the meat from grasslands would increase forest area by 10%. However, the grasslands would be at carrying capacity at 70%; and feeding more animals on grasslands would require clearing forests to create grasslands. Fewer animals can be fed per hectare on grasslands than by feeding them agriculture crops, so it would be counterproductive to convert forests to grassland instead of farm land.

The analysis shows that only part of the existing meat consumption can be produced from existing grasslands, so increases in meat demand will necessitate some forest clearing even if all grasslands are grazed to produce meat. More vegetarian diets could reduce farmland needs, although a mixture of plant crops would be needed to ensure healthy diets (Chapter 20).

Table 22.2 *Amounts of agriculture, forest, and grassland land under current and alternative scenarios of meat and vegetarian protein production [1, 14–16]*

	Agriculture area				Grassland area	
	Total	Direct food	Animal feed	Forest	Total	Used
Current meat consumption and feed distribution						
1.	11%	4%	7%	27%	23%	4%
Current meat amount, feed redistributed						
2. No meat fed on range	19%	4%	14%	19%	23%	0%
3. 70% meat fed on range	8%	4%	4%	30%	23%	23%
4. All meat fed on range	4%	4%	0%	24%	32%	32%
All people shift to 100% vegetarian diets						
5. All grain diet	6%	6%	0%	32%	23%	0%
6. All roots and tubers diet	9%	9%	0%	29%	23%	0%

22.5.2 Organic Farming

A proposed alternative to intensive agriculture is organic farming, which does not use artificial pesticides or fertilizers. A major concern is whether organic farming can produce comparable yields to intensive farming. Different studies comparing organic and other farming show mixed results, with some showing as little as half of the production with organic farming [21–23], others showing comparable production [17–19], and others showing mixed but optimistic production [20].

The results seem largely based on the crops grown, conditions and climate of growth, methods of treatment, and other factors. As experience and techniques of organic farming improve, it may become an increasingly viable alternative. Even if less food is produced per hectare, energy, fertilizer, pesticide, and irrigation water savings through organic farming could make it worthwhile. And, the lessons learned from organic farming could be used to modify intensive farming.

22.5.3 Other Measures: Rooftop and Urban Gardens, Waste Reduction

Urban, rooftop, and backyard gardens offer some alternative food production methods [24, 25]. Also, the human population may benignly contract to a lower level if the fertility rate declines (Figure 5.13).

Food waste can be reduced, although the extent of waste is uncertain and the means of reducing it are diffuse [26]. Reductions have been and will be accomplished by incrementally addressing different steps in the supply chain – harvesting, storing, processing, marketing, and disposing [27]. Much agriculture and food waste can be used to make biofuels and to feed animals such as pigs and goats.

Less food waste leads to less needed food production and has been accomplished by preserving foods using cold storage in transport and warehouses, canning processes, freeze drying, and preservatives as well as by transporting waste from all processing stages to the feedlot – from farms, processing plants, grocery stores, restaurants, and residential garbage disposals.

22.5.4 *Future Agriculture*

The dramatic social changes accompanied by the many innovations described above have overcome the fear of world famine [28]. The innovations have also enabled more crops to grow on each farmed area so that much area could return to more biodiverse systems (Chapter 5). However, they have had negative social consequences for cities and rural areas. Now the small rural populations are commonly below the critical mass to provide many amenities (Figure 4.6) so few people want to live there, leading to a cascading collapse of rural populations and infrastructures, including law enforcement.

The innovations have also led to issues of harmful pesticides such as DDT and PCBs (Chapter 7). They have increased fertilizer use which can pollute aquifers and rivers, salinate soils, and demand more water use to flush the chemicals. They have increased energy use in making nitrogen fertilizer. They are leading to scarcities of other fertilizers (Chapter 27). They are also reducing biodiversity, including harming populations of pollinators, at an extreme.

The issues that led to the current situation are not the same issues of concern today. World food supply is a concern but no longer a crisis; however, agriculture chemical pollution, irrigation water consumption, energy consumption, and a depauperate rural infrastructure are strongly emerging issues.

Current intensive agriculture could begin incorporating organic farming principles as technologies reduce the needs for extensive chemicals and irrigation, the chemicals become more benign, organic techniques prove efficient and effective, and site-specific actions done by trained specialists using modern techniques and technologies substitute for bulky, broad-scale applications. The general principles of organic farming are already becoming more prevalent, but chemicals will probably continue to be used sparingly as needed under similar practices as Integrated Pest Management (IPM), discussed above.

The principle is to use more information and precise actions instead of large-scale uniform actions and large amounts of energy, irrigation water, and chemicals. Intercropping can add nitrogen from legumes and reduce artificial fertilizers, can increase soil organic matter, can spread the seasonal tending, and can diversify the harvest. The diversity and longer seasons mean greater possibilities of full-time work and fewer pests. Staggering planting times and varieties of the same crop can further aid pest management.

The diversity of plants would lead to soil microbial diversity as well as insect and bird diversity. And, diverse croplands can serve better as corridors for migration among forest and range habitats (Chapter 16) as well as provide other ecosystem services [29].

The information-based farming trend would mean more skilled people could be employed at good wages for farming. In places, the seasonality of agriculture work could be complemented with work in forests, similarly using low-cost, small, highly technical equipment (Chapter 30). With critical masses of such rural jobs come indirect jobs and large enough communities to support diverse, pleasant lifestyles (Figure 4.6).

References

1. UN FAO. *FAOSTAT: Download Data.* (United Nations Food and Agriculture Organization, 2009 [cited Accessed 2012 September 16, 2012]). Available from: http://faostat3.fao.org/download/.
2. UN FAO. *FAO Statistical Yearbook.* (United Nations Food and Agriculture Organization, 2009 [Accessed October 8, 2014]). Available from: www.fao.org/docrep/014/am079m/PDF/am079m00a.pdf.
3. UN FAO. *FAO–Food Security Indicators.* (United Nations Food and Agriculture Organization, 2013 [Accessed June 25, 2016]). Available from: Food-Security-Statistics@FAO.org.
4. UN FAO. *Global Forest Resources Assessment 2015.* (United Nations Food and Agriculture Organization, 2015).
5. B. Kerremans. What Went Wrong in Cancun? A Principal-Agent View on the EU's Rationale Towards the Doha Development Round. *European Foreign Affairs Review.* 2004;9(3):363–93.
6. UN FAO. *A Handbook on Food Balance Sheet.* (United Nations Food and Agriculture Organization, 2001).
7. C. Delgado. Rising Demand for Meat and Milk in Developing Countries: Implications for Grasslands-Based Livestock Production, in McGilloway D. A., editor, *Grassland: A Global Resource.* (Wagingen Academic Publishers, 2005): pp. 29–39.
8. J. N. Galloway, M. Burke, G. E. Bradford, et al. International Trade in Meat: The Tip of the Pork Chop. *AMBIO: A Journal of the Human Environment.* 2007;36(8):622–9.
9. J. A. Foley, N. Ramankutty, K. A. Brauman, et al. Solutions for a Cultivated Planet. *Nature.* 2011;478(7369):337–42.
10. D. A. McGilloway. *Grassland: A Global Resource.* (Wageningen Academic Publishers, 2005).
11. E. J. Jacobo, A. M. Rodríguez, N. Bartoloni, V. A. Deregibus. Rotational Grazing Effects on Rangeland Vegetation at a Farm Scale. *Rangeland Ecology and Management.* 2006;59 (3):249–57.
12. C. Jones. Building New Topsoil. *Stipa Native Grasses Forum;* 3 May 2002; Armidale, Australia, 2002.
13. C. White. *Revolution on the Range: The Rise of a New Ranch in the American West.* (Island Press, 2012).
14. M. Kempen, T. Kraenzlein, editors. *Energy Use in Agriculture: A Modeling Approach to Evaluate Energy Reduction Policies.* 107th Seminar of the European Association; (The European Association, 2008).
15. D. Nijdam, T. Rood, H. Westhoek. The Price of Protein: Review of Land Use and Carbon Footprints from Life Cycle Assessments of Animal Food Products and Their Substitutes. *Food Policy.* 2012;37(6):760–70.
16. J. Wilkinson. Re-Defining Efficiency of Feed Use by Livestock. *Animal.* 2011;5(07):1014–22.
17. E. Holt-Gimenez, A. Shattuck, M. Altieri, H. Herren, S. Gliessman. We Already Grow Enough Food for 10 Billion People … And Still Can't End Hunger. *Journal of Sustainable Agriculture.* 2012;36(6):595–8.

18. L. C. Ponisio, L. K. M'Gonigle, K. C. Mace, et al., editors. Diversification Practices Reduce Organic to Conventional Yield Gap. *Proceedings of the Royal Society B*; 2015.

19. P. Ramesh, N. Panwar, A. Singh, et al. Status of Organic Farming in India. *Current Science*. 2010:1190–4.

20. V. Seufert, N. Ramankutty, J. A. Foley. Comparing the Yields of Organic and Conventional Agriculture. *Nature*. 2012;485(10 May):229–32.

21. D. J. Connor. Organically Grown Crops Do Not a Cropping System Make and nor Can Organic Agriculture Nearly Feed the World. *Field Crops Research*. 2013;144:145–7.

22. V. Smil. *Enriching the Earth: Fritz Haber, Carl Bosch and the Transformation of World Food Production*. (The MIT Press, 2001).

23. D. Melnychuk. Opportunities for Organic Agriculture in Ukraine. *Invited Seminar by Rector D. Melnychuk, National University of Life and Environmental Sciences of Ukraine*; January 21. (Yale University School of Forestry and Environmental Studies, 2012).

24. M. CoDyre, E. D. Fraser, K. Landman. How Does Your Garden Grow? An Empirical Evaluation of the Costs and Potential of Urban Gardening. *Urban Forestry & Urban Greening*. 2015;14 (1):72–9.

25. F. Orsini, D. Gasperi, L. Marchetti, et al. Exploring the Production Capacity of Rooftop Gardens (RTGS) in Urban Agriculture: The Potential Impact on Food and Nutrition Security, Biodiversity and Other Ecosystem Services in the City of Bologna. *Food Security*. 2014;6(6):781–92.

26. J. Parfitt, M. Barthel, S. Macnaughton. Food Waste within Food Supply Chains: Quantification and Potential for Change to 2050. *Philosophical Transactions of the Royal Society B: Biological Sciences*. 2010;365(1554):3065–81.

27. S. Lundie, G. M. Peters. Life Cycle Assessment of Food Waste Management Options. *Journal of Cleaner Production*. 2005;13(3):275–86.

28. P. R. Ehrlich, A. H. Ehrlich, G. C. Daily. Food Security, Population and Environment. *Population and Development Review*. 1993:1–32.

29. C. Kremen, A. Miles. Ecosystem Services in Biologically Diversified Versus Conventional Farming Systems: Benefits, Externalities, and Trade-Offs. *Ecology and Society*. 2012;17(4):40.

Part VIII

Energy

23

Energy Sources, the Energy Cycle, Exergy

23.1 Defining and Measuring Energy

Defining energy is difficult:

in the brief span of 20 years, energy was invented, defined and established as a cornerstone, first of physics, then of all science. We don't know what energy is ... but as a now robust scientific concept we can describe it in precise mathematical terms, and as a commodity we can measure, market, regulate, and tax it.

(H. C. Von Baeyer, 2004[1])

For this book, energy will be considered something that enables a particle, object, or system of objects to behave counter to such forces as gravity and atomic (nuclear) attractions and repulsions. As examples:

• An electron can absorb energy that enables it to move farther from its atomic nucleus.
• When an atomic nucleus splits or two nuclei combine, energy is released because the energy is no longer used to hold the nuclei together or apart.
• When a crystal such as ice absorbs energy, it is able to overcome the intermolecular attractions and lose its crystalline structure – forming water in the case of ice.
• As a liquid absorbs energy, the molecules can move faster and thus push farther apart. The liquid thus expands (or increases its pressure if it is constrained) and can even change to a gas. The gas can further absorb energy and expand, heat, and/or increase its pressure.
• A material will "absorb" energy if it is moved upward to a place where it can fall downward, attracted by the Earth's gravity. A book, for example, will absorb energy when placed upon a shelf, a rock contains energy if it is on a mountain ledge, and water contains energy if it is flowing down a river.

Energy can exist in many "forms": mechanical, thermal (heat), electromagnetic radiation (wave/particle such as light and ultraviolet), and chemical. It can change from one form to another and move elsewhere. The total energy in the Universe is almost constant (First Law of Thermodynamics: The Law of Conservation of Energy), with the exception that very small amounts of matter can be converted to very large amounts of energy in nuclear reactions.

The total amount of energy throughout the Earth is approximately in a steady state [2–8]. New energy arrives as radiation and gravity from the Sun and Moon, and the Earth radiates energy away through its atmosphere.

People make use of energy both to survive and to enjoy life. Through evolution, trial and error, science, and technologies, people have developed various ways to utilize energy between the times it enters and leaves the Earth. People do not consume the energy. They simply take it in one form and return it in another. People actually use very little of the energy that flows throughout the Earth. Some energy forms can create adverse consequences; and human ingenuity is constantly trying to overcome these consequences by using different forms or mitigating the consequences.

Energy is measured in different units, although consistent systems are being agreed upon [9]. Energy was described in very small units by research scientists doing experiments and in very large units by industry. Small units include British Thermal Units (BTUs), calories, joules, and foot pounds. Large units include barrels (for oil), or barrels of oil equivalent, metric or English tons (for coal), cubic meters or feet (for gas), and gallons or liters (for petroleum) [10].

The smaller units have been applied to larger scales by new names or prefixes. New names include "therm" which is 100 thousand BTUs; and "quad," one quadrillion BTUs. Prefixes include kilo (10^3), tera (10^{12}), mega (10^6), peta (10^{15}), giga (10^9), exa (10^{18}). For example, a gigajoule (gj) is 10^9 joules.

These measures and conversions complicate analyses and lead to mistakes. Various conversion tools are available on the internet to covert among units and sizes.

Energy will be expressed in BTUs and quads in this book. As a perspective, gasoline contains about 30 thousand BTUs/liter [10]. People consume about 10- to 11 thousand BTUs of food energy per day (2.5- to 3 thousand Calories; Chapter 20).

"Energy intensity" is commonly used as the energy within a unit of a specific substance (Table 8.2a) or required to perform a specific operation.

"Energy efficiency" is the inverse of energy intensity when quantifying the amount of operations performed by a specific amount of energy.

"Electricity" is referred to as energy by some authorities and not others. It is an energy carrier or a secondary source of energy. It carries energy through a slow flow of electrons but is not an independent source of energy similar to fossil fuels or renewable energy. In this book, the term "electric energy" will be used.

23.2 Potential and Kinetic Energy

Somewhere from nanoseconds to millions of years after energy has been absorbed, it is released; for example, energy is released when a crystal forms, water cools, a rock falls off a cliff, or an electron moves back toward the atom's nucleus. The released energy does not disappear.

An object can contain energy that can be released, known as "potential energy." Water in a reservoir has potential energy before it flows downhill. When it flows downhill, this potential energy is converted to "kinetic energy." Kinetic energy can power a water mill (for grinding flour or sawing lumber) or a hydroelectric turbine.

When an electron moves to a shell more distant from the nucleus – such as through photosynthesis – it absorbs kinetic energy. In the case of photosynthesis, the kinetic energy is in the form of radiation from the sun. The electron can be held in a more distant "shell" by combining it with other atoms to form molecules – organic molecules in the case of photosynthesis. These molecules have potential energy that is again released when the molecules are broken and the electrons return to closer shells, such as when an animal or microbe digests the molecule (respiration) or the molecule burns (combustion).

Some objects absorb more potential energy than others. Water, for example, absorbs much energy compared to the rate at which its temperature increases (heat capacity, Chapters 7 and 17).

Potential energy is often held by an object with a "threshold" – a relatively small amount of energy needed to stimulate the object to release its potential energy. A large rock resting on a mountain above a cliff has a lot of potential energy that would be released if it fell off the cliff. It would take relatively little energy – a threshold (leverage, Chapter 3) – to push the rock over the cliff's edge. Similarly, organic matter such as wood and petroleum contain a large amount of potential energy in their chemical bonds; but they almost never release this energy spontaneously, although relatively little energy such as the heat of a match can stimulate the greater release of energy of wood or petroleum burning.

23.3 Electricity and Exergy: Energy Quality

Some forms of energy are higher quality and thus capable of accomplishing work that other forms cannot. This quality is referred to as "exergy," or "absolute energy efficiency." Heat is usually considered low quality energy (microwave radiation, for example; Table 23.1) and less effective at some tasks than high quality energy such as X-rays. High quality energy can be transformed into a lot of heat; however, it is difficult to transform low quality energy such as heat into higher qualities. Heat is often considered a "degenerate" form of energy.

Electricity can carry energy of high exergy, but some energy is converted to low exergy (heat) energy in the transmission ("line loss") and does not reach the destination. Even more high exergy energy is lost to heat when high exergy fossil fuel is burned to generate low exergy heat to make steam that turns a turbine to generate electricity (high exergy). Only about 30% of the fossil fuel energy is recovered as electrical energy (Figure 24.3) in such power plants. The heat is usually intentionally dissipated in cooling towers, although it is increasingly recovered for steam heat as "combined heat and power" (CHP; also known as "cogeneration").

Electricity generation from fossil fuels began when the fossil fuels were cheap and plentiful, and the wasted fossil fuels were largely ignored. Large amounts of fossil fuel and carbon dioxide emissions can be saved by generating electricity in other ways (Chapter 25) and matching energy sources to tasks: "By thermodynamically matching sources to tasks, we can avoid the enormous waste of using high quality energy for low

Table 23.1 *Energy radiation wavelengths and comparative amount of energy per photon (calculated from conversion tables)*

Wavelength name	Length of wave	Number of energy particles (photons) needed to equal one gamma-ray photon
Radio	100.0 mm	5,000,000,000,000
Microwave	50.0 mm	100,000,000,000
Infrared	0.5 mm	68,000,000
Optical	0.0005 mm	25,000,000
UV	0.0002 mm	976,000
X-ray	0.000005 mm	1,000
Gamma-ray	0.00000001 mm	1

quality tasks and minimize the growing social and economic costs of energy production" [11] as cited in [12].

23.4 Energy Cycle into and out of the Earth

For energy considerations, the Earth includes the surrounding atmosphere as well as the liquid and solid sphere. The Earth has four sources of energy – the Sun's radiation, the heat generated from the Earth's interior, radioactive decay, and the Moon's gravity that generates tides. Solar energy is by far the largest of the four (Figure 23.1).

A total of 3.64 million quads of solar energy reach the Earth's outer atmosphere each year, with most of it arriving from the Sun [2–8]. The Sun's energy is primarily short-wave radiation (Figure 23.2). Of this incoming energy, about 5% is reflected by the Earth's atmosphere, 20% is reflected by clouds, and 5% is reflected by the Earth's solid and liquid surfaces. The rest is converted to other forms of energy and circulates among terrestrial, aquatic, and atmospheric Earth. Eventually it reradiates into space as long wave (heat) radiation (Chapter 7).

Between its entering and leaving the Earth's atmosphere, energy circulates through many processes and provides more energy for use and reuse than the amount entering and leaving each year. Without this circulation, the Earth would be almost as cold and sterile as Mars, where much less energy circulation occurs.

Parts of the Earth's energy circulates through potential energy that is held in storage for various lengths of time, such as in fossil fuels, gravity, biomass, or water vapor suspended in the air. This potential energy storage – as well as the internal circulation of energy – means that there was slightly more energy entering the Earth than leaving at some past time. The release of potential energy from burning fossil fuels during the past 120 years has circulated 20 thousand quads of energy more than has entered the Earth during this time (Figure 24.2). These 20 thousand quads are equivalent to the energy received from the sun in two days.

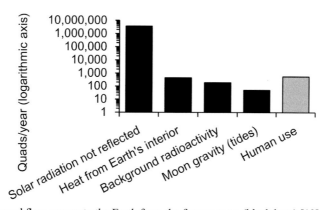

Figure 23.1 Annual flow energy to the Earth from the four sources (black bars) [13], compared with annual human energy use (gray) which comes largely from energy previously stored within the Earth as fossil fuels [14]. Note logarithmic scale.

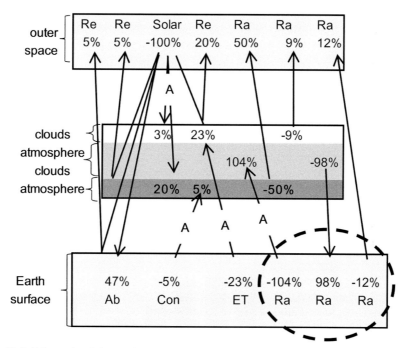

Figure 23.2 Schematic of the Earth's solar energy budget [2–8]. "Solar" = solar radiation, "Re" = reflection, "Ra" = radiation, "A" = absorption, "C" = convection, "ET" = evaporation and transpiration. Dashed circle area is discussed in the text.

Increased atmospheric greenhouse gases (Chapter 8) are absorbing and reradiating more long wave radiation in all directions, including back to the Earth (dashed circle of Figure 23.2). With fewer greenhouse gases, more long-wave radiation would simply exit the planet instead and the Earth would probably not be getting warmer (Figure 8.6).

23.5 Energy Flows Within the Earth

The circulating energy can be grouped into "pools" (Figure 23.3), similar to the ones in Odum et al. [13]. Kinetic "incoming energy" enters or emerges from the Earth (Figure 23.1) and moves among the different pools before reradiating beyond the atmosphere (Figure 23.2).

Much of the Earth's energy is used to warm the Earth, to stimulate the atmosphere and water, and to grow biological organisms (Figure 23.3). Ocean tides, currents, waves, winds, and rising air and water vapor are derived from the Sun differentially heating the Earth (Chapter 7). The energy can move into and out of the many pools of Figure 23.4 in a myriad of pathways.

Most energy that people currently use comes from fossil fuels (Figures 23.3–5), with lesser amounts from biomass, renewable energy (hydro-, solar-, and wind-power), nuclear energy, and a small proportion from food consumption. The total energy in all existing fossil fuels is nearly one-half of the incoming energy to the Earth for one year.

The annual incoming energy is about five thousand times as much as people use each year, and the total kinetic energy circulating in a year is about 12 thousand times as much (Figure 23.3). By tapping several of these energy pools for human use, it will be possible to provide many alternative sources of energy in the future.

Energy can move from one place to another in different ways, with some energy forms moving more readily and rapidly through some materials than others.

- Conduction is the flow of energy through an object. Some materials conduct some types of energy from places of higher energy to lower energy. Steel commonly conducts heat energy more rapidly than dry wood.
- Energy can be transported as a battery or fuel in a container. It can be transported heat and state changes of water (and other liquids; Chapter 7). Water can move in solid and liquid

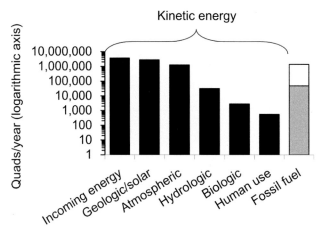

Figure 23.3 Major pools that the Earth's energy circulates through and global carbon storage [13]. Gray = fossil fuel "reserves"; white = total fossil fuel carbon storage.

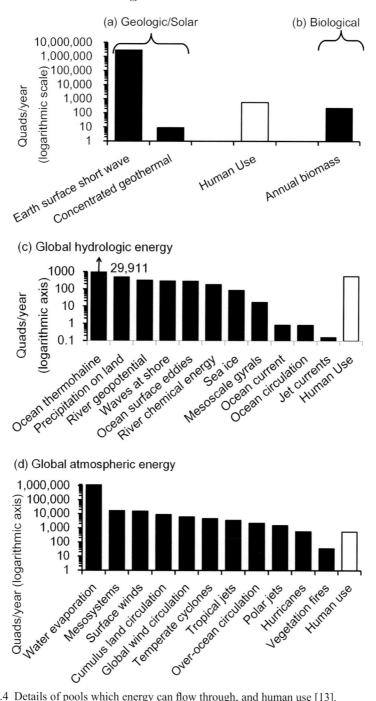

Figure 23.4 Details of pools which energy can flow through, and human use [13].

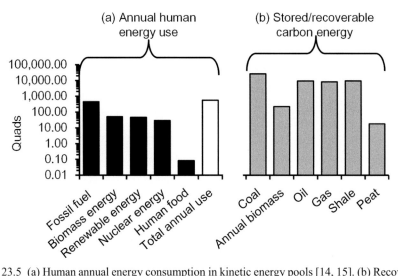

Figure 23.5 (a) Human annual energy consumption in kinetic energy pools [14, 15]. (b) Recoverable energy in carbon storage in various potential energy pools [14, 16–22].

states in water and in all three states in the air. Water and air currents and turbulent eddies move the water, which can release (or absorb) energy at a distant place.

- Energy also moves as radiation as "waves/particles" that can move through vacuums, such as the space between the Sun and the Earth. Some characteristics of radiant energy are best expressed if it is considered a wave; and other characteristics, by considering it as a particle. Radiant energy moves in waves of different wavelengths, or frequencies. Each wavelength emits energy "packets" – photons – of different amounts of energy, with shorter wavelengths having greater energy (Table 23.1). The greater energy (exergy) in shorter wavelengths meant that only 64 kg of uranium 235 emitting short wave radiation was needed to generate the devastating long wave (heat) energy of the Hiroshima bomb [23]. Radiant energy of a given wavelength striking a material can be transmitted, reflected, or absorbed. If absorbed, it can then be conducted, transported, or emitted as radiation.

- Transmission is the ability to pass through an object unchanged. Visible light is transmitted (passes) through the atmosphere, but most shorter wavelengths are blocked. Visible wavelengths can also be transmitted through clear glass, but long wave (heat) energy is not. Calcium bones and metals block X-rays while wood and animal flesh transmit them, allowing doctors to make "X-ray pictures" through people. If ultraviolet light were not blocked by the atmosphere, life on Earth would suffer dramatically or be annihilated; but the transmission of visible light is not only convenient for visibility but also uniquely suitable for photosynthesis.

- Reflection is casting back some radiant energy in the same wavelengths in which it strikes an object's surface. Radiation striking a material can also be reflected in only some wavelengths. Green leaves, for example, reflect energy in the green color wavelength, since this wavelength is not absorbed during photosynthesis. Also, color only indicates reflection of energy in the visible wavelengths. Materials such as silver-backed mirrors, wood, plexiglass, and gold may reflect or transmit most visible energy and appear cool, but absorb infrared energy and so be warmer than expected. Metals such as aluminum foil most completely reflect radiation in all wavelengths and thus prevent radiant energy from passing through.

- Absorption is taking energy into an object. Different materials absorb energy at different rates, in different wavelengths, and with different heat capacities (Chapter 7). A photon in the visible light wavelength can stimulate an electron to absorb the energy and jump to a "higher" shell farther from the nucleus, such as during photosynthesis in chlorophyll molecules and energy production in photovoltaic silicon molecules. Longer energy wavelengths do not have the concentrations of energy (exergy) to cause such changes, but can cause the molecules to vibrate and create heat. Absorbing materials and reflecting materials can be used on different sides of a wall to control the energy entering and leaving a building.

- Emission, or reradiation, is the release, or sending out, of energy from an object. An object can emit energy as radiation in specific wavelengths that are characteristic of the material. An "ideal" emission material – one that emits energy in all wavelengths – emits increasingly more energy at shorter wavelengths as it becomes hotter.

These variations in movement rates and forms in different materials generate the world's climate patterns (Chapter 7). They also allow energy to be managed for a variety of uses that are still being realized.

References

1. H. C. Von Baeyer. *Information: The New Language of Science* (Harvard University Press, 2004).
2. US National Aeronautics and Space Administration. *Climate and Earth's Energy Budget* (United States National Aeronautics and Space Administration, Earth Observatory, 2016 [Accessed March 14, 2016]). Available from: http://earthobservatory.nasa.gov/Features/EnergyBalance/.
3. M. G. Bosilovich, F. R. Robertson, J. Chen. Global Energy and Water Budgets in MERRA. *Journal of Climate*. 2011;24(22):5721–39.
4. J. T. Fasullo, K. E. Trenberth. The Annual Cycle of the Energy Budget. Part I: Global Mean and Land-Ocean Exchanges. *Journal of Climate*. 2008;21(10):2297–312.
5. J. T. Kiehl, K. E. Trenberth. Earth's Annual Global Mean Energy Budget. *Bulletin of the American Meteorological Society*. 1997;78(2):197–208.
6. R. Lindsey. *Climate and Earth's Energy Budget. Earth Observatory*, 2009 (Accessed May 17, 2017). Available from: https://earthobservatory.nasa.gov/Features/EnergyBalance/.
7. N. G. Loeb, B. A. Wielicki, D. R. Doelling, et al. Toward Optimal Closure of the Earth's Top-of-Atmosphere Radiation Budget. *Journal of Climate*. 2009;22(3):748–66.
8. K. E. Trenberth, J. T. Fasullo, J. Kiehl. Earth's Global Energy Budget. *Bulletin of the American Meteorological Society*. 2009;90(3):311–23.
9. J. Monteith. Consistency and Convenience in the Choice of Units for Agricultural Science. *Experimental Agriculture*. 1984;20(02):105–17.
10. A. Grubler, T. B. Johansson, L. Muncada, et al. Chapter 1 Energy Primer. In *Global Energy Assessment – Toward a Sustainable Future* (Cambridge University Press, 2012): pp. 99–150.

11. R. Torrie. *Half Life: Nuclear Power and Future Society: A Report to the Royal Commission on Electric Power Planning* (Coalition for Nuclear Responsibility, 1981).
12. Canadian Architect. Exergy. *Canadian Architect*, 2017 (Accessed July 18, 2017). Available from: www.canadianarchitect.com/asf/perspectives_sustainibility/measures_of_sustainablity/measures_of_sustainablity_exergy.htm.
13. H. Odum, M. Brown, S. Williams. *Handbook of Energy Evaluations Folios 1–4* (Center for Environmental Policy, University of Florida, 2000).
14. US Energy Information Administration. *International Energy Outlook (2004–2009)* (US Department of Energy, 2004–2009, DOE/EIA-0484, 2009).
15. M. Parikka. Global Biomass Fuel Resources. *Biomass and Bioenergy*. 2004;27(6):613–20.
16. US Energy Information Administration. *Shale Oil and Shale Gas Resources Are Globally Abundant* (US Energy Information Administration, 2014 [Accessed April 26, 2016]). Available from: www.eia.gov/todayinenergy/detail.cfm?id=14431.
17. British Petroleum. *BP Statistical Reviews of World Energy, June 2013–2016* (Accessed August 30, 2017). Available from: www.bp.com/statisticalreview.
18. R. Slade, R. G. Saunders, A. Bauen. *Energy from Biomass: The Size of the Global Resource* (Imperial College Centre for Energy Policy and Technology UK Energy Research Centre, 2011).
19. International Energy Agency. *Resources to Reserves 2013* (OECD, 2013).
20. UN FAO. *Global Forest Resources Assessment 2015* (United Nations, 2015).
21. World Energy Council. *Annexes*. 2013 (Accessed August 28, 2017). Available from: www.worldenergy.org/wp-content/uploads/2013/09/WER_2013_Annexes.pdf.
22. M. Xuehui, H. Jinming. Peat and Peatlands. *Coal, Oil Shale, Natural Bitumen, Heavy Oil and Peat*. 2009;2.
23. Atomic Bomb Museum. The First Atomic Bombs: "Little Boy" and "Fat Man" (Atomic Bomb Museum, 2006 [Accessed April 30, 2016]). Available from: www.atomicbombmuseum.org/2_firstbombs.shtml.

24

Conserving Energy and Renewable Energy

24.1 Human Use of Energy

People's pre-*Homo sapiens* ancestors first consumed just the energy produced from their food – currently about 10- to 11 thousand BTUs/(person day) (2.5- to 3 thousand Calories; Chapter 20). This energy ultimately came from photosynthesizing plants. People later used fire as additional energy. Now, each person in the developed world consumes over 500 thousand BTUs/day on average through heating, cooking, transportation, manufacturing, communicating, and other activities. The world average is about 165 thousand BTUs/(person day) (Figure 24.1) [1].

24.1.1 Early Energy Uses

Energy sources other than food developed slowly and opportunistically until the past several hundred years. As the advantages of extra energy became appreciated, people have been stimulated to develop new technologies, exploit new energy sources, use energy in different ways, and change their lifestyles and cultures to accommodate their newly discovered opportunities.

People first used fire for protection, managing vegetation, heating and cooking, and visibility at night. They later added wind, river currents, and domesticated animal energy for lifting and transporting. They used fire to make clay pottery and melt metal. They increased light by burning animal fats or petroleum from surface pools. They burned wood, shrubs, animal dung, dried peat, charcoal, and sometimes coal. Charcoal enabled wood energy to be transported farther and to create hotter fires for processing metals. Wind and water-powered mills were developed for grinding and other mechanical functions. "Greek fire" and gunpowder chemical energies were used in warfare. Energy sources were often not used sustainably. Wood was often depleted from a region and more or less suitable substitutes or conservation methods were found [7, 8]. Depleted forests in Europe led to the first conception, analyses, and practices of sustainability over 300 years ago (Chapter 29) [9].

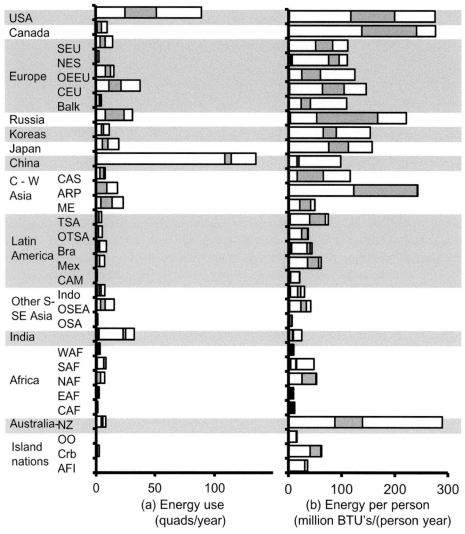

Figure 24.1 (a) Total and (b) per capita fossil fuel and wood energy consumption by country groups [2–6]. From left: black = wood; white = coal; gray = gas; right white = petroleum.

24.1.2 Industrial Revolution and Energy

Replacement of human and animal energy with other energy sources accelerated during the "industrial revolution" of the seventeenth and eighteenth centuries; and the amount of energy used in all functions increased – especially in heating, metallurgy, lighting, lifting, communicating, and transporting. The steam engine rapidly replaced water power in manufacturing, animals in ground transportation, and wind in water transportation. Increasing uses demanded more energy than wood sources could supply, so coal use

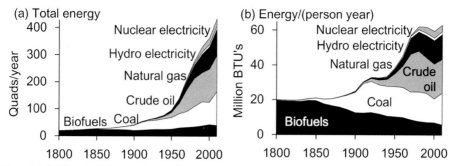

Figure 24.2 Historical changes in people's sources of (a) world energy and (b) energy/capita [11–13].

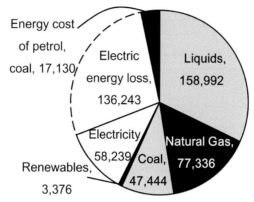

Figure 24.3 Quads of energy generated by source, including energy delivered to people, energy lost when generating electricity from fossil fuels, and energy consumed (lost) to process petroleum and coal, by source [16].

increased in the nineteenth century (Figure 24.2). Whales were exploited unsustainably for lamp oil until kerosene from fossil fuels replaced it [10]. As the nineteenth century closed, the internal combustion engine and gasoline had been perfected enough to enable the automobile, further revolutionizing transportation. The internal combustion engine also enabled the first airplanes, which were further advanced by jet engines.

24.1.3 Current Energy Uses

Fossil fuel has now become people's dominant energy source (Figures 24.2–3), used either directly or to generate electricity [14]. About 78% of the energy consumed in the world is from fossil fuels – oil, coal, and gas. Since 1880, burning has released 25 thousand quads of kinetic energy and about 550 billion tons of carbon that were tied up in fossil fuels [15]. Total consumption of any of the fossil fuels does not seem to have peaked and declined yet.

Wood has been used extremely inefficiently as a fuel in open fires and fireplaces. It contributes little to global energy consumption (Figure 24.1) even though half of the world's wood harvest is directly burned for energy (Chapter 28). New technologies are making wood more useful in energy conservation in other ways, as will be discussed.

Electricity generation began extensively in the late nineteenth century, coming from hydroelectric power and fossil fuels. Except for photovoltaic cells, most energy sources – from fossil fuels, solar heat, hydropower, or nuclear fission – generate electricity by turning a turbine. Electricity concentrates the energy as short-wave radiation that is used in metal processing, information analysis, communication, and lighting. The ability of electricity to be transmitted has enabled its use in mechanical, heating (thermal energy), transportation, and chemical functions. Currently, over 80% of the world's population has access to electricity [1], although some areas are subject to inconsistent supplies – blackouts and brownouts.

Fossil fuels and renewables are both used directly for energy or converted to electricity before use. Generation of electricity from fossil fuels creates large losses of energy; and transmission of electricity from any source creates a much smaller line loss. Globally, about 27% of the world's total human energy generation is lost to heat by burning fossil fuels to generate electricity (Figure 24.3). A similar proportion of carbon dioxide is emitted to the atmosphere.

24.1.4 Future Energy Concerns

Fossil fuels took hundreds of millions of years to develop (Figures 7.3–4). Conditions suitable for their formation may no longer exist. They are treated like minerals, in which the geologic system that produced them is exploited rather than preserved or modified (Chapter 27). Unlike elemental minerals, fossil fuels cannot be sustained by recycling in an anthropogenic system. When or before they are depleted, other energy sources and conservation will need to replace them if people's living conditions and the environment are to be sustained or improved.

Unused fossil fuels (and other minerals, Chapter 26) are measured as "reserves" and "resources." Reserves are sources that are technically, economically, and legally feasible to extract at present. Resources are potentially valuable deposits, whether or not they can presently be extracted. Recent estimates of the world's reserves are 34,000 quads: 9,500 quads of oil, 6,800 quads of gas, and 17,700 quads of coal [6]. This is an increase in oil reserve estimates by about 2,000 quads and a reduction in coal reserves by 10,000 quads [6]. Fossil fuel use and projection estimates vary between authorities and with time; but the world consumed about 434 quads of fossil fuel in 2010 and is projected to consume nearly 500 quads in 2020 and 565 quads in 2030 [6]. If the world fossil fuel consumption stabilized in 2030 and new fossil fuel reserves are not found or developed, the world would deplete its current reserves in 2075; other sources suggest depletion will occur about 2200. The reserves may increase or decline for technological, economic, environmental, and/or political reasons [6]. It is highly likely that fossil fuels' costs will increase as they are extracted from increasingly less accessible sources. It may be appropriate to restrict them to uses where few substitutes exist and to plan for future generations to obtain most energy from renewable sources.

Fossil fuel energy is also being criticized for its effects on global warming (Chapter 8) as well as other values (Chapter 25). Consequently, people are expanding renewable energy technologies such as hydroelectric and nuclear fission energies that have been used for decades; developing new technologies such as wind, solar, and tidal energies; and examining others such as nuclear fusion.

The energy and atmospheric pollution – CO_2, SO_2, and N_2O – produced per unit of energy varies widely with fossil fuels and wood (Table 8.2). Natural gas has sometimes been referred to as a "clean fuel" because it has fewer pollutants, and coals are referred to as "dirty" because they have more. These terms are relative since all produce atmospheric carbon dioxide when burned.

24.2 Future Energy Infrastructures

The past 10 thousand years have had a relatively benign environment (Chapter 9) with relatively stable sea levels, precipitation patterns, and temperature cycles. Consequently, large, stable infrastructures could be built and maintained – harbors, irrigation systems, pipelines, dams, transmission lines, and others for centralized development and use of resources. Future climates will probably be more dynamic; consequently, future energy and other resources may need more flexible infrastructures that can respond to changing conditions. This flexibility can be achieved with more small-scale technologies – such as handheld communication tools that, together with satellites and balloons, replace telephone lines, newspapers, and radios.

Most energy production facilities have been concentrated in small areas (Figure 24.4) [17]. However, small-scale high technologies such as photovoltaic (PV) cells are making less concentrated facilities available, similar to the "soft path solutions" in new water innovations (Chapter 19) [18]. Other opportunities are also emerging for using more diffuse energy through wood construction and biofuels.

Some concentrations of energy in cities and industrial areas will still be needed. New technologies such as Concentrated Solar Power (CSP) may replace fossil fuel energy there.

24.3 Emerging Renewable Technologies

The rest of this chapter first describes some promising renewable energy sources. Then, it discusses ways to store and transport energy, followed by ways to conserve it. Finally, it describes the primary sectors that use energy – industry, residences, commercial establishments, and transportation. Chapter 25 describes how renewable energy and conservation could be used in appropriate sectors to curtail fossil fuel use.

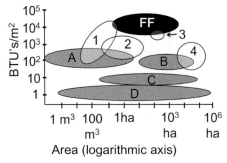

Figure 24.4 Area occupied and intensity of energy production and consumption methods [17]. Black with "FF" = fossil fuel energy production [17]. Gray = other energy production: A = solar PV; B = solar towers (CSP), tidal, hydropower; C = wind, D = photosynthesis. Clear ovals = energy consumption: 1 = residential and commercial buildings; 2 = industry; 3 = steel mills and refineries; 4 = cities.

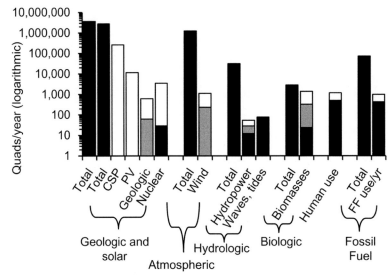

Figure 24.5 Current, short-, and long-term opportunities for renewable energy from various sources compared with theoretical total energy in each category [19, 20]. Note logarithmic scale (Black = total current and current; gray = short-term potential; white = long-term potential).

Each energy source has characteristics of production, transport, and storage that makes it fit for certain uses. Innovations in renewable energy generation are developing quite rapidly with many creative techniques (Figure 24.5) [17, 19]. The amount that each technology is being used is also changing rapidly as breakthroughs occur. The various techniques use energy from different pools (Figure 23.3). If only part of the energy were tapped from each pool (Figure 24.5), together they would generate more renewable energy than the world is projected to need for 2060. Consequently, these projections are optimistic, but certainly attainable (Chapter 25).

24.3.1 Nuclear Energy

Nuclear energy is discussed here although its classification as "renewable" is controversial. The explicit study of energy enabled the understanding and application of nuclear energy, primarily used in energy generation and medical applications but also as a weapon threat. The early promise after World War II of inexpensive, safe, small-scale nuclear energy generators did not materialize. Nevertheless, 6% of the world's power comes from nuclear energy. France obtained 75% of its power from nuclear reactors and was considered to be the exemplar of safe, efficient energy [21]. It is now reducing its dependence on these reactors for reasons of cost, safety, and technical expertise [22].

Worldwide, nuclear reactors have had many accidents [23, 24]; disposals of their wastes have created issues (Figure 17.3) [25]; and there are no international "rapid response" policies for accidents when radioactive pollutants go beyond national boundaries [26, 27]. The accidents and issues are primarily caused by people's difficulty in designing, constructing, and managing highly complicated systems with little allowance for error. The accidents and disposal issues have rendered large areas dangerous, unsustainable, and sometimes uninhabitable for many centuries [26]. On the other hand, the actual and potential deaths are fewer than caused by traffic accidents [27]; but there is still uncertainty in the long-term harm of increased background radiation. A further problem is that nuclear reactors are now being constructed in countries with less technical expertise and discipline than France and Japan, where the Fukushima nuclear accident occurred [28].

24.3.2 Hydroelectric Power

A turbine powered by flowing water first generated electricity in about 1880. Hydroelectric power can be generated from dams and from "run of the river" turbines, the latter being more sensitive to daily and seasonal stream flows.

Hydropower currently represents 16% of the world's electricity production [29] and 2.5% of total energy production. The largest producers of hydroelectricity are China, Brazil, Canada, United States, Russia, Norway, and India in descending order [30]. About two-thirds of the economically feasible hydroelectric power has not yet been developed, with untapped opportunities in Latin America, Central Africa, India, and China [30]. New technologies can be retrofitted into dams and make hydroelectric power more efficient while mitigating adverse effects to fish migration.

Hydroelectric power has long been criticized for high investment costs, effects on wildlife and fish habitats and migration patterns, and displacements of local people (Chapters 18–19) [30].

24.3.3 Biomass

Liquid biofuels – ethanol and biodiesel made from organic matter – come from three sources: woody (cellulosic) biomass from trees, shrubs, grasses, and other plants;

Table 24.1 *Energy obtained from bioethanol made from different residues obtained from management and harvest of all forests in the world [39, 40]*

Biofuel feedstock	Volume of bio-fuels (liters/ha)	Growth (quads/ [million ha years])	Forest land treated (million ha)	Energy produced (quads)
Lumber residues	335	0.007	3,445	29
Forest thinning	475	0.010	3,445	36
Mill sawdust	294	0.006	3,445	22

Table 24.2 *Percentage of world's forest land displaced to produce fifty quads of energy from various biofuel feedstocks [35, 39–42]. "B" = biodiesel; others = bioethanol*

Biofuel feedstock	Volume of biofuels (liters/ha)	Growth (quads/ [million ha years])	Land area needed (million ha)	Percentage of world forest area needed	Energy produced (quads)
Oil palm (B)	4,736	0.103	484	14%	50
Sugar cane	5,476	0.119	419	12%	50
Sugar beet	5,060	0.110	453	13%	50
Cassava	2,070	0.045	1,108	32%	50
Maize	1,995	0.044	1,149	33%	50
Rice	1,806	0.039	1,269	37%	50
Cassava	1,480	0.032	1,549	45%	50
Wheat	952	0.021	2,408	70%	50
Soybean (B)	552	0.012	3,445	100%	41
Sorghum	494	0.011	5,105	148%	55

agriculture crops grown specifically for biofuels; and waste from agriculture and elsewhere that can be refined. Cellulosic ethanol production from wood scrap appears promising but needs technical improvements [31]; however, other cellulose-derived fuels appear promising for jet fuel [32–34].

Efficient wood stoves such as pellet stoves are proving useful for generating heat more efficiently. Liquid biofuels have a unique niche in replacing fossil fuels in airplanes where lightweight, concentrated forms of energy are needed [32, 33]. World liquid biofuel energy production is projected to be 50 quads per year by 2020 [35, 36] – 10% of the world's total energy consumed in 2010.

Table 24.1 shows that thinnings (Chapter 29), wood residues, or mill sawdust from managing all of the world's forests could provide between 22 and 36 quads per year.

Table 24.2 shows that, depending on the crop used, between 14% and all of the world's forests would need to be cleared to provide agriculture land to grow the targeted 50 quads of biofuels. In 2012, the United States produced 49% of the world's biofuel energy, 1.9 quads; and Brazil produced 24%, 0.9 quads [37]. No other country produced more than 3.6%.

Devoting agriculture land to biofuels threatens biodiversity if more forests or other lands are cleared or maintained in agriculture (Chapters 9, 16). The United States is especially wasteful of agriculture land and irrigation water where it grows relatively inefficient corn (maize) for biofuel (Table 24.2). Biofuels from agriculture may have a niche by using the surplus grain produced during productive years (Figure 22.4).

Biofuels from waste may also be promising, but some may be better used to feed pigs, goats, and other animals to reduce the cropland needed for animal food (Chapter 22). The biomass energy contribution from food waste and residues is estimated at 15 to 28 quads, but can possibly be increased to between 55 and 115 quads if not used to feed animals [38, 39].

24.3.4 Wind Energy

Wind turbines for generating electricity are becoming popular despite some objections over their aesthetics. Bird deaths caused by collisions with wind turbines [43] create a trade-off in their use that needs to be addressed. Wind energy is effective where the wind blows consistently, although not necessarily constantly.

24.3.5 Solar Energy

The sun's radiant energy can be used directly in many ways:

1. Passive domestic solar hot water heaters.
2. Solar space heating using double wall panels that heat air between them before being circulated into rooms ("envelope houses").
3. Photovoltaic (PV) cells that can be laid on flat surfaces and generate electricity without a turbine. The electricity can be used locally and immediately, saved in a battery, or fed to a grid.
4. Concentrated solar power (CSP) that can use mirrors to concentrate sun rays and heat a liquid that generates electricity through a steam turbine.

Except for CSP, these systems are only effective when the sun shines unless there are energy storage systems. CSP uses a salt brine to retain heat at night and generate steam-powered electricity constantly.

24.4 Energy Conversion, Transport, and Storage

The world's geographic regions vary in their potentials to provide energy from different renewable sources (Figure 24.6). Consequently, many technologies can be used. There will be needs to convert, transport, and store energy. Figure 24.6 shows CSP to be most promising for large-scale energy based on recent studies [17, 19]. Subsequent advances in battery storage are enabling PV solar energy to extend to places previously assumed best for CSP. Many such technologies use relatively rare elements whose availability may limit

(a) Concentrated solar power (CSP)

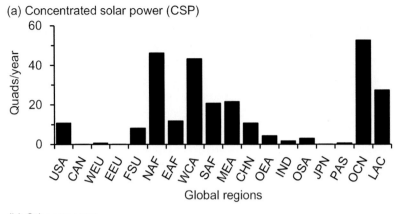

Global regions

(b) Other energy

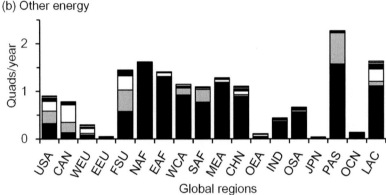

Global regions

Figure 24.6 Potentially available (a) CSP and (b) other renewable energy sources by global regions [17, 19].[A] (b) From bottom: gray = concentrated PV cells; black = nuclear power; white = offshore and land wind; upper gray = geothermal; black = decentralized PV cells; top white = other energy (crops, hydropower, crop residues, animal waste, wood waste).
[A] USA = United States, CAN = Canada, WEU = Western Europe incl. Turkey, EEU = Central and Eastern Europe, FSU = Former Soviet Union, NAF = Northern Africa, EAF = Eastern Africa, WCA = Western and Central Africa, SAF = Southern Africa, MEE = Middle East, CHN = China, OEA = other East Asia, IND = India, OSA = other South Asia, JPN = Japan, PAS = other Pacific Asia, OCN = Australia, New Zealand, and other Oceania, LAC = Latin America and Caribbean.

how widely a technology is used – such as the need for platinum in fuel cells (Chapter 27). With advanced knowledge and planning, a diversity of technologies, research into substitutes, and recycling can help overcome such shortages.

24.4.1 Conversion

Energy is produced in different concentrations and generates heat at different intensities (Table 8.2). Airplanes need concentrated energy from fossil fuels or biofuels, discussed earlier. Many space and water heating facilities are already powered by direct solar radiation, using insulated glass panels that transmit solar radiation into a room or chamber but do not transmit heat radiation out. Ground transportation vehicles can use nonfossil fuel energy sources to convert energy to electricity – and powering the cars, trains, and buses through tracks, batteries, or overhead cables. Electric automobile stations, similar to current gas stations, enable electric car batteries to be recharged – or even exchanged. Turbine electricity plants can be run by Concentrated Solar Power (CSP). When retrofitted to cogenerate steam as well as electricity ("Combined Heat and Power," CHP), the efficiency of electrical plants can be doubled. The cooling can also be used to desalinate water, as will be discussed.

24.4.2 Transport

Metal industries and other intensive energy users are commonly located close to energy sources to save transportation costs. Energy is also transported both to areas of existing industrial infrastructures and to further conversion and distribution places where it is again transported to end users. The transportation network can be quite complicated – a mixture of rails, ships, barges, trucks, and pipelines (Figure 4.4).

Energy transportation leads to energy losses. Transportation costs vary by carriers (Figures 24.11–12); distances transported; infrastructure costs of rails, highways, pipelines, docks, and other facilities; probability/costs of accidents; and various taxes. Transport of energy also has dangers of accidents and sabotage.

Electricity generated by fossil fuels is widely used to transport energy. Generating and transmitting it long distances, distributing it to many small-volume users, and tapping into it illegally cause efficiency losses. Isolated facilities may more efficiently use alternative and/or local energy to generate electricity, such as wind, PV solar cells, or wind-powered generators.

24.4.3 Storage

Many sources do not generate energy uniformly throughout the day and year. Users' energy consumptions often vary also. Solar heat or electricity generators, for example, function less on cloudy days and at night. Residences and commercial areas use less energy at night, and commuters have peak hours of transportation energy consumption in mornings and evenings.

Less energy is wasted in electricity generation and direct fuel use if motors are run and fuels are burned at a constant rate, which occurs in systems where excess energy generated during times of low consumption ("off-peak") is stored and then used during high energy consumption ("peak") periods. Fossil fuels store energy until needed. Hydroelectric "pump storage" dams pump water uphill to reservoirs using excess energy of off-peak hours and release it to flow downhill and generate electricity as needed. Insulation and thermal/hygrothermal mass systems, discussed below, help homes retain the heat (or cold in air-conditioned houses) and so help store energy. Batteries that are more efficient, longer lasting, and cheaper and that can take more discharging and recharging are also being developed. Hydrogen fuel cells that store energy and convert it into electricity are being developed.

The issue of irregular demand for energy is also being overcome by coordination so that some energy-intensive activities are done during off-peak hours. Computer controls that coordinate activities in factories and homes are also helping reduce energy usage fluctuations as well as amounts.

24.5 Avoiding Energy Use

Equally important as changing the sources of energy used is reducing energy use, especially if it does not impose on people's lifestyles. Reducing energy use in a non-imposing way can take a variety of forms in addition to not using fossil fuels to generate electricity, discussed earlier.

24.5.1 Lifestyles

Single-family, suburban home lifestyles common in the United States in the twentieth century are very energy-intensive (Chapter 4) [44].

An increasingly common alternative is living in mid-rise apartment buildings (Figure 4.7), where shops, work, schools, and commuter systems are closer – often within walking distance – because of the higher concentration of people. Less time is consumed commuting, driving to shops, and maintaining the residence and its grounds; one's time is spent in other ways – a lifestyle decision (Chapter 4).

24.5.2 Solid Wood Use and Lifecycle Analyses

Innovative "mass timber" designs are allowing mid-rise buildings, bridges, soccer stadium roofs, and many other structures to be constructed from wood. Solid, cross laminated timbers (CLT; "mass timber") replace steel-reinforced concrete slabs in floors, walls, ceilings, posts, beams, and stairs (Figure 24.7). The wood construction is lighter, faster, and as durable. It is more energy-efficient (Figure 24.8) because steel, concrete, and brick require much more fossil fuel energy to produce, haul, and assemble than wood [45]. "Mass timber" designs using CLT can be especially energy-saving in mid-rise buildings (10 to 20 stories). Such mass timber construction in buildings is more fireproof and earthquake resistant than traditional, reinforced concrete and steel buildings.

Lifecycle energy consumption in buildings can be divided into "cradle-to-gate" and "operational" components. Heating and cooling – major operational components – depend partly on the insulation and partly on the modification of day/night temperature fluctuations, with high fluctuations requiring more energy. These fluctuations can be moderated by insulation and by either thermal mass and hygrothermal mass [46]. Thermal mass uses dense concrete, stone, or a similar product to absorb and release heat. Hygrothermal mass uses uncovered wood to absorb and release water and so uses changes in water's states

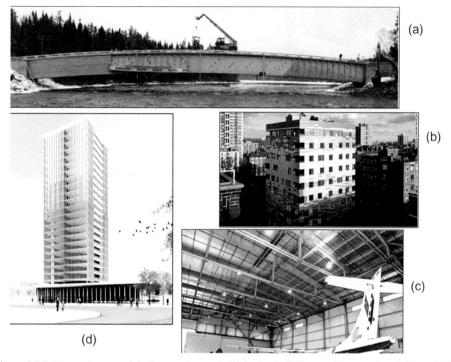

Figure 24.7 Innovative wood designs and uses in high-rise buildings and heavy-duty bridges [45]. Reproduced (open access) from Taylor and Francis Group Ltd.

between vapor in the air to liquid in the wood to release and absorb heat energy (Chapter 7) [46, 47]. Uncovered wood similarly moderates humidity. As a general rule, mid-rise, multifamily buildings made from mass timber are much more efficient in saving energy in construction (cradle-to-gate) and use (operational) than single-family houses [44].

24.5.3 Insulation

Historically, more energy has been used during operational components of heating and cooling a building during its life time than during constructing it; however, the proportions of energy used vary with climate, use, and building materials. About 50% of the energy consumed by residential and commercial buildings is for heating and cooling [48]. Technologies already exist for avoiding most of this energy with better insulation. With better insulation and moderation of daily temperature fluctuations, the initial construction energy (cradle-to-gate) becomes more important in the building's total energy budget [44]. Case-specific analyses are needed to determine if an old building should be retrofitted or replaced to improve its energy efficiency [49].

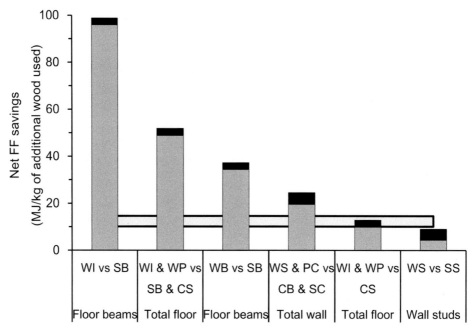

Figure 24.8 Fossil fuel (FF) savings from targeted construction components made of wood compared to steel or concrete and wood energy range [horizontal bar] [45]. See also Figure 28.6. Black = FF energy displaced by wood energy; gray = FF avoided; horizontal bar = FF avoided if wood is directly burned. WI = engineered wood I-beam; SB = steel beam; WP = plywood; CS = concrete slab; WB = solid wood beam; WS = wood studs; PC = interior and exterior plywood sheathing; CB = concrete block; SC = stucco; SS = steel stud. Reproduced (open access) from Taylor and Francis Group Ltd.

24.6 World's Energy Uses

The world's current (2010) energy use can be divided into four sectors (Figure 24.9): industrial, commercial, residential, and transportation.

24.6.1 Industrial Energy

Figure 24.10 shows the primary uses of industrial energy. Unlike other sectors, industrial products are shipped globally. Consequently, industrial energy consumption cannot easily be assigned to specific countries. A high production of building products is expected at least through 2050 as the world triples its infrastructure to meet the maturing population (Figure 5.12) [50]. Iron and steel, cement, and similar products have traditionally been used for such construction, but are very energy intensive to produce compared to wood construction (Figure 24.8).

Industrial energy is also consumed in extracting and processing coal and petroleum (Figure 24.10), adding an additional energy impact to fossil fuels use. Petrochemicals are another large consumer of energy for production of important fertilizers, organic chemicals,

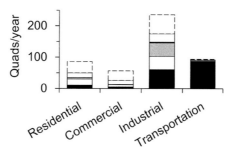

Figure 24.9 World energy use by sector and fuel type [16]. From bottom: black = liquids; white = natural gas; gray = coal; upper black = renewables; upper white = electricity; top white with dashed border = electric energy loss.

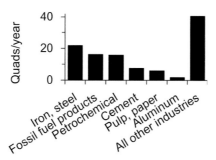

Figure 24.10 Global industrial energy uses [36, 48].

and plastic plumbing and irrigation materials as well as frivolous plastic shopping bags and toy wrappers. Many chemical and plastic products or their substitutes can be made without fossil fuels (Chapter 28). Paper is also a large consumer of fossil fuel energy, even though it generates about one-half of its energy needs from wood waste.

Industries have commonly located around inexpensive sources of energy or raw materials – coal deposits, hydroelectric facilities, or iron deposits. Concentrated Solar Power (CSP) facilities (Figure 24.6) in deserts [19, 51] could make them new industrial centers, as has been discussed (Chapter 18).

24.6.2 Commercial and Residential Energy

Commercial and residential energy per capita is consumed largely in developed countries and especially in those with abundant fossil fuel reserves (Figure 24.1). As other countries develop, their energy consumption will also rise. Newly developed countries may have newer, energy efficient technological installations. Most developing countries are in warm climates, so less energy will be needed for space heating, although some savings may be offset by space cooling needs.

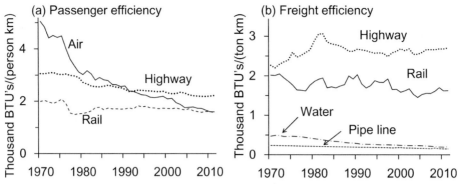

Figure 24.11 Changes in energy intensity (inverse of energy efficiency) of (a) USA passenger and (b) USA freight transportation [52].

Energy used residentially and commercially often comes from fossil fuel-generated electricity and is inefficient. Space heating consumes about 50% of the household and commercial energy, appliances about 20%, and water heating about 15% in developed countries [48]. Both space heating and cooling energy can be saved by innovative home construction including better insulation; and both electricity and fossil fuels can be saved by passive solar water heaters.

24.6.3 Transportation Energy

Locations of major transportation routes are in Figure 4.4. Linking train, highway, and/or pipeline routes across Africa, South America, and central Asia would dramatically change the energy costs of moving people and freight throughout the world.

In the United States, 2.5 times as much energy was expended directly on transporting people than transporting freight in 2011, compared to 3.1 times as much in 1970 [52]. Data for indirect expenditures on road, railroad, airport, or canal construction and maintenance were not available.

Transporting people is generally becoming more fuel-efficient, based on US data (Figure 24.11) [52]. Automobiles are becoming more energy-efficient and are changing from fossil fuels to electric energy. If the electricity is produced renewably, the change will reduce nonrenewable fossil fuel consumption. Commercial airplanes have become much more energy efficient and now rival rail and highway travel in efficiency, especially over long distances so that high takeoff energy is balanced by low inflight energy.

Efficiencies vary within each transportation category (Figure 24.12). Private airplanes that carry few people are very fuel-consuming per person, while many-passenger commercial airplanes are not. Light rail and transit buses are quite inefficient, probably because they carry few passengers during off-peak hours and because they make many stops and starts that take energy. Intercity and school buses are very fuel efficient, rail systems are intermediate, and personal vehicles are quite inefficient.

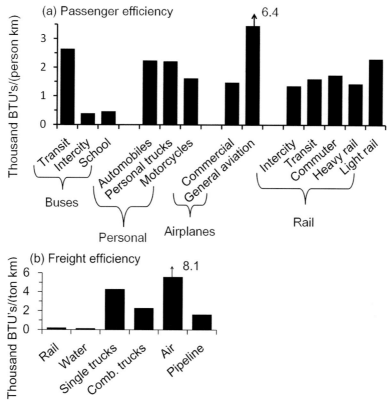

Figure 24.12 Energy intensities (inverse of efficiencies) within transportation categories for (a) passenger and (b) freight transportation [52, 53].

Efficiency improvements in freight transport have had less impact than in passenger transport. Freight efficiency has changed relatively little in the United States in the past 40 years (Figure 24.11b). Shipping freight by rail and water are extremely energy-efficient, while air and trucks are inefficient. Trailer trucks ("combination trucks") are much more efficient than single unit trucks (Figure 24.12b). Pipelines are more efficient than trucks for moving oil and gas.

The United States is the largest energy consumer for transportation because it is large and populous and has long had access to inexpensive fossil fuels. Its many citizens own many large cars and trucks and spend much time driving long distances for work or leisure. They often live in single-family houses, driving to work and shopping. The poor bus and rail system is contributing to both this commuter traffic and to shipping by trucks on the same highways, further adding to congestion as well as to the danger of driving small cars.

Recent innovations are expected to reverse the fossil fuel intensity of automobiles but keep the flexibility they provide. Innovations in electric automobiles and batteries may remove the automobile dependence on fossil fuels if the batteries are charged and recharged by renewable energy. Innovations in driverless cars and programmed travel may allow cars traveling to similar destinations to link up as "trains" and thus save energy. At the same time, mass transportation is expanding and becoming more popular as traffic congestion increases and new generations enjoy less solitary conditions – or watching their cellular telephones instead of the road.

Air transportation is so much a part of current lifestyles that it probably will not disappear, although the glamor of being a member of the "jet set" has disappeared.

References

1. The World Bank. *World Development Indicators*. (The World Bank, 2016 [Accessed May 8, 2016]). Available from: http://data.worldbank.org/indicator/EG.USE.PCAP.KG.OE.
2. US Central Intelligence Agency. *Country Comparison*. (US Central Intelligence Agency, 2016 [Accessed June 24, 2016]). Available from: www.cia.gov/library/publications/the-world-factbook/index.html.
3. US Energy Information Administration. *International Energy Statistics*. (US Department of Energy, 2013–2015 [Accessed July 18, 2016]). Available from: www.eia.gov/cfapps/ipdbpro ject/IEDIndex3.cfm?tid=5&pid=54&aid=4.
4. UN FAO. *FLUDE: The Forest Land Use Data Explorer*. (United Nations Food and Agriculture Organization, 2015 [Accessed May 25, 2016]). Available from: www.fao.org/forest-resources-assessment/explore-data/en/.
5. UN FAO. *Global Forest Resources Assessment 2015*. (United Nations Food and Agriculture Organization, 2015).
6. British Petroleum. *BP Statistical Reviews of World Energy*, June 2013–2016 (Accessed August 30, 2017). Available from: www.bp.com/statisticalreview.
7. J. Perlin. *A Forest Journey: The Story of Wood and Civilization*. (The Countryman Press, 2005).
8. A. Mather. The Transition from Deforestation to Reforestation in Europe. *Agricultural Technologies and Tropical Deforestation*. 2001:35–52.
9. F. Schmithüsen. Three Hundred Years of Applied Sustainability in Forestry. *Unasylva*. 2013;64 (1):3–11.
10. S. J. V. Geology. *How the Oil Industry Saved the Whales*. (San Joaquin Valley Geology, 2015 [Accessed November 26, 2016]). Available from: www.sjvgeology.org/history/whales.html.
11. US Census Bureau. *Historical Estimates of the World Population*. (US Dept. of Commerce, 2016 [Accessed July 1, 2016]). Available from: www.census.gov/population/international/data/world pop/table_history.php.
12. U. C. Bureau. *Total Midyear Population of World: 1950–2050*. (US Dept. of Commerce, 2016 [Accessed July1, 2016]). Available from: www.census.gov/population/international/data/world pop/table_population.php.
13. V. Smil. *Energy Transitions: History, Requirements, Prospects*. (ABC-CLIO, 2010).
14. J. R. Craig, D. J. Vaughan, B. J. Skinner. *Resources of the Earth: Origin, Use, and Environmental Impacts*, fourth edition. (Prentice Hall, 2011).
15. IPCC. *Climate Change 2014: Synthesis Report*. (Intergovernmental Panel on Climate Change, 2014).
16. US Energy Information Administration. *International Energy Annual 2006*. (USA EIA. 2006).
17. A. Grubler, X. Bai, T. Buettner, et al. in Chapter 18 – Urban Energy Systems. *Global Energy Assessment – Toward a Sustainable Future*. (Cambridge University Press, 2012): pp. 1307–400.
18. P. H. Gleick. Global Freshwater Resources: Soft-Path Solutions for the 21st Century. *Science*. 2003;302(5650):1524–8.

19. H. H. Rogner, R. F. Aguilera, C. L. Archer, et al. in Energy Resources and Potentials. *Global Energy Assessment – Toward a Sustainable Future.* (Cambridge University Press, 2012): pp. 425–512.

20. H. Odum, M. Brown, S. Williams. *Handbook of Energy Evaluations Folios 1–4.* (Center for Environmental Policy, University of Florida, 2000).

21. A. Grubler. The Costs of the French Nuclear Scale-Up: A Case of Negative Learning by Doing. *Energy Policy.* 2010;38(9):5174–88.

22. U. Irfan. France Loses Enthusiasm for Nuclear Power. *Scientific American.* 2015.

23. E. J. Bromet. Lessons Learned from Radiation Disasters. *World Psychiatry.* 2011;10(2):83–4.

24. S. Raju. Estimating the Frequency of Nuclear Accidents. *Science and Global Security.* 2016;24 (1):37–62.

25. P. Slovic, J. H. Flynn, M. Layman. *Perceived Risk, Trust, and the Politics of Nuclear Waste.* (National Emergency Training Center, 1991).

26. S. Zibtsev, C. D. Oliver, J. G. Goldammer, et al. Wildlife Risk Reduction from Forests Contaminated by Radionuclides: A Case Study of the Chernobyl Nuclear Power Plant Exclusion Zone. *The International Wildland Fire Conference*; May 9–13, 2011.

27. A. Hohl, A. Niccolai, C. D. Oliver, et al. The Human Health Effects of Radioactive Smoke from a Catastrophic Wildfire in the Chernobyl Exclusion Zone: A Worst Case Scenario. *Earth Bioresources and Quality of Life.* 2011;1.

28. M. Holt, R. J. Campbell, M. B. Nikitin. *Fukushima Nuclear Disaster.* (Congressional Research Service, 2012).

29. International Energy Agency. *IEA Report Sets Course for Doubling Hydroelectricity Output by 2050.* (OECD, 2016 [Accessed May 1, 2016]). Available from: www.iea.org/topics/renewables/ subtopics/hydropower.

30. H. Perlman. *Hydroelectric Power Water Use.* (U.S. Geological Survey, U.S. Department of the Interior, 2016 [Accessed May 1, 2016]). Available from: http://water.usgs.gov/edu/wuhy.html.

31. B. Lippke, M. E. Puettmann, L. Johnson, et al. Carbon Emission Reduction Impacts from Alternative Biofuels. *Forest Products Journal.* 2012;62(4):296.

32. J. R. Regalbuto. Cellulosic Biofuels – Got Gasoline? *Science.* 2009;325(5942):822–4.

33. P. Fairley. Introduction: Next Generation Biofuels. *Nature.* 2011;474(7352):S2–S5.

34. B. Lippke, R. Gustafson, R. Venditti, et al. Sustainable Biofuel Contributions to Carbon Mitigation and Energy Independence. *Forests.* 2011;2(4):861–74.

35. D. Rajagopal, D. Zilberman. *Review of Environmental, Economic and Policy Aspects of Biofuels.* Contract No.: 4341. (The World Bank Development Research Group, Sustainable Rural and Urban Development Team, 2007).

36. US Energy Information Administration. *International Energy Outlook (2004–2009).* (US Department of Energy, 2004–2009, DOE/EIA-0484(2009)).

37. US Energy Information Administration. *Total Biofuels Production.* (U.S. Department of Energy, Energy Information Agency, 2016 [Accessed July 22, 2016]). Available from: www.eia.gov/ cfapps/ipdbproject/iedindex3.cfm?tid=79&pid=79&aid=1&cid=regions&syid=2000 &eyid=2012&unit=TBPD.

38. R. Slade, R. G. Saunders, A. Bauen. *Energy from Biomass: The Size of the Global Resource.* (Imperial College Centre for Energy Policy and Technology UK Energy Research Centre, 2011).

39. J. Daystar, C. Reeb, R. Venditti, R. Gonzalez, M. E. Puettmann. Life-Cycle Assessment of Bioethanol from Pine Residues via Indirect Biomass Gasification to Mixed Alcohols. *Forest Products Journal.* 2012;62(4):314–25.

40. US Department of Energy. *Alternative Fuels Data Center.* (US Department of Energy, 2016 [Accessed April 15, 2016]). Available from: www.afdc.energy.gov/fuels/ethanol_feeedstocks .html.

41. UN FAO. *Biofuels: Prospects, Risks, and Opportunities.* (United Nations Food and Agriculture Organization, 2008).

42. R. L. Naylor, A. J. Liska, M. B. Burke, et al. The Ripple Effect: Biofuels, Food Security, and the Environment. in *Environment: Science and Policy for Sustainable Development.* 2007;49 (9):30–43.

43. W. P. Erickson, G. D. Johnson, M. D. Strickland, et al. *Avian Collisions with Wind Turbines: A Summary of Existing Studies and Comparisons to Other Sources of Avian Collision Mortality in the United States*. (RESOLVE, Inc., 2001).

44. E. Puurunen, A. Organschi. Multiplier Effect: High Performance Construction Assemblies and Urban Density in US Housing. In *Mitigating Climate Change*. (Springer, 2013): pp. 183–206.

45. C. D. Oliver, N. T. Nassar, B. R. Lippke, J. B. McCarter. Carbon, Fossil Fuel, and Biodiversity Mitigation with Wood and Forests. *Journal of Sustainable Forestry*. 2014;33(3):248–75.

46. K. Nore. Hygrothermal Mass: Potential Improvement of Indoor Climate Using Building Materials. In *Wood Forum Building Nordic*. June 16. (Treteknisk–Norwegian institute of wood technology, 2016).

47. I. Katavic, K. Nore, T. Aurlien. Measured Moisture Buffering and Latent Heat Capacities in Clt Test Houses. In *35th AIVC Conference: Ventilation and Airtightness in Transforming the Building Stock to High Performance*. September 24–25. (Energy in Building Communities Programme, 2014).

48. OECD International Energy Agency. *Worldwide Trends in Energy Use and Efficiency: Key Insights from IEA Indicator Analysis*. (OECD/IEA, 2008).

49. A. Saynajoki, J. Heinonen, S. Junnila. A Comparative Greenhouse Gas Analysis of Energy Efficiency Renovation and New Construction in a Residential Area–A Preliminary Study. In *19th Annual Pacific-Rim Real Estate Conference*; January 13–16, 2013.

50. K. C. Seto, B. Güneralp, L. R. Hutyra. Global Forecasts of Urban Expansion to 2030 and Direct Impacts on Biodiversity and Carbon Pools. *Proceedings of the National Academy of Sciences*. 2012;109(40):16083–8.

51. T. M. Pavlović, I. S. Radonjić, D. D. Milosavljević, L. S. Pantić. A Review of Concentrating Solar Power Plants in the World and Their Potential Use in Serbia. *Renewable and Sustainable Energy Reviews*. 2012;16(6):3891–902.

52. US Department of Energy. *Energy Intensity Indicators Data: Transportation Sector*. (US Department of Energy, 2016 [Accessed January 13, 2017]). Available from: https://energy.gov/eere/analysis/downloads/energy-intensity-indicators-data.

53. S. C. Davis, S. W. Diegel, R. G. Boundy. *Transportation Energy Data Book*. Laboratory ORN, editor. (US Dept of Energy, 2008).

25

Future Energy: Reducing Fossil Fuel Use

25.1 Issues Generated by Fossil Fuels

Just as DDT was beneficial by saving people from malaria, fossil fuel has been beneficial by curtailing whale oil use [1] and showing the possibilities of life with abundant energy. Like DDT, fossil fuels' time may be passing (Chapter 8). It may be time to appreciate what fossil fuels have taught about abundant energy but move to less harmful energy sources.

Some fossil fuels may be needed for specialized chemicals and technologies; however, these need only be a small fraction of the current consumption.

Most regions produce fossil fuels, about two-thirds import them, and about one-half export them (Figure 25.1).

The global trade in fossil fuels is much greater than for any agricultural crop (Figure 25.2 compared to Figure 22.5a). Except for the ruling elite, people in some oil-rich countries can be quite poor because the elite do not depend on the people's taxes for wealth and so do not enhance their well-being (Chapter 6). The elite get wealth from oil exports [4].

The "true" cost of fossil fuels, including health and environmental damages, has been estimated at over five trillion US dollars [5]. Two recent wars – the Kuwait and Iraq Wars – would probably not have occurred without people's dependence on fossil fuel energy [6]. The money spent on these two wars could have been spent in more productive ways (Figure 25.3).

Fossil fuels are not evenly distributed and so are subject to political, economic, environmental, and technological challenges in their availability, with places such as Western Europe being dependent on Russia and elsewhere for natural gas (Figure 25.4).

A challenge is to find ways for everyone to achieve a decent well-being but transform away from fossil fuels. This task can be accomplished. There are many indications that the transformation has already begun. The COP meetings (Conferences of the Parties, United Nations Framework Convention on Climate Change) since the Rio Earth Summit of 1992 have been instituting policies that encourage alternatives to fossil fuels [10]. Many people are taking personal interests in avoiding fossil fuels – such as purchasing automobiles that consume less fossil fuel [2] and using more photovoltaic and other renewable energy. New technologies are providing renewable energy to substitute for fossil fuels (Chapter 24). There is a gradual alignment of individual actions; changing technologies; and a few, consistent policies that are making alternatives to fossil fuels increasingly used.

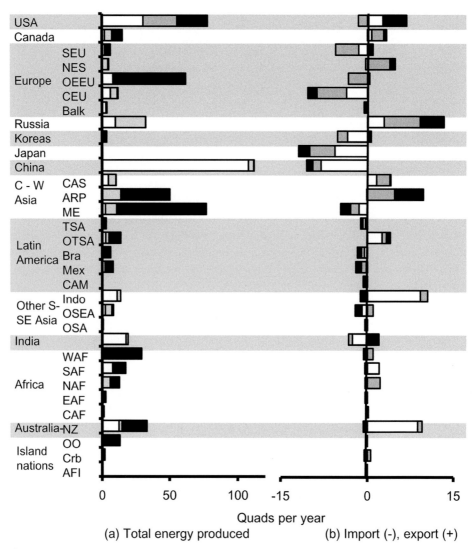

Figure 25.1 Total energy (a) production and (b) import (−) and export (+) by fuel types and country groups [2, 3]. White = coal; gray = gas; black = petroleum. Includes data from BP Statistical Review of World Energy.

25.2 Recent Trends in Energy Sources

Most of the world's current energy is from fossil fuels (Figure 25.5), with minor contributions from the long-tested renewables hydroelectricity and nuclear energy.

Innovative renewable energy technologies and investments in them have been growing at a very rapid rate since about 2006 (Figures 25.6–7, Table 25.1). One reference [12] reports hydroelectricity in 2006 at 33.6 quads, about three times as high as 10.3 quads reported

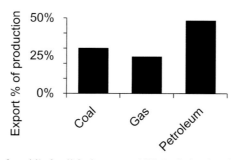

Figure 25.2 Percentage of world's fossil fuels exported [2]. Includes data from BP Statistical Review of World Energy.

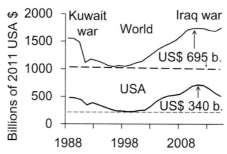

Figure 25.3 US and global military expenditures (2011 US$) in relation to recent Middle Eastern wars [7]. Dashed lines show base, non-war military expenditures. Data used with permission of Stockholm International Peace Research Institute.

elsewhere. (The calculation showing 33.6 quads may include electricity generation energy loss that is avoided.) Renewable energy output is still much less than fossil fuel energy because the renewable base was initially so small. Decline in nuclear energy use actually masks the progress in other renewables. If growth rates of the other renewable energies continue and were combined with other appropriate actions, there could soon be a substantial reduction of fossil fuel use.

25.3 Scenario to Reduce Fossil Fuel Use

The discussion below will suggest how fossil fuels could be eliminated with specific, broad measures. The analysis is simplistic compared to others [14–16]; however, it avoids the confusions of incorporating time delays and costs and simply shows the scales of impact possible. The 2006 energy consumption level [2] will be used with the presumption that all measures could be implemented instantaneously and completely. It also presumes that eliminating energy needs in one sector allows available renewable energy to be shifted to remaining sectors and replace their dependency on fossil fuel.

In 2006 (Figure 25.8a, Table 25.2), the world consumed about 471 quads, with approximately 408 being fossil fuels and 63 being renewables – primarily hydroelectricity and

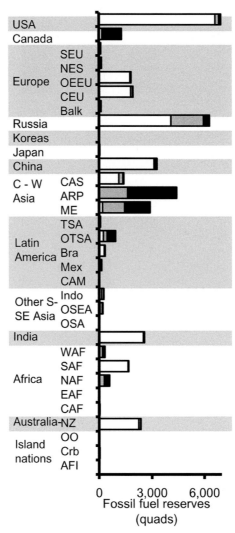

Figure 25.4 Quads of currently known fossil fuel reserves by country groups [8, 9]. White = coal; gray = gas; black = petroleum. Includes data from BP Statistical Review of World Energy.

nuclear energy. Fossil fuels can be eliminated while maintaining and increasing people's lifestyles by avoiding energy waste and so using just half as much total energy – 257 quads. The world is already using 63 quads of existing renewables. Consequently, only 194 quads of additional renewables will be needed. The dramatic increase of renewable energy that will be needed has precedence when global fossil fuel energy production increased by 230 quads between 1950 and 1990 (Figure 24.2a).

The first measure (Figure 25.8b) would be to insulate home and commercial buildings better and to use passive solar hot water heaters. These actions could eliminate 50% of the

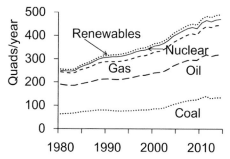

Figure 25.5 Change in world energy consumption from different sources [3, 11]. Creative Commons 3.0 Attributions License.

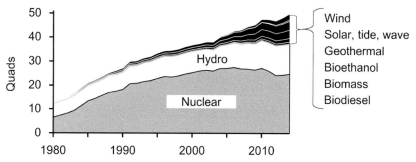

Figure 25.6 Changes in nonfossil fuel energy consumption [3, 11]. Creative Commons 3.0 Attributions License.

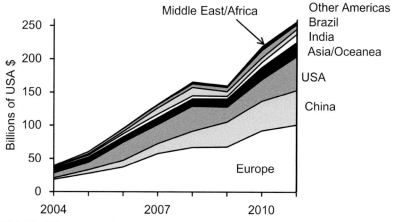

Figure 25.7 World investment in renewable energy by regions, 2004–11 [13].

energy expended in homes and commercial enterprises (Chapter 24) without requiring renewable energy substitutes. If the eliminated energy were from fossil fuels, total world energy use would lower to about 400 quads, with 337 quads being fossil fuel and 63 being existing renewables (Table 25.2).

Table 25.1 *Changes in fossil fuel and renewable fuels uses, 2006–14 [3, 11]*

	2006 (quads)	2006 (% total energy)	2014 (quads)	2014 (% total energy)	Change (quads)	Change (% 2006 energy)
					Change: 2006–13	
Hydro	10.3	2.4%	12.6	2.5%	2	22%
Nuclear	27.5	6.3%	24.8	5.0%	−3	−10%
Biodiesel	0.2	0.0%	0.8	0.2%	1	306%
Biomass waste electricity	0.8	0.2%	1.5	0.3%	1	85%
Fuel ethanol	1.6	0.4%	3.5	0.7%	2	123%
Geothermal	1.8	0.4%	3.8	0.8%	2	114%
Solar/tide/wave electricity	0.0	0.0%	0.6	0.1%	1	3,145%
Wind electricity	0.4	0.1%	2.4	0.5%	2	431%
Coal	115.5	26.5%	134.0	27.1%	19	16%
Gas	105.5	24.2%	126.1	25.5%	21	20%
Oil	171.9	39.5%	184.9	37.4%	13	8%
Total	435.5		495.0		59	

Table 25.2 *Quads of energy and energy change by source with each step in Figure 25.8 (existing renewables include existing nuclear and hydroelectric operating in 2006)*

	Fossil fuels[B] Total	Fossil fuels[B] Change	Current renewables	Additional renewables Total	Additional renewables Change	Total energy Total	Total energy Change
a Baseline	408		63	0	0	471	0
b Insulation/solar water heating	337	−71	63	0	0	400	−71
c Replace fossil fuel electricity	244	−93	63	29	+29	336	−64
d Wood construction	165	−79	63	29	0	257	−79
e Wood residue biofuel	136	−29	63	58	+29	257	0
f Additional renewables	0	−136	63	194	+136	257	0

[B] Includes 60% conversion and line loss when converting to electricity.

A second measure (Figure 25.8c) would be to change fossil fuel electric plants to renewable electricity. This would save three times as much fossil fuel energy as would be needed in substitute renewables, because 60% to 70% of the fossil fuel energy used to generate electricity is lost as heat and other inefficiencies (Figure 24.3). Fossil fuels also

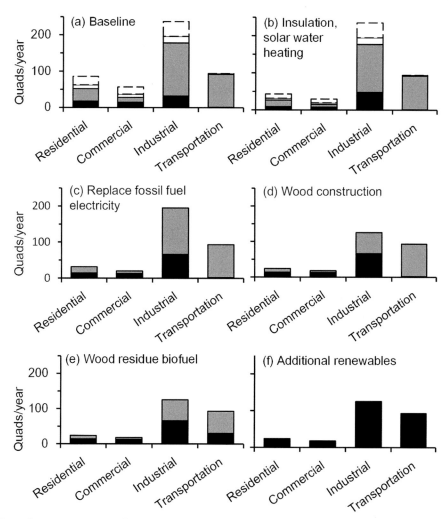

Figure 25.8 Measures to eliminate fossil fuel consumption with little change in lifestyles (also Table 25.2) (black = renewable energy consumption; gray = consumption of fossil fuels not converted to electricity; white = fossil fuels converted to electricity [dashed border = energy lost converting to electricity; solid border = electricity consumed]).

take about 4% additional energy to extract and refine; therefore, whenever fossil fuel energy is avoided – electric or other – 104% of the directly avoided energy is saved. By replacing all fossil fuel electric power plant outputs with new renewable energy yet to be installed, 93 quads of fossil fuel energy would be saved. The result would be a total energy consumption of about 336 quads, with 244 quads being fossil fuel and 29 quads being new renewables – plus the 63 quads of existing renewables (Table 25.2).

Third (Figure 25.8d), 79 quads of fossil fuel could be directly avoided by using innovate wood construction techniques to construct mid-rise buildings, bridges, and other structures that are currently built from steel, concrete, or brick (Figures 24.7–8) [17, 18]. The wood used would equal the excess that is growing in the world's forests every year; and if more wood were needed, intensive plantations could provide it. This additional measure would result in a total energy use of 257 quads – 165 being fossil fuel and 92 being new and existing renewables (Table 25.2).

Fourth (Figure 25.8e), the scrap-wood generated while harvesting and processing this construction wood could be made into bioethanol (Table 24.1), replacing 29 quads of fossil fuels. The additional measure would maintain a cumulative energy use of 257 quads, but only 136 quads would be fossil fuels.

Finally, the various renewable and energy-saving technologies could eliminate the additional 136 quads of fossil fuel energy (Figure 25.8f). Locating industrial areas in deserts and using Concentrated Solar Power (CSP) have been discussed (Chapter 18). Rapidly changing electric automobile innovations may dramatically reduce fossil fuel needs for transportation.

The measures above do not need to be done sequentially. The first and third measures (Figures 25.8b, d) could be combined and implemented without waiting for renewable energy capacity to develop.

Even if not completely implemented, partial implementation of the measures described would greatly contribute to fossil fuel emissions reductions.

25.4 Other Emerging Issues with Fossil Fuels

Social pressures to eliminate fossil fuel use are emerging not only because of its contributions global warming (Chapters 6–8) and wars, but also because of its price instability (Figure 25.9).

Break-even prices of fossil fuels varied between US$ 1.53 and US$ 9.46 per million BTUs (February 2016) [19], indicating that fossil fuel production should have been unprofitable in certain times and locations. Nearly half of the producer costs in Figure 25.9 are for infrastructures. Once an infrastructure's capital cost is paid, its only cost is maintenance. Consequently, fossil fuel will continue to be produced if the non-infrastructure costs (operational expenditures) are below the oil price as long as the infrastructure remains intact. These operational expenditures ranged from US$ 0.86 to US$ 5.53 per million BTUs [19].

From an economic perspective, fossil fuel companies probably make little profits when energy prices are low; but the low prices probably discourage investments in renewable energy. Alternatively, high energy prices financially benefit fossil fuel energy companies but invite competition from renewable energy and conservation. The fluctuating prices create a confusing investment environment and indicate that shifts to renewable energy may need policies that reduce this confusion [20, 21].

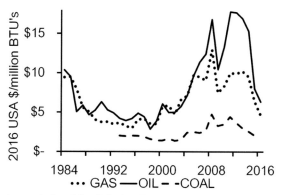

Figure 25.9 Change in non-inflated fossil fuels prices [2]. Includes data from BP Statistical Review of World Energy.

A policy shift to renewable energy could allow fossil fuel investments to be phased out as their infrastructure declines. Pipelines are generally owned by independent, publicly traded companies. An opportunity might exist to convert many oil and gas pipelines to water pipelines (although new pipes may need to be installed within the rights-of-way), similar to the recent water pipeline from Turkey to Cyprus [22]. This conversion would provide alternative markets to the pipeline companies, reduce their dependence on fossil fuel consumption, and enable water to be moved to regions becoming increasingly arid.

25.5 Renewable Energy as Economic Stimuli

Renewable energy and its associated changes in labor needs and lifestyles could be economic stimuli – social drivers that form new and different economic sectors (Chapter 6). It can be a wise business decision as well as a moral commitment [23]. China and South Korea, for example, are investing heavily in environmentally friendly activities as social and economic stimuli [24]. Morocco is heavily investing in solar energy to take advantage of its desert.

Alternative energy, demographics, and other changes can alter places that have advantages in providing resources and other values. The result will be shifts in occupations, cultures, and centers for various activities. The challenge is to attract people to do new, fulfilling things.

Information and communication are becoming more diffuse and decentralized, enabling people to live in new ways that add selective elements of an energy-dependent world. For example, beekeepers and yak herders in Western China retain their transhumance lifestyles in tents and yurts but already have televisions, electric lights, and even computers powered by solar panels and batteries. People in the Himalayas already use small hydroelectric power systems for cooking, heating, charging cell phones, and internet communication; but they walk for many days to reach automotive transportation.

Both pressures to conserve energy and new opportunities to use it cause changes and trade-offs in lifestyles. People travel more frequently by airplane, but dress more informally. A change in lifestyles so that wealth is displayed in more energy-conservative ways could be done by people being highly recognized for donating money to hospitals, schools, universities, museums, or the arts instead of owning luxurious homes or jets.

Much of the type and amount of future energy used will depend on how developing countries build their infrastructures. Developed countries' infrastructures are based on lifestyles that were partly shaped by historical energy/technology opportunities and limitations – cheap gasoline, steel, and concrete; no concern about fossil fuel consumption; uncrowded highways; high resale values of homes; stationary telephones; and bulky (paper) information systems. These have led to high use of automobiles and trucks at the expense of mass transit; single-family houses with long commutes; and access to information only in homes or offices – and hence spending much time there.

Many of these opportunities and limitations may no longer apply; and countries that do not already have an infrastructure in place may develop different lifestyles with energy and technology infrastructures to match. Countries may prefer to develop many smaller cities to avoid people migrating to a few megacities.

References

1. San Joaquin Valley Geology. *How the Oil Industry Saved the Whales*. (San Joaquin Valley Geology, 2015 [Accessed, November 26 2016]). Available from: www.sjvgeology.org/history/whales.html.
2. British Petroleum. *BP Statistical Reviews of World Energy*, June 2013–2016 (Accessed August 30, 2017). Available from: www.bp.com/statisticalreview.
3. US Energy Information Administration. *International Energy Statistics*. (US Department of Energy; 2013–2015 [Accessed July 18, 2016]). Available from: www.eia.gov/cfapps/ipdbproject/IEDIndex3.cfm?tid=5&pid=54&aid=4.
4. P. Bond. *Looting Africa: The Economics of Exploitation*. (Zed Books Ltd, 2006).
5. A. Wernick. *IMF: "True Cost" of Fossil Fuels Is $ 5.3 Trillion a Year*. (Public Radio International, 2015 [Accessed December 13, 2016]). Available from: www.pri.org/stories/2015–06–07/imf-true-cost-fossil-fuels-53-trillion-year.
6. J. R. Craig, D. J. Vaughan, B. J. Skinner. *Resources of the Earth: Origin, Use, and Environmental Impacts*, fourth edition. (Prentice Hall, 2011).
7. SIPRI. *Sipri Military Expenditure Database*. 2015 (Accessed August 12, 2017). Available from: www.sipri.org/research/armaments/milex/milex_database.
8. C. Rühl. *BP Statistical Review of World Energy*. (British Petroleum, 2008).
9. US Energy Information Administration. *International Energy Outlook (2004–2009)*. (US Department of Energy, 2004–2009, DOE/EIA-0484[2009]).
10. B. Dawson, M. Spannagle. *The Complete Guide to Climate Change*. (Routledge, 2008).
11. The Shift Project. *Historical Energy Consumption Statistics*. 2015 (Accessed May 6, 2016). Available from: www.tsp-data-portal.org/Energy-Consumption-Statistics#tspQvChart.
12. USEIA. *International Energy Annual 2006*. (USA Energy Information Administration, 2006).

13. E. Macguire. *Who's Funding the Green Energy Revolution?* (CNN [Cable News Network], 2012 [Accessed August 9, 2017]). Available from: www.cnn.com/2012/06/12/world/renewables-finance-unep/index.html.

14. J. H. Williams, B. Haley, F. Kahrl, et al. *Pathways to Deep Decarbonization in the United States.* Energy and Environmental Economics, Inc. (Institute for Sustainable Development and International Relations, 2014).

15. The White House. *United States Mid-Century Strategy for Deep Decarbonization.* (The White House, 2016).

16. IDDRI (Institute for Sustainable Development and International Relations). *DDPP: Deep Decarbonization Pathways Project.* (Sustainable Development Solutions Network: A Global Initiative for the United Nations), 2016 [Accessed August 30, 2017]). Available from: www .deepdecarbonization.org.

17. C. D. Oliver, N. T. Nassar, B. R. Lippke, et al. Biodiversity Mitigation with Wood and Forests. *Journal of Sustainable Forestry.* 2014;33(3):248–75.

18. B. Lippke, R. Gustafson, R. Venditti, et al. Sustainable Biofuel Contributions to Carbon Mitigation and Energy Independence. *Forests.* 2011;2(4):861–74.

19. R. Patterson. *Oil Production Is Going to Drop and Oil Prices Are Likely to Increase.* (Peak Oil Barrell, 2016 [Accessed April 16, 2016]). Available from: http://peakoilbarrel.com/oil-production-is-going-to-drop/.

20. A. Saxena, V. Sharma, S. Rachuri, H. Joshi. Climate Change and Business Strategies: The Case of Automobile and Associated Ancillary Sectors in Madhya Pradesh. *IUP Journal of Business Strategy.* 2015;12(3):46.

21. K. Gillingham, J. Sweeney. Barriers to Implementing Low-Carbon Technologies. *Climate Change Economics.* 2012;3(04):1250019.

22. E. Gies. Northern Cyprus Sees Hope in Water Pipeline. *The New York Times.* 2013.

23. D. Esty, A. Winston. *Green to Gold: How Smart Companies Use Environmental Strategy to Innovate, Create Value, and Build Competitive Advantage.* (John Wiley & Sons, 2009).

24. E. Von Weizsacker, K. Hargroves, M. Smith, C. Desha, P. Stasinopoulos. *Factor Five: Transforming the Global Economy through 80% Improvements in Resource Productivity.* (Earthscan/Routledge, 2009).

Part IX

Minerals

26

Rocks and Mineral Properties, Mining

26.1 The Earth's Mineral Composition

"Mineral" has several definitions [1]. Chapters 26 and 27 discuss minerals as nonliving materials found within the Earth that are usually inorganic and solid. They can be pure elements, covalent molecules, and ionic compounds (sometimes known as "ionic molecules"). Petroleum, bromine, and mercury are exceptions because they are liquids; coal, petroleum, and methane are organic; and methane, hydrogen, nitrogen, oxygen, fluorine, chlorine, and the noble gases are gaseous. Some minerals can be fabricated, but elements cannot be fabricated.

Geologically, "mineral" generally refers to a solid, nonhuman-fabricated, inorganic arrangement of atoms with a definite chemical composition and a characteristic crystal structure [2, 3]. A "mineraloid" has most mineral properties. "Rocks" in Chapters 26 and 27 are consolidated aggregates of one or more minerals – a slightly different meaning than in Chapter 10. Similar to fossil fuels (Chapter 24), "reserves" in geology are valuable mineral deposits that are technically, economically, and legally feasible to extract at present; whereas "resources" are potentially valuable deposits whether or not they can be extracted at present [4]. Mineral industry terms may also differ from agriculture and geology terms: "clay" in mineral industries can include "silt" in agriculture (Chapter 10).

Rock and mineral uses vary from needed elements in high technologies to bulk materials for heavy construction. Minerals – pure elements, molecules, or mixtures – can be found as parts of satellites, lasers, computers, and jewelry. Rocks gravels, sands, and stones, and cement (made from calcite) can be the basis for roads, bridges, harbors, airport runways, rocket platforms, and buildings.

It is uncertain if there are sufficient amounts of all elements in the Earth's land, air, and water to provide all foreseeable human needs. Some elements may simply not exist in sufficient quantity; others may be inaccessible; and others such as the "rare earth metals" are not rare but are generally found in such low concentrations that they are difficult to obtain.

Like water and energy (Chapters 17 and 23), mineral elements are essentially indestructible and so can be made available by circulation. However, the geologic recirculation of

409

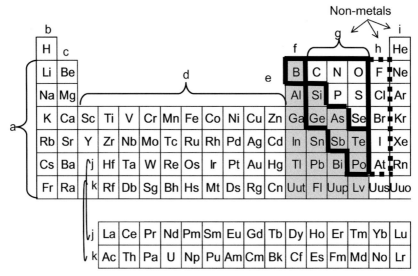

Figure 26.1 Periodic table of the elements, showing element symbols in boxes. a = alkali metals; b = hydrogen; c = alkali earth metals; d = transition metals; metalloids: e = "poor metals" gray not in boxes; f = "metalloids" gray in boxes; g = other nonmetals; h = halogens; i = noble gases; j = lanthanide; k = actinide [5]. Appendix II gives names, atomic numbers, and weights.

elements can take hundreds of millions of years for it to diffuse into the Earth's land, air, or water; and then concentrate – "fractionate" – and finally emerge to an accessible location [4]. And, the sizes of these concentrations can be quite small. Minerals that are valuable as molecular structures, such as vermiculite and mica, can be destroyed and possibly fabricated rather than just circulated.

The Earth contains about 100 elements (Figure 26.1), but relatively few are abundant. The elements are classified as metals and nonmetals depending on their properties and behaviors, although classifications vary. Hydrogen has properties of both metals and nonmetals. Fewer than 20 nonmetals and about 80 metals exist, but the nonmetals greatly outweigh metals in the Earth's crust and atmosphere (Figure 26.2, and Table 8.1). Metals can be divided into seven groups (Figure 26.1): alkali metals (6 elements), alkaline earth (6 elements), transition (29 elements), heavy transition (8 elements), lanthanide (15 elements), actinide (15 elements), and metalloids (18–19 elements). Metalloids are sometimes divided into poor metals (11 elements) and metalloids (7 elements). The nonmetals are divided into halogens (5 elements), noble gases (6 elements), and others (6 elements). Elements with atomic numbers greater than 94 (Appendix II) are highly radioactive and have limited uses.

Except for nonreactive noble gases, nonmetals combine with each other and the approximately 80 different metals and metal-like elements to create approximately 1,500 different groups of minerals, with variations within a group.

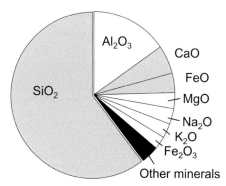

Figure 26.2 Major mineral proportions of the terrestrial Earth's crust [3]. (Varies slightly from [6].)

26.2 Metal and Nonmetal Elements

Both metals and nonmetals vary in properties, behavior, size, weight, and abundance. Many properties and variations can be understood from Figure 26.1 and some generalizations can be made [7–10]. Metals and nonmetals can be valuable in their elemental forms but also are used in mixtures of other metals and nonmetals. Homogenous mixing two or more metals or metals and nonmetals can form an "alloy."

A common property of most pure metals is their lustrousness; they lose their luster if oxidized and generally form basic oxides. Metals are malleable and flexible; strong in compression, tension, and torque; and hard. They are solid at room temperature except for mercury, which is liquid. They are also able to form alloys, conduct electricity, and catalyze.

Metals are often classified by their properties and rarities, with different authorities varying in their classifications [8, 9]. Increasingly, rare metals are used for low-volume, specialized, high-value uses [10].

26.2.1 Precious Metals

Precious metals are rare, naturally occurring, of high market value, and with low corrosiveness and high luster: gold, silver, platinum, palladium, and others of the platinum group (iridium, osmium, rhodium, and ruthenium). Other classifications include any metal as "precious" as long as it is rare. Precious metals are used for industrial, high technology, cosmetic, wealth display, investment, and adornments for religious institutions [8]; however, their rarity means they are not used in bulk amounts. Gold and silver are obtained in relatively small amounts from quite a few places (Figures 27.1, 4).

26.2.2 Base Metals

Metals that are not "precious" are generally "base." Depending on the classification system, base metals are either common, inexpensive metals; ones that corrode or oxidize relatively easily, such as iron, nickel, lead, and zinc; the four nonprecious and nonferrous metals, copper, lead, nickel, and zinc; or a range of metals including iron, aluminum, copper, lead, molybdenum, cobalt, manganese, tin, chromium, zinc, and nickel [8]. These metals are found in many places and are mined where they are concentrated (Chapter 27). They are used in large amounts for their strength and durability – in buildings, engines, vehicle bodies, plows, chains, pipes for conducting fluids, electrical wires, and modern industrial applications.

26.2.3 Noble Metals

Noble metals are different elements from "noble gases" (Figure 26.1). Metals generally considered "noble" are gold, silver, platinum, palladium, osmium, iridium, rhodium, and ruthenium – and sometimes mercury or rhenium. "Noble" and "precious" metals are not completely the same set of metals in some classifications. Noble metals tend to resist corrosion and oxidation in moist air [8]. They tend also to be valuable because of their rarity. Some noble metals are currently obtained from only a few places in the world (Chapter 27).

26.2.4 Rare Earth Elements, Lanthanides

These are the 15 elements of the actinide series plus the transition metals scandium and yttrium (Figure 26.1). They are sometimes divided into three groups [11]:

- Scandium and yttrium are in a group by themselves;
- Light rare earth elements: lanthanium, cerium, praseodymium, neodymium, promethium, and samarium;
- Heavy rare earth elements: europium, gadolinium, terbium, dysprosium, holmium, erbium, thulium, ytterbium, and lutetium.

Rare earth elements are useful in modern metal alloys, reflective and refractive devices (such as lenses and lasers), catalysts, X-ray devices, and for many other highly technical needs [8, 12]. They are relatively ubiquitous but not found in high concentrations and so are difficult to obtain.

Elements of the lanthanide series had once been included as rare earth elements, but generally are not at present. Uranium, of the lanthanide series, is obtained from moderately many places, but primarily from Canada, Australia and New Zealand, Central Asia, and Southern and Western Africa.

26.2.5 Nonmetal Elements

Nonmetals have variable properties but generally the opposite of metals. They are not lustrous or malleable, are good insulators, and most are less dense than most metals. Their melting and boiling points are lower than metals. At room temperature, many nonmetals (hydrogen, nitrogen, oxygen, fluorine, chlorine, and the noble gases) are gases; one (bromine) is a liquid; the remaining are solids. Nonmetals in the gaseous state exist as single atoms or diatomic molecules.

They can form large, complex covalent molecules or molecular radical anions because they tend to share or take electrons rather or give them up as metals do. Consequently, they are found in very many ionic compounds, generally as anions or parts of anions in association with metal cations. Nonmetals form acidic oxides. (A "radical" is a term meaning an anion or cation, either from individual atoms or covalently bonded clusters.)

Including the noble gases, twenty-one nonmetals exist. Noble gas atoms are inert; they do not "react" (form ionic or covalent bonds) with other atoms. Four nonmetal atoms are polyatomic – C, P, S, Se; they can form long chains of covalent molecules or molecular radicals. Six nonmetals and hydrogen are diatomic – H, N, O, F, Cl, Br, and I; they are commonly found as two atoms of the same element covalently bonded. All diatomic nonmetals except bromine and iodine are gaseous molecules at room temperature.

26.3 Mineral Chemical Structures

Minerals sometimes contain impurities and structure variations. Minerals occur in several forms [2, 3, 13], described below.

26.3.1 Amorphous Substances

These vary in their composite elements or their noncrystalline (amorphous) structure. Pure elements such as gold, lead, and silver can be found as rare deposits of amalgamated (but not bonded) atoms; usually they are "transition metals" that do not have an extreme propensity to give up electrons. Obsidian (volcanic glass), amber (petrified resin), and opal contain somewhat varying elements and amorphous structures.

26.3.2 Van der Waals Bonds

A few minerals such as graphite (pure carbon) and pure sulfur contain these weak bonds that allow such things as slippage of graphite layers.

26.3.3 Covalent Molecules

Elements can also occur in nature as covalently bonded molecules. Except for noble gases, many of the other nonmetals as well as many metalloids, poor metals, and transition metals can form covalent crystals. The halogen elements can occur in diatomic molecules of the single element as well as in ionic compounds (halides).

26.3.4 Pure Crystals

Elements such as carbon have such a balance of electron needs/offerings (four of each) that they can form long chains and crystals of covalent bonds of atoms of just this element.

26.3.5 Ionic Compounds

Most metals are found as ionically bonded compounds. Each compound has a metal cation – or element that is behaving as a metal (such as silica in quartz crystals) – and either an individual nonmetal anion or a covalently bonded cluster of atoms behaving as the anion radical.

Mineral ionic compounds are commonly polarized, with a "positive" pole on the cation's side and a negative pole on the anion's side. These enable such minerals to dissolve in polarized solutions such as water, to mix with other polarized minerals, and to form crystals. The charged end of the compound is attracted to oppositely charged ends of other polarized compounds, creating solutions and crystals (Figure 26.3). Solutions are mixtures of mineral compounds that are irregularly attracted to or attached to other compounds of the same or different minerals in a haphazard way; the solutions can be solid or liquid. Crystals are minerals that are attached to the same minerals in a regular fashion that creates a geometric structure. Crystalloids are partial crystals, such as Figure 26.3. Mixed crystals occur if compounds of one mineral begin forming a crystal and either switch to another analog or another mineral and so create a small, intimate mixture of crystals within a deposit. Two compounds can begin forming crystals apart from each other, but the crystals "grow" together, forming various interfaces (Figure 26.3).

Irregularities in crystalloids cause variations that can make them weak, attractive, or colored differently. Some causes of variations described below refer to Figure 26.3 [2, 3]:

1. Cleavage planes: A single crystal type can converge from different directions of growth – either because they begin as separate crystals or because some cation/anion combinations become attached at an irregular angle. The junction of these two growths is known as a "cleavage plane." A cleavage plane can also occur when two different crystal types grow together.
2. Cluster Proxy: A group of atoms, such as a covalently bonded anion radical, can substitute for another group. This cluster substitution can also occur among cations.
3. Omission: Sometimes a cation or anion radical is omitted from the matrix when a crystal or solid solution is forming, resulting in a permanent omission and an imbalance of electron charges. The charge imbalances are sometimes resolved by having omissions of ions of the opposite charge nearby.
4. Interstitial stuffing: An atom, ion, compound, or molecule can also be inserted at an irregular interval within a crystal or mixture, especially if the inserted entity is small relative to the overall matrix. Sometimes these entities are charged and are balanced by omissions or stuffings of oppositely charged entities.

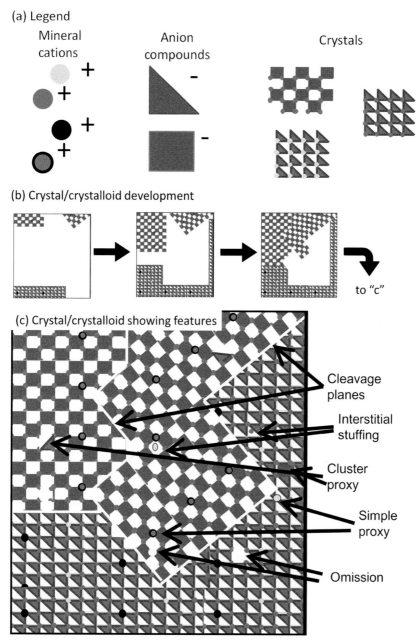

Figure 26.3 Schematic development of a saturated solution into a mixture of crystals, mixed crystals, crystalloids, and solid solutions. (A black and white version of this figure will appear in some formats. For the color version, please refer to the plate section. Color plate 11.)

5. Simple Proxy: A single atom (anion or cation) can substitute for another one in a crystal lattice. This substitution commonly occurs among metal cations, but can also occur among anions.

The different anion radicals can combine with scores of different cations and create over 8,000 currently named mineral species. Many minerals have traditional names – such as "salt" for "sodium chloride." Efforts are underway to reduce confusion and duplication by systematizing mineral naming with suffixes such as "-ide," "-oxygenate," and "-hydrogenate." Crystals and ionic radicals sometimes have polymorphs, different chemicals with the same proportion of atoms configured in different ways. Graphite and diamonds are polymorphs of pure carbon.

Relatively few of the 8,000 minerals comprise most of the Earth's crust. The most common mineral groups are [2, 3]:

- Oxygenates: This most common group contains a metal cation and nonmetal atoms covalently bonded to one or more oxygens as an anion radical. The oxygenates can be further classified by the nonmetal atoms in the radical.
- Silicates: The most common mineral/metalloid, they include feldspars, pyroxenes, quartz, and micas.
- Carbonates: Carbonates include calcium carbonate and dolomite.
- Phosphates: They include apatites.
- Sulfates: Sulfates include gypsum – calcium sulfate.
- Halides: Halogen atoms (F, Cl, Br, I, and At; sometimes H is listed) form individual, nonmetal anions and link with a (usually single) cation to form binary compounds. Salt (NaCl), iron fluoride, hydrogen chloride, and others.
- Sulfides and Oxides: Sulfur and oxygen atoms can form anions similar to halogens and can link individually or in groups to a metal cation to form minerals such as galena (PbS) or magnetite (Fe_3O_4).
- Hydrous Oxides: Hydrogen and oxygen can form an ionic radical that links to metal cations, forming minerals such as iron or aluminum hydroxides.

26.4 Minerals and Elements Within

Valuable elements and compounds are obtained from several forms of rocks and minerals (Figure 26.1). In cases such as salt, marble, and quartz, the molecular or ionic compound structure is valuable and is saved. In other cases, such as aluminum, mercury, silver, uranium, and copper, the molecules or compounds are destroyed to obtain the valued element. Silver and copper can be found in pure, elemental forms; or they can be extracted from compounds. Others can only be extracted from compounds. Bulk rocks are primarily valued for road and harbor construction purposes in their intact conditions.

26.5 Mineral Formation and Concentration Systems

The formation of elements in the universe and their coalescing into the Earth and other planets takes 10- to over 100 million years (Chapter 11) [4, 14, 15].

Most elements and many ionic radicals and crystals can be found in some amounts in nearly all geologic formations, but generally in concentrations too low for extraction. Fortunately, they gradually concentrate – "fractionate" – as minerals in various forms through geologic processes [2, 3, 13, 16] and become available to people after 30- to 3,000 million years, often after an additional period of resurfacing from deep within the Earth [4]. These concentrated deposits are often found in specific landforms. When found, the fractionated minerals usually occur as rocks and solid mixtures with a range of chemical compositions (mineral compounds) and different degrees of crystallization. Relatively rarely, elements are found as pure, amorphous deposits. Variations occur within a solid mineral solution in several ways (Figure 26.3), based on processes occurring during formation and solidification and differences in sizes of composite atoms and resulting ions.

26.5.1 Igneous Processes

Mineral formation can be a considered a cyclic process. Beginning with igneous rock formation (Chapter 10), rocks then weather, decompose, and form sedimentary rocks which then metamorphose and continue weathering. During the process, some rocks remelt and reform igneous rocks. The process varies in time, with very old igneous rocks still remaining from 4.4 billion years ago in Australia [17] and igneous basalt flows currently occurring in Iceland (Chapter 11).

Molten rock (magma) "intrudes" into layers within the Earth's crust or flows on the surface and both cools and contracts, forming igneous rocks (Chapter 10). The magma's cooling and pressure reduction first allow molecules and mineral compounds to form. Which molecules and compounds form depends partly on the initial conditions – which elements are present – and partly on how rapidly the magma declines in pressure and temperature, based on the magma's proximity to the surface. Minerals tend to segregate based on melting points, sizes, shapes, and polarization configurations. The processes occur in a continuum of pressure and temperature in different parts of the new igneous rock. The result is solid solutions and crystals of minerals with varying concentrations and irregularities. Very highly metamorphosed rock can behave similarly. Other events lead to further changes.

The area surrounding an igneous intrusion or beneath a surface flow cools the adjacent magma more rapidly than the inner magma cools. Consequently, minerals with low melting points tend to solidify and concentrate there. Very large crystals known as pegmatites can form [3]. In addition, the heat and pressure of the molten rock can push atoms and minerals into adjacent, older rocks and heat them and change their chemical composition.

Table 26.1 *Examples of valued elements, minerals, and rocks obtained from various forms of intact and decomposed minerals [3]*

Name	Chemical formula	Compound name
Pure amorphous material		
Gold	Au	
Copper	Cu	
Silver	Ag	
Intact crystal		
Diamond	C	
quartz	SiO_2	
Intact ionic compound		
Salt	NaCl	Sodium chloride
Marble	$CaCO_3$	Calcium carbonate
Decomposition of ionic compound		
Iron	Fe_2O_3	Oron oxide
Iron	Fe_3O_4	Magnetite
Copper	CuS	Copper sulfide
Mercury	HgS	Cinnabar
Silver	Ag_2S	Argentite
Aluminum	$Al_2O_3*nH_2O$	Bauxite
Uranium	U_3O_8	Pitchblende
Intact rock		
Granite	mineral mixture	
Pumice	mineral mixture	
Basalt	mineral mixture	

Most minerals contain oxygen, silica, aluminum, hydrogen, and iron as anionic or cationic parts (Table 26.1) – the major components of the Earth's surface (Figure 26.2). Rocks vary in chemical composition, but generally contain relatively common metals such as aluminum, iron, calcium, and magnesium. Atoms of less common metals are incorporated within the crystals and give the crystals color. However, other, "poor metals" such as gold and lead can be concentrated between crystals, along the periphery of a formation, and elsewhere. Cracks form in the cooling and contracting rocks. These cracks – veins, fissures, and seams – fill with solutions of metals, elements, and minerals. These solutions can form crystals and crystalloids such as quartz, diamonds, graphite, or sulfur; amorphous solid solutions or noncrystalline molecules such as gold or lead; or solid solutions. Metals and other elements can also be found in various forms within the original rocks as large or small crystals, noncrystalline molecules, cations within the crystals, amorphous substances, or impurities. Such concentrations within rocks are known as "lodes" [1].

Some areas can have local concentrations of radioactive elements; others, concentrations of poisons such as arsenic; others, seeps of methane; and others, hot water (hot springs) often accompanied by sulfur deposits.

26.5.2 Decomposition of Rocks

Rocks decompose over time with chemical and physical weathering (Chapter 10) [2, 3]. Chemically and physically less resistant minerals such as feldspars dissolve and release cations and anionic radicals to the soil solution. The dissolved material and radicals can either wash away in ground water or remain deeper in the soil (Chapter 17). In either case, they can reform as clays such as illite and montmorillinite or other compounds. More resistant minerals such as silicates and some heavier metals such as gold remain within the soil or are washed short distances downstream and form silts, sands, and gravels of different sizes (Chapter 12), depending on the sizes of crystals and compounds formed within the parent rock. Metal deposits within such gravels are known as "placers" [18]. The high density of metals such as gold causes them to sink when washed by the flowing water or by intentional mining.

Decomposition of rocks can also expose seams of large crystals, crystalloids (Figure 11.3b), and metal particles.

26.5.3 Recomposition of Sedimentary Rocks

The resulting rocks, sand, silt, and clay move with gravity, wind, and water and become redeposited on land, rivers, lakes, or oceans. They commonly form coastal plains that at times are submerged beneath oceans. The particles sort themselves by sizes into layers typical of sediments – and later sedimentary rocks (Chapters 11–12).

Three types of minerals constitute these sedimentary layers:

• Fragmented segments of the resistant silicate and similar materials, often sands or silts;
• Reformed minerals from parent materials, such as clays (as previously described);
• Completely new minerals formed by combining minerals that are added from different sources. For example, carbonates from marl, coral, and other marine life [19] can form compounds with metals, as well as evaporites such as gypsum and halites. Coal also forms layers where organic sediments accumulate in lagoons and similar places and do not decompose. Volcanic ash (tephra) and mudflows can add layers of other compounds.

The resulting sedimentary rocks have the different minerals relatively segregated both into layers and into geographic regions, based on climate, topography, proximity to oceans, and surrounding eroding landscape at the time of formation. Cracks also form in sedimentary layers which create veins and seams that eventually fill with mineral crystals, crystalloids, solid solutions, and noncrystalline molecules similar to seams in igneous formations (Figure 11.3b).

26.5.4 Metamorphosis of Sedimentary and/or Igneous Rocks

Minerals in sedimentary deposits alter their chemical composition because of time, pressure from the overburden, chemical and physical actions of organic soil leachates, and other physical and chemical weathering processes (Figure 8.4).

They still commonly retain their segregation of materials (Figure 11.4), but crystalline structures change and chemical reformations of minerals continue. Quartz sandstones and shales generally become harder as the particles fuse more. Limestone – calcium carbonate – can change to marble.

Veins and seams continue to form as the rocks crack during folding; and veins commonly appear at interfaces between rock strata – such as between a sandstone and shale layer. Rock layers weather, erode, and dissolve at different rates, creating cavities that are often lined and/or filled with newly precipitating minerals and metals.

26.5.5 Liquids and Gases Within Rocks

Liquids and gases useful as minerals can be found within rocks. Aquifers have already been discussed (Chapter 17); the minerals dissolved in the water can make it mild, poor tasting, undrinkable, and/or poisonous [20]. Aquifer water can also absorb methane in coal beds and release it as pressure is reduced when the water surfaces; this behavior has enabled "fracking" – fracturing methane-containing geologic formations to release and capture the methane. Buried organic matter transforms to coal, petroleum, and methane. Methane eventually rises to the surface and dissipates, so some fossil fuel geologic formations contain more than others. Both petroleum and methane can be trapped beneath impervious rock layers and so remain for a long time. And, the methane and coal can catch fire and burn for decades.

26.6 Mining

Types of mineral deposits found are often characteristic of specific landforms. Mineral seams, veins, and layers are commonly mined at concentrations of a desired mineral.

Bulk formations – or thick layers of rock – are mined if the entire formation is valuable, such as coal for fuel; phosphates, potash, and nitrates for fertilizers; marble, sand, limestone, granite, or tuff for building; and granite or basalt for gravel. Liquid or gaseous deposits of fossil fuels are often extracted in bulk through wells. Low concentrations of a desired mineral within a formation are sometimes extracted by modifying the entire formation.

Minerals from placer deposits are often extracted by washing the stream gravel of the precious minerals.

The type of mine depends on the type and location of the target mineral deposits; needs for extraction and transport; and environmental, health, and safety considerations [1, 10, 21]. Many mines have environmental concerns. They expose chemicals in a "reduced" (nonoxidized) state to air (oxygen) and can generate acids that pollute water and air and kill plants and animals, including fish. They also can eliminate overlying vegetation. Many of these issues can be mitigated to various degrees.

A large proportion of the cost of minerals is the extracting and transporting processes. Consequently, mineral prices often fluctuate with energy prices [22].

26.6.1 Shaft

Shaft mines, often resembling caves, are among the oldest mines [21]. They follow the targeted seam or layer as it extends underground. They require relatively little machinery to remove unwanted material; however, they are both dangerous and unhealthy. They can collapse when the tunnels or removed seams or layers leave underground openings; and the air inside can harm the workers. Shaft mines can leave "tailings" – unwanted material – at their surface openings. And, they can be dangerous for people falling into them – or if they collapse with age or when heavy buildings are constructed above them.

Wells are variations of shaft mines but generally have fewer hazards because people do not enter them.

26.6.2 Open Pit

Various open pits have been used where a seam, layer, or bulk material is near the surface [21]. Undesired overlying material ("overburden") is removed and the desired mineral is then taken. Open pits have fewer health and safety concerns than shaft mining; however, they generally require heavy machinery only available in the past century or, previously, inexpensive labor. Open pits for removing stones and gravel are referred to as "quarries." Removal of overburden along a continuous front to extract a layer such as coal is referred to as "strip mining." Laws and techniques are being instituted in many places to replace the overburden and restore vegetation after the minerals have been extracted.

26.6.3 Hydraulic

Hydraulic mining has been used to extract low concentrations of a valuable mineral such as gold from bulk formations. A water body is dammed or diverted to erode a large hill or mountain. The valuable mineral is collected at the bottom where it concentrates because it is heavy, while the water and eroded material washes downstream. This mining has been outlawed in many places because of its destruction to downstream houses and farms, its diversion of water from other uses, and its destruction to mountains [23].

26.6.4 Placer

Placer mining removes the desired minerals from stream beds. Placer mining can temporarily create sediment in the streams, can eliminate aquatic species from the area and downstream, and can alter the stream or river channel. Mining gravel in streams for making roads or concrete can remove gravel used by spawning fish and disrupt wetland and aquatic habitats.

26.6.5 Fracking

Oil shale can be mined to generate natural gas by drilling vertical and horizontal holes within the bulk material and injecting pressurized water to fracture the formation and release the gas [24, 25]. The process, known as "fracking," requires large amounts of water, can subject the area to earthquakes, and leave it quite deteriorated. Fracking also requires mining of sand and increasing the infrastructure of transportation. Although the process is quite old, its widespread implementation is new; and the environmental effects are only recently being understood and mitigated [26–28].

26.6.6 Acid Leach

Acid leaching is used to extract low concentrations of gold or other valuable minerals from bulk formations [29]. A large lake with an impervious bottom and containing acids is created. Bulk material is dumped into the lake and the acid separates the valuable mineral, which sinks to the bottom and is recovered. These mines create environmental problems if they leak or if waterfowl land in the toxic lake. They are also problematic to restore afterwards.

References

1. US Bureau of Mines. *Dictionary of Mining, Mineral, and Related Terms*. (US National Archives and Records Administration, 2016 [Accessed May 21, 2016]). Available from: http://webharvest.gov/peth04/20041015011634/imcg.wr.usgs.gov/dmmrt/
2. B. J. Skinner, S. C. Porter, J. Park. *Dynamic Earth*. (Wiley, 2004).
3. C. C. Plummer, D. McGeary, D. H. Carlson. *Physical Geology*, ninth edition. (McGraw-Hill, 2003).
4. P. D. Sackett. Elemental Cycles in the Anthropocene: Mining Aboveground. *Geological Society of America Special Papers*. 2016;520:SPE520–11.
5. Los Alamos National Laboratory. *Periodic Table of the Elements*. (US Los Alamos National Laboratory Chemistry Division, 2016 [Accessed December 10, 2016]). Available from: http://periodic.lanl.gov/images/periodictable.pdf.
6. W. F. McDonough. Ch. 1. The Composition of the Earth, in Teisseyre R., Majewski E., editors, *Earthquake Thermodynamics and Phase Transformations in the Earth's Interior*. (Academic Press, 2000): pp. 5–24.
7. A. M. Helmenstine. *Metals Versus Nonmetals*. (about.com, 2016 [updated February 2, 2016; Accessed May 19, 2016]). Available from: http://chemistry.about.com/od/periodictableelements/a/Metals-And-Nonmetals.htm.
8. C. R. Hammond. Section 4: The Elements, in Lide D. R., editor, *Handbook of Chemistry and Physics*. 89 ed. (CRC Press/Taylor and Francis, 2009).
9. R. L. Virta. *Minerals Yearbook*. (US Department of the Interior, US Geological Survey, 2006).
10. American Geological Institute. *Dictionary of Mining, Mineral, and Related Terms*, second edition. 1997.
11. European Commission. *Report on Critical Raw Materials for the EU*. 2014.

12. Mission 2016. *Categorizing the Strategic Elements*. (Massachusetts Institute of Technology, 2016 [Accessed May 14, 2016]). Available from: http://web.mit.edu/12.000/www/m2016/final website/elements/index.html.
13. J. S. Monroe, R. Wicander. *Physical Geology*. (Brooks/Cole, 2001).
14. W. D. Arnett, J. N. Bahcall, R. P. Kirshner, S. E. Woosley. Supernova 1987a. *Annual Review of Astronomy and Astrophysics*. 1989;27:629–700.
15. S. Taylor, S. McLennan, R. Armstrong, J. Tarney. The Composition and Evolution of the Continental Crust: Rare Earth Element Evidence from Sedimentary Rocks [and Discussion]. *Philosophical Transactions of the Royal Society of London A: Mathematical, Physical and Engineering Sciences*. 1981;301(1461):381–99.
16. W. K. Hamblin, E. H. Christiansen. *The Earth's Dynamic Systems*, tenth edition. (Prentice Hall, 2001).
17. J. W. Valley, A. J. Cavosie, T. Ushikubo, et al. Hadean Age for a Post-Magma-Ocean Zircon Confirmed by Atom-Probe Tomography. *Nature Geoscience*. 2014;7(3):219–23.
18. E. H. Macdonald. *Alluvial Mining: The Geology, Technology, and Economics of Placers*. (CRC Press, 1983).
19. Amethyst Galleries. *Limestone*. (Amethyst Galleries, 1995 [Accessed May 14, 2016]). Available from: www.galleries.com/rocks/limestone.htm.
20. D. K. Nordstrom. Worldwide Occurrences of Arsenic in Ground Water. *Science*. 2002;296 (5576):2143–5.
21. H. L. Hartman, J. M. Mutmansky. *Introductory Mining Engineering*. (John Wiley & Sons, 2002).
22. J. Kooroshy, C. Meindersma, M. Rademaker, et al. Scarcity of Minerals. A Strategic Security Issue. *The Hague, Centre for Strategic Studies*. 2009(02–01):10.
23. F. H. Nichols, J. E. Cloern, S. N. Luoma, D. H. Peterson. The Modification of an Estuary. *Science*. 1986;231(4738):567–73.
24. D. Healy. *Hydraulic Fracturing or "Fracking": A Short Summary of Current Knowledge and Potential Environmental Impacts*. (Environment Production Agency of Ireland, 2012).
25. L. H. Huffman, G. S. Knoke, inventors. The United States of America as Represented by the United States Department of Energy, assignee. *Hydraulic Mining Method*. (United States of America patent US4536035 A. 1985).
26. K. O. Maloney, S. Baruch-Mordo, L. A. Patterson, et al. Unconventional Oil and Gas Spills: Materials, Volumes, and Risks to Surface Waters in Four States of the US. *Science of the Total Environment*. 2017;581:369–77.
27. T. Myers. Potential Contaminant Pathways from Hydraulically Fractured Shale to Aquifers. *Groundwater*. 2012;50(6):872–82.
28. J. E. Saiers, E. Barth. Potential Contaminant Pathways from Hydraulically Fractured Shale Aquifers. *Groundwater*. 2012;50(6):826–8.
29. J. Marsden, I. House. *The Chemistry of Gold Extraction*. (The Society for Mining, Metallurgy, and Exploration Inc., 2006).

27

Rocks and Minerals: Production, Use, Distribution

27.1 Volumes of Minerals Produced

Minerals were probably first used by people's ancestors as bulk rocks for tools and weapons – both to hit while holding and to throw. Later, they became the blades on spears, knives, and arrows; and the differentiation of rock types became important – flint, obsidian, and quartz made better cutting tools and arrowheads than limestone, marble, or granite.

Gems and other minerals were used for jewelry; flint and quartz were used for starting fires; and rocks improved fireplaces and heated water. Stones were also used for building, and fresh dirt of specific textures and low stickiness was used to line floors of caves and other dwellings.

Later, but still thousands of years ago, specific textures of clays were identified as better for pottery; and lime was used to make a cement. Pottery kilns led to metalworking, which demanded an understanding of minerals, ores, and mineral seams. Over time, other minerals were identified and used with increasing precision – salt for food preservation; lime, saltpeter, and guano for fertilization; and different metals for plowing, weapons, and precision clocks. Glass production led to magnification that allowed people to understand more about minerals in small rocks and the larger universe. Bulk rocks, gravel, and sand have continued to be used for roads, buildings, and harbors for thousands of years.

Mineral uses will probably increase as well as change as new technologies render needs for some materials obsolete or as substitutes are found. New technologies have made flint arrowheads less important and crystals for radios less in demand but have increased the need for tungsten, talc, lithium, and sulfuric acid.

The amount of various rocks, metals, and minerals presently produced varies over one billion-fold from coal at one extreme to diamonds at the other (Figure 27.1).

27.2 Locations of Minerals

Bulk minerals such as gravel, coal, quarry stone, and cement are usually produced close to where they are used because of high transportation costs. Many such as aluminum and steel

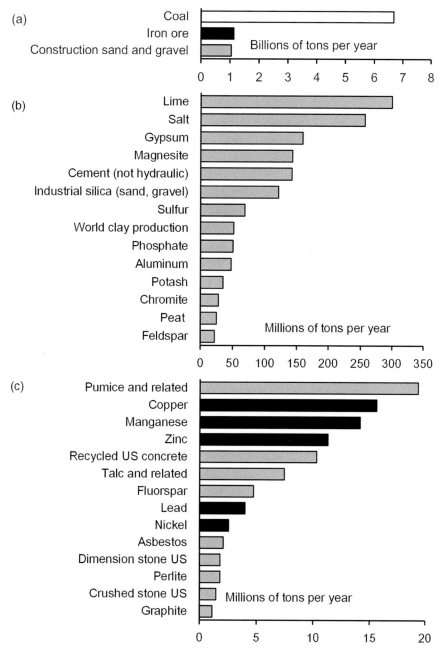

Figure 27.1 Mineral, rock, and similar material annual production [1]. Black = metals; white = organic; gray = other.

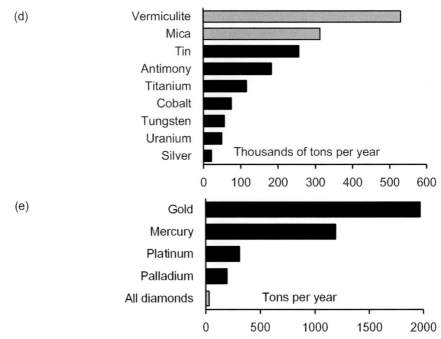

Figure 27.1 (cont.)

are refined near energy sources such as coal deposits or hydroelectric facilities. Countries with rapidly developing infrastructures use many bulk minerals such as cement, bauxite for aluminum, and iron ore for steel (Figures 27.2, 9). Australia produces large amounts of bulk minerals that it ships to rapidly developing countries in Asia. Other minerals are produced in the few places where they are found [2].

China produced 45% of the world's aluminum, followed by 8% from Russia [2]. Chile produced the most copper, 33%, followed by China, 8%. China produced 45% of the world's iron and steel, with Australia (17%) and Brazil (13%) next. China produced most of the world's lead, 54%; Australia was second with 13%; and the United States was third with 7% [2].

Some base metals such as zinc, manganese, and tin are only found and mined in certain countries with specific geologic formations, while copper, chromium, and nickel are obtained throughout much of the world (Figure 27.3).

Silver and gold are found in many places, but nowhere in such large quantities that the market can be monopolized. Gold and silver are rare enough to command a high value (Figure 27.4). Alternatively, diamonds are only found in a few places which can monopolize their distribution and so keep them rare and valuable.

Rare elements that have recently become essential for advanced technologies are often only supplied from a few places in the world (Figure 27.5).

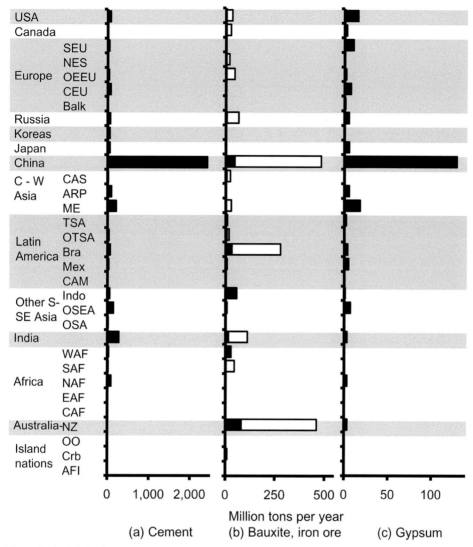

Figure 27.2 Origins by country groups of (a) cement, (b) bauxite for aluminum (black) and iron ore (white), and (c) gypsum [2].

27.3 Forms and Uses of Minerals

"Ores," useful minerals, are sometimes used in the forms they are found in and are sometimes converted to other forms. For example, lead is occasionally found in elemental form, but is more commonly made from lead sulfide. Talc is used in its mined, noncrystalline form; but calcite needs to be processed to make cement. Mica is used in its pure, crystalline forms.

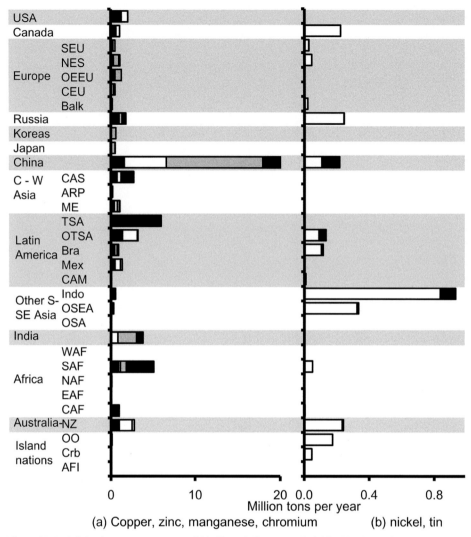

Figure 27.3 Origins by country groups of (a) (from left) copper (left black), zinc (white), manganese (gray), chromium (right black); (b) nickel (white) and tin (black) [2].

27.3.1 Gases

Mineral gases largely consist of noble gases, other elemental gases, and organic gases. Noble gases are generally used for cooling in refrigerants and superconducting magnets because of their low melting and boiling points. They replace air in technical procedures such as gas chromatographs, medical procedures, and arc welding because of their non-reactivity. Lasers use them because of their tendency to glow. Argon can be an insulator, and helium has been lifting dirigibles since hydrogen proved too dangerous.

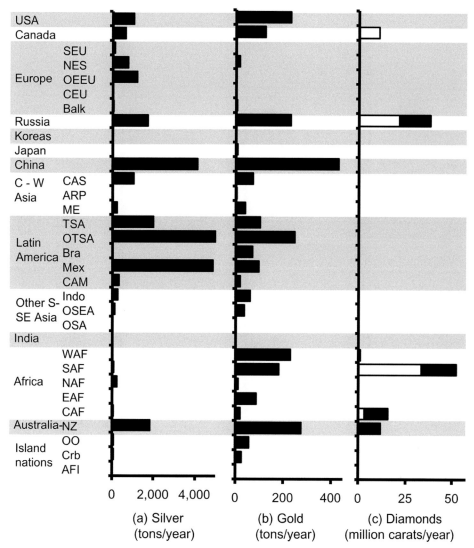

Figure 27.4 Origins by country groups of (a) silver, (b) gold, and (c) gemstone (white) and industrial (black) diamonds [2].

Elemental gases that are not noble include hydrogen, oxygen, and nitrogen. Diatomic hydrogen and oxygen are highly flammable and can be dangerous. Properly controlled, hydrogen looks promising in fuel cells to store energy similarly to batteries. Oxygen is used to enrich air for various chemical and ignition uses. Diatomic nitrogen makes up most of the atmosphere (Table 8.1) and is quite unreactive. It is used in cooling and freezing and to make nitrogen fertilizer.

Many gaseous molecules of multiple elements exist, including ammonium, carbon dioxide, methane, and other organics.

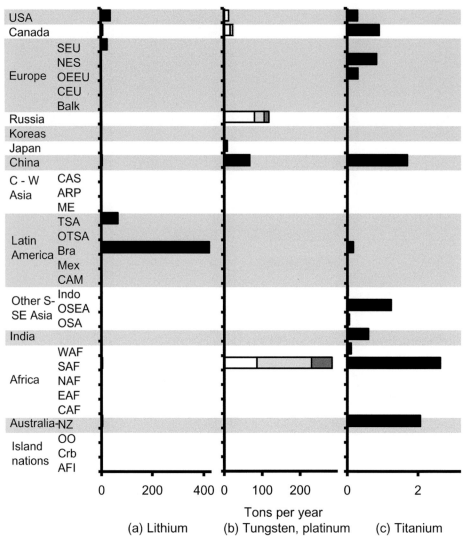

Figure 27.5 Origins by country groups of (a) lithium; (b) tungsten (black), palladium (white), platinum (light gray), other platinum group (dark gray); (c) titanium [2].

27.3.2 Liquids

Mercury has been used in thermometers, in dental fillings, to recover gold, as a chemical catalyst, and in electrical switches. It evaporates quite readily. Recent understanding of its poisonous nature has led to greater precautions in its use.

Bromine has been used as a fire retardant, in dyes, and in agriculture chemicals including insecticides. Its toxicity is leading to its being phased out or used more cautiously.

Fossil fuels, water, and other mineral liquids have been discussed elsewhere in other chapters.

27.3.3 Metals

Metals were discussed in general in Chapter 26. Each metal has properties that give it different roles, especially in the highly technological sectors ranging from circuit boards to airplanes [3]. Some metals do not have substitutes if they become unavailable [4]. The dependence of modern society on these limited, irreplaceable metals is a sustainability concern, discussed toward the end of this chapter.

27.3.4 Crystals

Mineral crystals can be composed of either a single element or several elements. Common properties of most crystals are their physical hardness, sharp edges, beauty, bright colors, and consistent planar angles. Other properties that make them useful are their uniformity; vibratory ability; and light reflectivity, refractivity, dispersivity, and transmissibility [5].

Crystals have historically been mined in seams or similar places. Mined crystals have different sizes and shapes, with different irregularities that reduce their precision and value for industrial purposes. Industrial crystals such as quartz and diamonds are now also artificially made. Artificial production ensures quality, consistency, and purity but does not provide aesthetically valued irregularities (Figure 26.3).

Crystals can be roughly classified into covalent crystals, metallic crystals, and ionic crystals [5]. Many crystals are gradations among classes and show intermediate properties. Aesthetically, crystals are appreciated in jewelry, chandeliers, and other places that sparkle in light. Industrial diamond crystals are commonly used in cutting because of their hardness. Other crystals concentrate, distort, reflect, or refract light, such as in lasers. Crystals are also used for their consistent vibratory resonance for frequency-sensitive purposes.

27.3.5 Bulk Rocks

Rocks, containing many minerals, are used where weight and strength are needed [1, 6, 7]. Important properties of rocks for these uses are local abundance and accessibility, durability, inertness, and hardness. Igneous and metamorphic rocks are more useful as bulk than weak rocks such as sandstone, limestone, and pumice.

Some common bulk sizes and uses are described below:

- Rocks of very large sizes are used to construct jetties into oceans; and progressively smaller rocks are used for terracing, river bank stabilizing, and beds beneath roads and railroads.
- Crushed stones of progressively smaller diameters are generally layered on road and railroad beds to form a stable platform.

- Construction gravel is used on roadbeds, but also as a component of concrete.
- Construction sand is used as a component of concrete and mortar.
- Other sand can be used for road beds.

   Rocks and gravels are either obtained from existing deposits such as in glacial outwash plains or river beds or by blasting rock formations using dynamite and then further crushing the fragmented stones mechanically.

   Recent innovations are creating ways to reduce the amount of gravel used, such as adding cement directly to soil to form a hard road bed or adding a layer of heavy plastic within the rock and gravel bed layers. "Fillers" are replacing some sand and gravel in concrete. These innovations reduce the energy costs of extracting and transporting sands and gravels.

27.3.6 Quarry Stone (Dimension Stone)

Quarry stone is used for aesthetic, strength, and durability purposes in buildings, bridges, and other structures; in statues, outdoor tables and benches; and in gravestones and commemorative plaques [6, 7]. A variety of rocks can be used, but commonly granites and metamorphic limestones, marbles, sandstones, and slates are used.

27.3.7 Ornaments

Both metals and crystals are used for jewelry; ornaments such as door knobs and handles, chandeliers, lamps, and other light fixtures in building and vehicles; serving dishes and utensils, pens, and other ornaments and decorations; and religious decorations. Most high-quality ornamental metal objects are made of precious metals.

27.3.8 Energy

Some minerals generate energy in conventional and unconventional ways. Radioactive elements such as uranium generate energy used in nuclear reactors. Coal generates energy when burned (Chapter 23). Rocks and soil on hills and mountains contain potential energy that is only realized when they fall downward in landslides, rockslide, and soil slumps (Chapter 23). It is theoretically possible to capture and utilize this energy of hillside rocks (Chapter 23).

27.3.9 Currency and Intrinsic Value

Gold has a strange and confusing value that has lasted for thousands of years. It is hoarded in times of insecurity but has little use except that it is beautiful and desired by other people. Until recently gold was used by countries to assure the value of their currency.

27.3.10 Other Materials Processed from Minerals

Different raw materials and chemicals can be extracted from minerals, utilizing them and sometimes altering existing compounds or covalent radicals.

The strong acids and bases, a variety of elements, many mineral shapes, and polarized charges give a wide opportunity for different chemical properties to be exploited.

Various physical and chemical means are used to process the raw materials, from simple extraction and shipping in the case of rock salt for roadways to complex chemical activities for various acid productions.

Some of the materials and chemicals utilized or manufactured from minerals include [8]:

- Fertilizers: Deposits of calcium nitrate, gypsum, and minerals containing phosphate and potassium (potash) have been mined and both directly applied as fertilizers and further processed to produce concentrated fertilizers. In addition, sulfur and lime are applied to change the soil's acidity (Chapter 21). Concentrated mineral deposits used for fertilizer are often mined until they are depleted; then, deposits elsewhere in the world are mined [1, 9]. As nitrate deposits were exhausted, techniques to produce nitrogen fertilizer from atmospheric nitrogen were developed (Chapter 21) [9, 10]. Phosphorus is an important fertilizer that is mined as inorganic phosphate rock and guano – bird feces that can accumulate to large deposits over millennia [11]. Potassium is also an important mineral mined as potash for fertilizer. The presence of phosphates and potash in only certain places in the world (Figure 21.8) leads to sustainability concerns for these fertilizers [12].
- Gypsum board: In addition to fertilizer, the gypsum mineral is also used to make "sheetrock," or "wallboard," used as the inside walls of many buildings.
- Sulfuric and other acids are produced in large quantities from mineral deposits of sulfur. Sulfuric acid is one of the most largely produced and widely used chemicals in industrial societies. It was once a discharged waste product of extracting copper from ores; however, its recovery has proved beneficial both by not polluting the environment and by providing financial and material benefits [13].
- Asbestos is a general name for several minerals that have common properties: strength, flexibility, chemical and thermal stability, and low electrical and thermal conductance. It is used as a thermal and electrical insulator, for roofing, and for friction products. Concerns of its cancer-causing properties have emerged.
- Vermiculite, when expanded by heat, becomes a lightweight, inert, odorless, and fire-resistant substance. It is used as a filler in plaster and concrete because of its fire resistant and thermal insulation properties. Vermiculite also is used as a "carrier" to absorb and apply fertilizers and pesticides.
- Common salt is the mineral ionic compound of sodium chlorine. It is used almost universally as a food flavoring and meat preservative, especially in the absence of freezers. It is used in winter to melt ice on roadways. Salt is also used to make polyvinyl chloride and caustic soda, a chemical important in pulp/paper manufacturing. Salt can be extracted from desert deposits or by drying seawater.

- Cement is a raw chemical made by heating the minerals calcium carbonate and clay (or other sources of the silica oxide mineral), resulting in calcium silicates. When crushed and wetted, it will harden to an extremely strong, slightly flexible bonding material. Concretes and mortars are then made by adding various gravel and/or sand fillers to the wetted cements. Concretes and mortars are generally about 15% cement by weight; but this concentration varies by end use. High-quality sand deposits suitable for making concrete are limited; substitutes are being developed [14].
- Clays of different characteristics are used for ceramic tiles, absorbents, foundry sand bonding agents, bricks, Portland cement, and drilling mud.
- Glass is made from firing the silica and sodium and calcium oxide minerals.
- Mica is mined as large mineral crystals that can be split into transparent sheets. It is used for windows in furnaces and other places that endure temperature extremes.

27.4 Mineral Production

With the development of highly technological societies, properties specific to individual elements are becoming understood and utilized. Specific properties include ability or inability to conduct, catalyze, reflect, refract, and react.

27.4.1 Increasing Volumes and Varieties Consumed

Increasing amounts and varieties of minerals are being consumed (Figure 27.6).

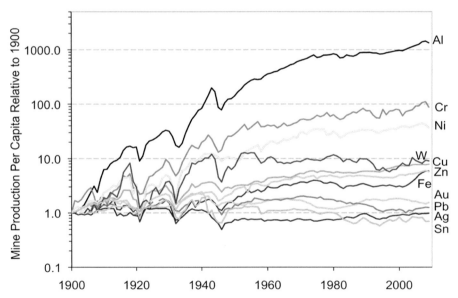

Figure 27.6 Change in per capital production of many metals during past century [15]. Vertical axis is logarithmic scale. Copyright (2007) The American Chemical Society.

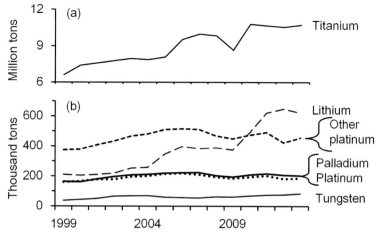

Figure 27.7 Global production of specialized metals. "Other platinum" = other platinum group metals: iridium, osmium, rhodium, ruthenium. Scale changes for titanium. [2].

Use of relatively rare, little-used minerals generally increases as innovative technologies new to both the developed and developing world find need of them (Figure 27.7). The rise of lithium production is largely because of its use in batteries.

Global productions of aluminum and copper have been increasing, although their mining rates have declined because recycling provides some of the production (Figure 27.8) [16]. Much of the production is to provide infrastructures for rapidly developing countries. Developed countries such as the United States, the United Kingdom, and France seem to have saturated their per capita consumption of iron [17]. Lead mining has actually declined because of both recycling and declining use [18].

Global cement production has risen sharply with the economic development and associated infrastructure construction in China and India (Figures 27.9–10) [8]. China produced 57% of the world's cement, India produced 15%, and no other country produced more than 2% in 2015. Concrete consumes about 5% of the world's fossil fuel energy [19]; however, the increased concrete production in China and India may not continue if its use declines once the countries' infrastructures are developed and their populations stabilize (Chapter 5). Other countries may show a temporary increase in cement use as they develop, unless alternatives to cement-based infrastructures emerge (leapfrog technologies). Wood as an alternative to concrete and steel construction may help (Chapters 24 and 28).

27.4.2 Impending Mineral Shortages

Shortages of some metals within the next few decades are already being projected [20]; it will become technically or economically unfeasible to extract any more of these from raw materials. Copper, zinc, platinum, and phosphorus would become very scarce and

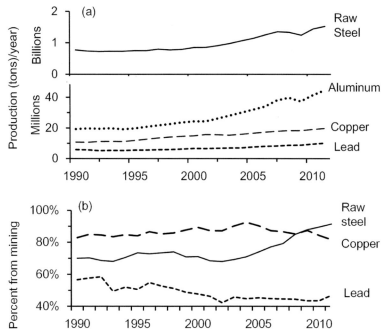

Figure 27.8 (a) Global production of lead, copper, aluminum, and raw steel [18]. Aluminum includes remelted scrap; copper is solely from mines. Scale change for raw steel. (b) Percent of raw steel, copper, and lead that is from mining, different from recycling.

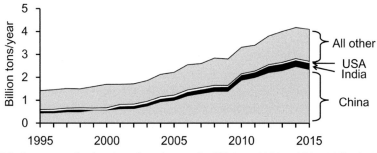

Figure 27.9 Cement production for selected countries [8] from 1995 to 2015. "All other" = rest of world [2].

prohibitively expensive if the whole world begins using as much as the developed world does at present [12]. Zinc is used in brakes and galvanizing steel and so dissipates into the atmosphere where it is difficult to recover. Platinum is used in fuel cells and would be in short supply if everyone used them. Phosphorus is mined for fertilizer (Figure 21.8), but the deposits are diminishing [12].

In addition to the raw material being technically or economically unavailable, minerals may be unavailable because their toxicity to people or the environment prohibits their use or

Figure 27.10 China is using much cement to build an infrastructure of highways and cities.

because political rivalries block their international trade [4, 21]. Such minerals of concern are known as "critical" minerals. The United States [22] and the European Commission [21] have developed lists of critical minerals (Table 27.1).

27.5 Sustaining Mineral Systems

Like water and energy, minerals are constantly circulating with geologic and anthropogenic processes (Table 27.2). Minerals have historically been discarded to junkyards and landfills or released to the Earth's waters or atmosphere. The human circulation of some minerals has actually exceeded the background geologic cycling rate (Table 27.2), as shown in the Human–Nature Elemental Dominance Index (Klee–Graedel Index) [23].

It will be difficult to sustain the supply of certain minerals in perpetuity; however, the supply can be extended far into the future by intentionally circulating them. The geologic system that circulates minerals is too slow to produce enough minerals; consequently, unlike the climate system that may best be preserved unaltered, the mineral system may best serve people by being altered. The geologic system for circulating minerals could continue to be used; but a second, currently nascent anthropogenic system of recycling could be greatly expanded through a combination of exploiting geologic deposits, recycling, conserving, substituting, and synthesizing (Figure 27.11). Then, an equitable annual distribution amount of each mineral can be calculated to ensure present and future generations are not unfairly burdened [25], similar to the "sustained yield" calculation of forest timber [26, 27].

27.5.1 Recycling and Recovery

Ores by definition contain sufficient concentrations of desired minerals to enable minerals to be extracted efficiently. As concentrated ores are depleted, less concentrated ores are

Table 27.1 *Number and proportion of weight of 54 important "raw materials" (RM) and European Union 20 "critical raw materials" (EUCRM) provided by major supplying countries [21]*

Country	RMs supplied	Proportion of RM weight	EUCRMs supplied	Proportion of EUCRM weight
	(Number out of 54)	(percent)	(Number out of 20[A])	(percent)
China	48	30%	18	49%
USA	36	10%	9	9%
Russia	42	5%	16	4%
Brazil	36	5%	11	6%
Australia	34	4%	10	1%
South Africa	26	4%	9	6%
Chile	18	3%	2	1%
Canada	30	3%	11	3%
India	30	3%	8	2%
Turkey	25	2%	7	3%
Japan	18	2%	2	1%
Total		71%		85%

[A] The European Commission list contains 20 minerals:

Antimony	Coking coal	Magnesite	Platinum-group metals
Beryllium	Fluorspar	Magnesium	Heavy Rare Earth elements
Borates	Gallium	Natural Graphite	Light Rare Earth elements
Chromium	Germanium	Niobium	Tungsten
Cobalt	Indium	Phosphate Rock	Silicon Metal

processed. Not uncommonly, previously discarded tailings and other recycling targets have sufficient concentrations of remaining minerals to be reprocessed [15].

Automobiles, plumbing, electric wiring, and batteries are being increasingly recycled. Minerals such as sulfur that were considered useless, discarded by-products of mining are being recovered and reused [13], as are nitrogen and other fertilizer minerals from municipal and industrial wastes, which are applied to agriculture crops [28].

Not all discarded objects can be recycled. The material must have sufficient mass and concentration to justify collection, be in a recoverable form, and be regulated appropriately [15]. Ideally, goods would be manufactured with the intent of recycling the mineral components when no longer needed.

Recycling is not a perfect solution for obtaining minerals in perpetuity because some material is lost to the atmosphere, water, soil, or other unrecoverable sinks during manufacturing and recycling.

Table 27.2 *The human influence on mineral flow compared to background flows for Aluminum, Copper, Iron, and Zinc [24]*

	Aluminum	Copper	Iron	Zinc
	(million tons per year)			
Background flows				
Net primary production	26.00	1.90	40.00	8.60
Sea spray	0.36	0.01	0.33	0.03
Denudation	1,200.00	1.10	690.00	2.70
Anthropogenic flows				
Fossil fuel burning	25.00	0.03	33.00	0.08
Biomass burning	0.19	0.01	0.37	0.09
Mining	30.00	16.00	520.00	8.40

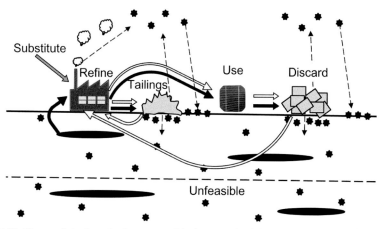

Figure 27.11 The exploited geologic system (black arrows) of deposits of a mineral (in black) and newly developed anthropogenic system (white arrows) of reuse.

27.5.2 Substitution

Sometimes a technology is developed that requires a rare, valuable, heavy, or otherwise awkward mineral. A technical refinement can then enable a more available mineral to substitute for the rare one. Consequently, the technology becomes more useable.

Aluminum has replaced steel in automobile engines and elsewhere to make them lighter. Graphite has replaced metals in many tools, also for lightness and strength. Plastic has replaced lead in automobile battery casings; and other metals are replacing tungsten in efficient light bulbs.

Elements such as lithium, zinc, and cadmium have excellent substitutes in many uses; consequently, these elements' uses can be concentrated to where there are no substitutes.

Other elements such as magnesium, manganese, and lead generally have only poor substitutes. (See [4] for complete lists.) No known substitutes exist for thallium in supercomputers or for germanium in fiber optic cables [25].

Substitution can also occur by technology transformation. For example, if an element needed in batteries is projected to be in short supply, then different batteries or energy systems not needing batteries can be promoted.

27.5.3 Synthesis

Once a mineral compound proves useful and rare, people also attempt to synthesize it. Quartz and diamond crystals are now synthesized, as is nitrogen fertilizer, as examples. Much natural gas is needed to synthesize nitrogen fertilizer, and people will probably be seeking other ways to produce it – or ways of recycling and conserving it (Chapter 21). Elements are generally not synthesized, except rarely in nuclear reactions.

27.5.4 Forecasting and Adjusting

The availability, needs, and substitutes can be forecast and updated to focus and adjust research and development in ways that will avoid and circumvent shortages.

In the long term, more efficient ways of extracting minerals at low concentrations may also help alleviate mineral shortages.

References

1. US Geological Survey. *Minerals Yearbook, Vol. III Area Reports–International.* (US Department of the Interior, 2006–2012).
2. US Geological Survey. *International Minerals Statistics and Information.* (US Department of the Interior, 2014 [Accessed October 14, 2016]). Available from: http://minerals.usgs.gov/minerals/pubs/country/.
3. T. Graedel, L. Erdmann. Will Metal Scarcity Impede Routine Industrial Use? *MRS Bulletin.* 2012;37(04):325–31.
4. T. E. Graedel, E. Harper, N. T. Nassar, B. K. Reck. On the Materials Basis of Modern Society. *Proceedings of the National Academy of Sciences.* 2015;112(20):6295–300.
5. A. M. Helmenstine. *Types of Crystals.* (about.com, 2016 [updated May 11, 2016; Accessed May 14, 2016]). Available from: http://chemistry.about.com/cs/growingcrystals/a/aa011104a.htm.
6. J. A. Howe. *The Geology of Building Stones.* (Routledge, 2016).
7. E. M. Winkler. *Stone: Properties, Durability in Man's Environment.* (Springer Science and Business Media, 2013).
8. US Geological Survey. *Minerals Information: Commodity Statistics and Information.* (US Department of the Interior, 2016 [updated March 24, 2016; Accessed May 14, 2016]). Available from: http://minerals.usgs.gov/minerals/pubs/commodity/.
9. G. J. Leigh. *The World's Greatest Fix: A History of Nitrogen and Agriculture.* (Oxford University Press, 2004).
10. J. R. Craig, D. J. Vaughan, B. J. Skinner. *Resources of the Earth: Origin, Use, and Environmental Impacts*, fourth edition. (Prentice Hall, 2011).

11. J. Hays. *Phosphates, Bird Guano, and Fertilizer.* 2013 (Accessed May 14, 2016). Available from: http://factsanddetails.com/world/cat51/sub326/item2401.html.

12. P. D. Sackett. Elemental Cycles in the Anthropocene: Mining Aboveground. *Geological Society of America Special Papers.* 2016;520:SPE520–11.

13. G. G. Hedgcock. Injury by Smelter Smoke in Southeastern Tennessee. *Journal of the Washington Academy of Sciences.* 1914;4.

14. G. Goss-Durant. *Footprints in the Sand.* (Palladian Publications Ltd., 2015 [Accessed December 11, 2016]). Available from: www.worldcement.com/special-reports/16062015/foot prints-in-the-sand-829/.

15. J. Johnson, E. Harper, R. Lifset, T. Graedel. Dining at the Periodic Table: Metals Concentrations as They Relate to Recycling. *Environmental Science and Technology.* 2007;41(5):1759–65.

16. O. Dzioubinsku, R. Chipman. *Trends in Consumption and Production: Selected Minerals.* Contract No.: 5. (UN, Economic and Social Affairs, 1999).

17. D. B. Müller, T. Wang, B. Duval. Patterns of Iron Use in Societal Evolution. *Environmental Science and Technology.* 2010;45(1):182–8.

18. G. R. Matos. *Historical Global Statistics for Mineral and Material Commodities.* (US Department of Interior, Geological Survey, 2015).

19. C. D. Oliver, N. T. Nassar, B. R. Lippke, J. B. McCarter. Carbon, Fossil Fuel, and Biodiversity Mitigation with Wood and Forests. *Journal of Sustainable Forestry.* 2014;33(3):248–75.

20. R. B. Gordon, M. Bertram, T. Graedel. Metal Stocks and Sustainability. *Proceedings of the National Academy of Sciences.* 2006;103(5):1209–14.

21. European Commission. *Report on Critical Raw Materials for the EU.* (Ad hoc Working Group on Defining Critical Raw Materials, 2014.)

22. T. Office of the Under Secretary of Defense for Acquisition, and Logistics. *Strategic and Critical Materials 2013 Report on Stockpile Requirements.* (US Department of Defense, 2013).

23. R. Klee, T. Graedel. Elemental Cycles: A Status Report on Human or Natural Dominance. *Annual Review of Environmental Resources.* 2004;29:69–107.

24. J. N. Rauch. Global Spatial Indexing of the Human Impact of Al, Cu, Fe, and Zn Mobilization. *Environmental Science and Technology.* 2010;44:5728–34.

25. T. E. Graedel, R. J. Klee. Getting Serious About Sustainability. *Environmental Science and Technology.* 2002;36(4):523–9.

26. L. Davis, K. Johnson, P. Bettinger, T. Howard. *Forest Management, fourth edition.* (McGraw-Hill, 2001).

27. F. Schmithusen. Three Hundred Years of Applied Sustainability in Forestry. *Unasylva.* 2013;64 (1):3–11.

28. R. Schulz, V. Römheld. Recycling of Municipal and Industrial Organic Wastes in Agriculture: Benefits, Limitations, and Means of Improvement. *Soil Science and Plant Nutrition.* 1997;43 (Supplement 1):1051–6.

Part X

Forests

28

Forest Commodity and Non-Commodity Values

28.1 An Underused Resource

The world's forests cover 4 billion hectares of land – 31% of the terrestrial area (Table 15.2); another 1.4 billion hectares are classified as "shrub/trees" and "forest fallow." Together they contain over 500 billion cubic meters of wood [1]. Despite their large extent, they have not lived up to their potential as contributors to people or the environment.

- Only 0.4% of the world's workforce is employed in forest harvesting and restoration [2].
- Only 20% of the wood that the world's forests grow is being harvested [3, 4].
- Forest-dwelling species are the largest endangered group (Figure 14.13).
- Forest commerce represents only 0.4% of the world gross domestic product (GDP) [5].
- Many forests are overcrowded and in danger of burning up from neglect and mismanagement.

Forests are generally considered a "residual use" and are confined to lands that are not needed for houses, cities, agriculture, and similar uses. A small fraction – about 0.2% per year – are being converted to agriculture and other uses (Figure 28.1) [3]; and another 0.5% are burning each year [3, 6, 7].

Forests offer an opportunity to increase employment, relieve crowded cities, reduce fossil fuels and carbon dioxide emissions, and provide more water and biodiversity if they are managed appropriately.

Forests still cover about 70% of the area they occupied before agriculture clearing (Chapter 15), and probably more area than 18 thousand years ago (Figure 9.5). Appropriately attended to, they can provide many more non-commodity and commodity values collectively known as "ecosystem services" and can help alleviate people's pressure on the world's environment and other resources.

28.2 Forest Values and Stand Structures

The array of values from forests is being recognized through "Criteria for Sustainable Forestry" developed collectively for many nations throughout the world [8]. The Criteria comprise about seven values [9–11]:

Figure 28.1 (a) Residual forest burned in Liberia after most valuable trees were removed. (b) Burned forests are often converted to marginal agricultural land, especially when city infrastructures collapse such as during war (Liberia, 2011).

1. conservation of biological diversity;
2. maintenance of ecosystem productive capacity;
3. maintenance of ecosystem health and vitality;
4. maintenance of soil and water resources;
5. contribution to global carbon cycles;
6. enhancement of long-term socioeconomic benefits;
7. enhancement of legal, institutional, and economic frameworks.

Efforts are gaining momentum to certify and incentivize appropriate forest management [12], to pay landowners for providing various ecosystem services [13–17], and to zone forests for different priorities.

A stand forms a structure and then changes to another through growth and/or disturbances (Figure 14.5b) [18]. An open structure can still be a forest if it will grow back to another forest structure. When, in each structure, a stand provides some commodity and non-commodity values but excludes others. Figure 28.2 shows generalized relations of values and structures, although exceptions occur.

28.3 Non-Commodity Values

Non-commodity values are extremely diverse, depending on both the nature of the forest and culture and the creativity of local people [19]. This chapter will describe several values and focus on a few.

28.3.1 Biodiversity

Forests contain more country-endemic endangered species than other cover types (Figure 14.13) but do not contain markedly more species except for tropical rainforests

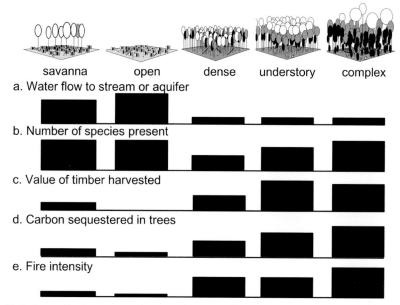

savanna · open · dense · understory · complex

a. Water flow to stream or aquifer

b. Number of species present

c. Value of timber harvested

d. Carbon sequestered in trees

e. Fire intensity

Figure 28.2 Generalized amounts of some values provided by each stand structure.

(Figure 15.3). Causes of this concentration of endangered species in forests are uncertain, but are probably attributable to imbalances of structures and fragmentation (Chapter 15). To correct these situations [20, 21], the coarse filter approach to maintaining biodiversity can be used by ensuring all structures are maintained within a landscape and maintaining corridors among landscapes (Figure 14.10; Chapter 16).

28.3.2 Water

Each stand structure can have contrasting effects on water flow depending on local conditions (Figure 17.6). Except where counteracted by fog drip or snowpack, reduced leaf covers in the open and savanna structures generally result in less evapotranspiration and more water flowing to groundwater and then slowly to streams or aquifers (Chapter 17). Periodic return to other stand structures maintains the soil's structure, keeping water percolating into the soil instead of flowing overland, with harmful consequences (Figures 17.7–8). Consequently, a diversity and constant changing of stand structures across a landscape can provide both water flow and percolation that prevents overland flow.

28.3.3 Fire Protection

Fires need appropriate fuel, weather, and an ignition source to burn. When fuels and weather are suitable, fires can ignite from a variety of sources, so blaming a large fire on

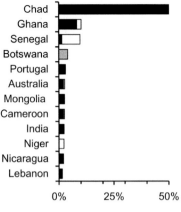

Figure 28.3 Percentage of forest covered annually by fires in most extensively burned countries [1] (black = wildfires; white = other fires; gray = unknown fire origin).

the cause of ignition ignores more controllable factors [22]. It is also difficult to control the weather; consequently, fires are best controlled by managing the fuels – living and dead trees and other organic matter. Fires are hotter, burn more deeply into the soil and wood, move through the forest much faster, and are much more difficult to stop with more fuels.

A diversity of structures across a landscape forces wildfires to burn less intensively, move more slowly, and so be more easily controlled and less harmful. Fires generally do not burn intensively in open and savanna structures because there is relatively little fuel and the large trees in a savanna are too far apart for a fire to move continuously through the upper canopy (Figures 14.5b, 14.10). More fuels exist in closed forests (dense understory and complex structures). Once a fire begins burning the upper canopy of closed structures, it becomes a "crown fire" and can burn very hot and travel rapidly for many kilometers.

Fuels accumulate as forests grow and frequent, mild fires are prevented [23]; timber is not harvested; or only the large trees are harvested. Fuels can be high immediately after a windstorm, logging operation, or similar disturbance until the fuels near the ground decompose. Of longer-lasting danger are crowded forests where many of the dead, dry trees remain standing and so do not rapidly decompose.

Forest wildfires of different periodicities and intensities occur nearly everywhere, from the tropics to boreal regions [22, 24, 25]. Such fires release carbon dioxide [26]; alter habitats [27]; and destroy human lives, timber, and buildings [28]. Forest fires cover about 20 million ha each year – about 0.2% of the world's forest area; locally, they can be more extensive [1]. If 3% of the forest land burns every year (Figure 28.3), each stand would be burned every 33 years on average; or more likely, some areas burn more frequently and others, less so.

28.3.4 Danger and Refuge

Forests can be places of danger or refuge [29]. People in conflict with the law often live, hide, and make whiskey in closed forests; and grow coca (for cocaine) and other illegal substances in openings [30]. Unexpected intruders are sometimes treated with hostility.

Human enemies and predator animals commonly hide in closed forests in some places and attack their enemies or prey in adjacent open forests, agriculture fields, pastures, or villages. Consequently, villages used to remove closed forest structures near their dwellings. Removing nearby closed structures also prevented hot fires from burning there and moving to adjacent dwellings, crops, or domestic animals.

Elsewhere, societies identify themselves positively with forests or other locations as a "sense of place." They respect certain locations as having traditional, mythological, historical, aesthetic, or sacred values [31, 32].

Forests have been claimed to have "existence value" to people who simply know that the forest exists, even if they never physically visit it [33, 34]. This value can create tensions if a large, faraway population asserts existence value and denies a forest's access to local people.

Forests are enjoyed for recreation where they are abundant (Figure 28.4) and contain a variety of structures that can be enjoyed in diverse ways [17, 35–41].

28.3.5 Carbon Sequestration and Other Climate Change Issues

A growing forest takes in carbon dioxide (CO_2), decomposes it, and makes cellulose, lignin, and other materials from the carbon (Figure 20.1). A forest growing from an open structure will accumulate carbon at a rate based on the species, detailed spacings of trees, and productivity of the stand – a reflection of its climate and soils (Figure 28.5a) [42, 43].

The trees become crowded as a stand grows, and some trees die, rot, and/or burn, in all cases returning CO_2 to the atmosphere [44]. The rate of net CO_2 intake of the whole stand declines as more trees die [45, 46], although some trees survive, grow larger, and take CO_2 from the atmosphere for many centuries [47]. Older stands generally contain more CO_2 (Figure 28.5a) but actually remove less CO_2 from the atmosphere than younger stands where trees are not dying (Figure 28.5a).

The stands in Figure 28.5 are 70 years old when their forest volumes are growing at their maximum rate and CO_2 is taken in at maximum efficiency (Figure 28.5b). This age is the "culmination of Mean Annual Increment" (MAI) [49]. If all stands were harvested and regrown to this age, the maximum amount of wood would be produced in the long term.

In a sustainable forest where equal volumes were harvested each year at culmination of mean annual increment and regrown, the average stand age, volume, and CO_2 sequestered would be half of those at MAI culmination. This means much less CO_2 would be sequestered in the managed forest than in the really old forest (Figure 28.5b). On the other hand, the cumulative CO_2 sequestered by using the harvested wood for construction and fuel *plus*

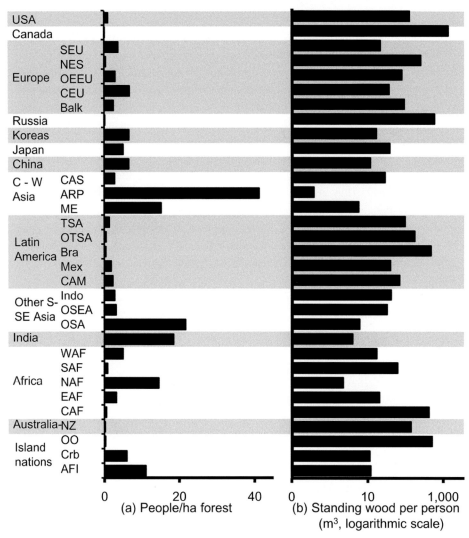

Figure 28.4 (a) Concentration by country groups of people per forest area and (b) standing forest volume per person [1].

the lesser CO_2 amount in the sustainably managed forest would exceed the CO_2 in the unharvested forest (Figure 29.5c) [4].

The best way that forests can sequester CO_2 is to use wood to replace steel, concrete, and bricks and thus avoid burning fossil fuel (Figures 24.8, 28.5b, 28.6) [4]. Carbon would be stored in the long-term slow domain (Figure 8.3) as unburned fossil fuel, and so prevented from entering the atmosphere as CO_2. Wood construction would also store some fast domain CO_2 in the preserved structural wood (Figure 28.6) [4, 50–52].

Table 28.1 *Legend for Figure 28.5*

(a)	Curved line = changes in CO_2 sequestered if forested ha grows to 160 years "UF" = CO_2 sequestered in the 160-year-old forest. Also in (b) and (c). Light gray = same as "UF"
(b)	Curved lines = changes in CO_2 sequestered in a forested ha repeatedly grown and harvested at 70 years. "HF" = average CO_2 sequestered in a sustained forest of equal areas harvested every 70 years. Also in (c). Light Gray = same as "HF"

(c)	"UF" and "HF" same as in (a) and (b).
	Curved lines = CO_2 sequestered in forest plus different harvested products and scrapwood burned for energy:
	    "F+IJ" = forest plus wood I-joists and scrapwood for energy
	    "F+SB" = forest plus solid wood beams and scrapwood for energy
	    "F+f" = forest plus all harvested wood burned for energy
	"Grays" = also total sequestered over time in alternatives.

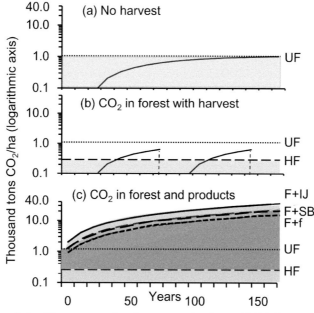

Figure 28.5 Cumulative CO_2 sequestration from preserving forest (a) VS (c) managing for different products [4, 48]. (b) compares CO_2 sequestered in forest with sustained harvests and no harvest. Table 28.1 gives legend.

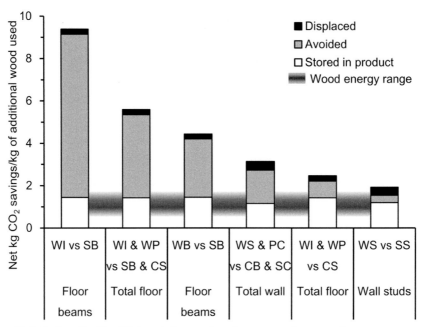

Figure 28.6 Carbon dioxide (CO_2) emissions savings from targeted construction components made of wood compared to steel or concrete and range (fuzzy horizontal bar) for wood energy [4]. Black = FF CO_2 saved by wood energy; gray = CO_2 avoided; white = CO_2-equivalent carbon stored FF in product. Abbreviations on horizontal axis same as Figure 24.8. Reproduced (open access) from Taylor and Francis Group Ltd.

Forests can also replace fossil fuels with wood unfit for construction as a fuel, either as pellets or as biofuel (Table 24.1). Wood burned in traditional, subsistence fireplaces sequesters little CO_2 until the harvested wood regrows [53]. Wood pellet stoves are more efficient than burning gas in the short and long term; and bioethanol from wood avoids fossil fuels in airplanes and other uses needing light weight, concentrated energy (Chapter 24).

Most degraded forests (Chapter 29) probably contained more volume and sequestered more CO_2 before they became degraded by improper harvesting [54]. Less CO_2 in degraded forests, or any harvested forest, does not mean forest management has been a net source of CO_2 emissions because much fossil fuel CO_2 emissions were prevented by using the harvested wood for fuel or building [55].

The recent IPCC report combined deforestation (land use change), swidden ("slash and burn") agriculture, and forest degradation and concluded they create a net increase in atmospheric CO_2 [56, 57]. More detailed examination of the report shows that forest degradation is recorded as a net CO_2 savings because of the generated wood products that avoid use of steel and concrete [53, 58, 59].

Trees also emit aerosols and methane that react with other chemicals and contribute to global warming, counteracting the CO_2-reducing effects of trees [60–62]. Closed evergreen forests at high latitudes also create a dark canopy that absorbs more sunlight for heat in winter than white snow or deciduous trees. Consequently, coniferous forests may actually increase atmospheric warming [63] compared to deciduous forests or forests that include open and savanna structures.

28.4 Non-Timber Forest Products

Non-timber forest products include decorative greenery and flowers, medicinal plants, foods and fragrances, fibers, mushrooms, pollinators and honey, saps and resins, among others [64–68]. Globally, they comprise 19% of the approximately 160 billion US dollars obtained annually from forests [69]. Biomass and fuelwood are sometimes considered in this category as well and contribute another 16%. These products are often locally concentrated and may form a large part of a forest community's income, especially in countries with a high PPPA (Purchasing Power Parity Advantage, Figure 6.4).

The presence and value of each non-timber forest product largely depends on the culture, climate, associated species, and stand structure. Most fruits, berries, and flowers (except orchids); pollinators; and honey are found in the open or savanna structures. Mushrooms are found in several structures but commonly in complex structures. Non-timber forest products such as ginseng (*Panax ginseng*), mushrooms, caffeine-bearing plants, gum from chicklet trees (*Manilkara zapota*), and sap from sugar maple (*Acer saccharum*) trees have become so popular that they are being cultivated. Brazil nuts (*Bertholletia excels*) are still collected from wild trees but have become a large commercial business. Some wild plants are being discovered to have valuable medicinal properties; searches for these plants – biological prospecting – may become lucrative.

Even if not a fulltime occupation, collecting and selling non-timber forest products can be an essential part of a rural family's livelihood, along with farming and seasonal labor. If any of these essential parts is lost, the family may not be able to survive in the rural area and so migrate to the city [70].

Non-timber forest products have commonly yielded low, subsistence incomes to villages, with "middlemen" purchasing them cheaply and selling them for high profits in cities. With internet and cell phone access, villagers are realizing and demanding non-timber forest product market values and so are returning more profits to their villages. The profits motivate them to collect more, but add a need to ensure their collection is sustainable.

28.5 Timber Products

A wood's use is based on both characteristics of the species and the way it is grown. Conifers (also known as "softwoods") are generally used for construction because they are light relative to their strength. Angiosperms (also known as "hardwoods" or "broadleafs")

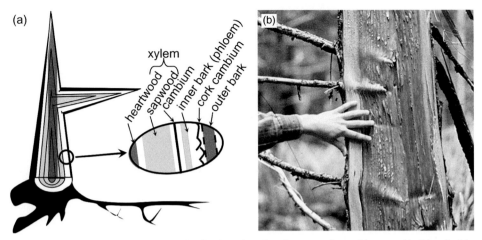

Figure 28.7 (left) Schematic tree longitudinal section showing annual wood layers added just inside the cambium, forming a tapered woody column up the stem. Dark, untapered column at tree center is juvenile wood. Other layers shown in oval enlargement; sapwood and heartwood annual wood growth layers are not shown here. (right) Split tree showing bark, sapwood, heartwood, and wood grain distortions caused by limbs that create knots.

are generally used for furniture, flooring, and paneling because they can be harder and heavier than conifers. Some hardwoods belie their name and can be quite soft and light, such as cottonwood (*Populus*) and kapok (*Ceiba*); conifers such as yew (*Taxus*) can be quite hard.

A tree forms wood each year by adding a thin layer of many wood cells to the outside of a tapered woody column inside its bark (Figure 28.7a) [71–74]. It also adds a very thin layer of material to the bark's inside. The layers consist largely of long, thin cells that are generally oriented parallel to the tree stem, thus creating strength and stability in this orientation (the wood "grain"). Wood shrinks and swells mostly perpendicular to the grain. Where a branch protrudes from the stem, wood is added both along the branch and around the branch forming a knot, a strong connection but a distortion in the parallel orientation of the grain (Figure 28.7b). Consequently, knots cause weaknesses and warping in lumber.

Between the wood (known as "xylem") and bark is the "cambium" – a single layer of cells that divides inward and outward to add wood and bark (Figure 28.7a). The additions inside stretch the outer bark, which splits in characteristic patterns for each species. Cells added to the xylem can be of different types and uses – structural, transport, and repair. Cell size and density can vary during the growing season, causing annual rings. The cells themselves have layered cell walls consisting of long-chain cellulose molecules and other very hard "lignins."

Wood is weak and often prone to warping when it is added to a young part of a tree – whether a young tree or a newly formed, high part of an old tree's stem. This wood is "juvenile wood" (Figure 28.7a). The tree's center also contains many problematic knots. After about 18 years, "mature" wood is formed that is stronger and less prone to warping. Wood that grows too rapidly can have problems with strength and warping, but wood that grows too slowly can also be weak. The outer wood of a tree, the "sapwood," contains some cells that remain alive. Inside the sapwood is non-living, often colored wood characteristic of the species – the "heart wood" (Figures 28.7a–b). The sapwood and mature heartwood are generally similar in strength and warping, but the heart wood is sometimes valued for beauty and/or rot resistance.

Wood is used by cutting its solid form to a desired shape, by cutting and regluing it, by decomposing it into its component chemicals, or by burning it as fuel. Wood can be cut into poles, timbers, and boards for structural and aesthetic uses. It can be cut and re-glued to avoid the defects and take advantage of the strength of wood parallel to its grain. Gluing wood layers in perpendicular directions can make strong and stiff plywood and cross-laminated timber (CLT, mass timber). Perpendicular orientations and strong glues enable traditionally weak woods (from juvenile wood and certain species) to be used in some construction.

Different species react to cutting, drying, and other processing differently; and tools and machines need to be adjusted accordingly. Sometimes species with similar processing characteristics can be grouped and milled together at a single machine adjustment.

Wood construction is excellent for withstanding earthquakes. It resists fires when built from large beams that char on the outside but stay strong beneath – either a single cut piece or pieces glued together. Most species rot if continually moist unless treated appropriately. Wood structures are light and require less massive foundations than concrete or brick structures; in fact, wood structures sometimes need to be anchored in windy areas.

Wood is also a matrix of long-chain hydrocarbons, pore spaces, and organic chemical bonds (lignin) that can readily be converted to plastic and microfibers. Wood has great, untapped potential for industrial chemical uses. It can be mechanically and chemically decomposed into its cellulose and lignin components to make papers of various strengths and qualities, plastics, nanofibers [75], plastics, aviation fuel [76, 77], and other materials.

Wood's potential for avoiding carbon dioxide emissions (Figures 28.5–6) has also been discussed in Chapter 9 (Figure 9.8). Its potential for avoiding fossil fuel use is discussed in Chapters 24–25 (Figures 24.1, 24.8, 25.8, Table 25.2).

References

1. UN FAO. *Global Forest Resource Assessment*. (United Nations Food and Agriculture Organization, 2010).
2. International Labour Organization. *Global Employment Trends: 2014. Supporting Data Sets*. (United Nations, International Labour Organization, 2014 [Accessed July 24, 2016]). Available from: www.ilo.org/global/research/global-reports/global-employment-trends/2014/WCMS_234879/lang–en/index.htm.
3. UN FAO. *Global Forest Resources Assessment 2015*. (United Nations, 2015).

4. C. D. Oliver, N. T. Nassar, B. R. Lippke, J. B. McCarter. Carbon, Fossil Fuel, and Biodiversity Mitigation with Wood and Forests. *Journal of Sustainable Forestry.* 2014;33(3):248–75.
5. International Labour Organization. *International Year of Forests 2011: What About Labour Aspects of Forestry?* (International Labour Organization, 2011).
6. J. Sundberg. NGO Landscapes in the Maya Biosphere Reserve, Guatemala. *Geographical Review.* 1998;88(3):388–412.
7. H. Powell. People Power and Sustainable Forestry Keep Deforestation at Bay in Guatemala. *Living Bird Magazine.* 2016.
8. C. D. Oliver. Sustainable Forestry: What Is It? How Do We Achieve It? *Journal of Forestry.* 2003;101(5):8–14.
9. R. Guldin, H. Kaiser. *National Report on Sustainable Forests–2003.* FS-766. (US Department of Agriculture, Forest Service, 2004).
10. N. E. Stork, T. Boyle, V. Dale, et al. *Criteria and Indicators for Assessing the Sustainability of Forest Management: Conservation of Biodiversity.* CIFOR Working Paper No. 17. 1997.
11. D. Wijewardana. Criteria and Indicators for Sustainable Forest Management: The Road Travelled and the Way Ahead. *Ecological Indicators.* 2008;8(2):115–22.
12. B. W. Cashore, G. Auld, D. Newsom. *Governing through Markets: Forest Certification and the Emergence of Non-State Authority.* (Yale University Press, 2004).
13. C. Bottorff. *Payments for Ecosystem Services: Cases from the Experience of US Communities.* (Key-log Economics, 2014).
14. G. C. Daily, S. Polasky, J. Goldstein, et al. Ecosystem Services in Decision Making: Time to Deliver. *Frontiers in Ecology and the Environment.* 2009;7(1):21–8.
15. J. Farley, R. Costanza. Payments for Ecosystem Services: From Local to Global. *Ecological Economics.* 2010;69(11):2060–8.
16. K. H. Redford, W. M. Adams. Payment for Ecosystem Services and the Challenge of Saving Nature. *Conservation Biology.* 2009;23(4):785–7.
17. G. Daily. *Nature's Services: Societal Dependence on Natural Ecosystems.* (Island Press, 1997).
18. M. C. Duguid, E. H. Morrell, E. Goodale, M. S. Ashton. Changes in Breeding Bird Abundance and Species Composition over a 20 year Chronosequence Following Shelterwood Harvests in Oak-Hardwood Forests. *Forest Ecology and Management.* 2016;376:221–30.
19. F. Montagnini, C. F. Jordan. *Tropical Forest Ecology: The Basis for Conservation and Management.* (Springer Science and Business Media, 2005).
20. N. Myers. Environmental Services of Biodiversity. *Proceedings of the National Academy of Sciences.* 1996;93(7):2764–9.
21. M. L. Oldfield. *The Value of Conserving Genetic Resources.* (Sinauer Associates Inc., 1984).
22. S. J. Pyne, P. L. Andrews, R. D. Laven. *Introduction to Wildland Fire.* (John Wiley and Sons, 1996).
23. H. D. Safford, D. A. Schmidt, C. H. Carlson. Effects of Fuel Treatments on Fire Severity in an Area of Wildland–Urban Interface, Angora Fire, Lake Tahoe Basin, California. *Forest Ecology and Management.* 2009;258(5):773–87.
24. M. A. Cochrane, M. D. Schulze. Forest Fires in the Brazilian Amazon. *Conservation Biology.* 1998;12(5):948–50.
25. R. L. Sanford, J. Saldarriaga, K. E. Clark, C. Uhl, R. Herrera. Amazon Rain-Forest Fires. *Science.* 1985;227(4682):53–5.
26. C. Wiedinmyer, J. Neff. Estimates of CO_2 from Fires in the United States: Implications for Carbon Management. *Carbon Balance and Management.* 2007;2(10).
27. A. C. Camp, C. D. Oliver, P. Hessburg, R. Everett. Predicting Late-Successional Fire Refugia Pre-Dating European Settlement in the Wenatchee Mountains. Forest Ecology and Management 95: 63–77. *Forest Ecology and Management.* 1997;95:63–77.
28. D. C. Morton, M. E. Roessing, A. E. Camp. *Assessing the Environmental, Social, and Economic Impacts of Wildfire.* (Global Institute of Sustainable Forestry, School of Forestry and Environmental Studies, Yale University, 2003).
29. U. Onon, translator. *Chinggis Khan: The Golden History of the Mongols.* Bradbury S, editor. (Folio Society, 1993).

30. L. M. Dávalos, A. C. Bejarano. Conservation in Conflict: Illegal Drugs Versus Habitat in the Americas. *State of the Wild 2008–2009: A Global Portrait of Wildlife, Wildlands, and Oceans.* (Island Press, 2008): pp. 218–25.

31. J. Hughes, M. S. Chandran. *Sacred Groves around the Earth: An Overview.* (Oxford and India Book House, 1998).

32. S. R. Kellert. Connecting with Creation: The Convergence of Nature, Religion, Science and Culture. *Journal for the Study of Religion, Nature and Culture.* 2007;1:25–37.

33. R. G. Walsh, J. B. Loomis, R. A. Gillman. Valuing Option, Existence, and Bequest Demands for Wilderness. *Land Economics.* 1984;60(1):14–29.

34. A. Randall, J. R. Stoll. Existence Value in a Total Valuation Framework. *Managing Air Quality and Scenic Resources at National Parks and Wilderness Areas.* (Westview Press, 1983): pp. 265–74.

35. C. Clayton, R. Mendelsohn. The Value of Watchable Wildlife: A Case Study of McNeil River. *Journal of Environmental Management.* 1993;39(2):101–6.

36. P. Maille, R. Mendelsohn. Valuing Ecotourism in Madagascar. *Journal of Environmental Management.* 1993;38(3):213–18.

37. G. Pinchot. A Primer of Forestry. *Farmers' Bulletin.* 1903;173.

38. T. Whelan. *Nature Tourism: Managing for the Environment.* (Island Press, 1991).

39. G. G. Whitney. *From Coastal Wilderness to Fruited Plain: A History of Environmental Change in Temperate North America from 1500 to the Present.* (Cambridge University Press, 1996).

40. R. Costanza, R. d'Arge, R. De Groot, et al. The Value of the World's Ecosystem Services and Natural Capital. *Nature.* 1997(387):253–60.

41. C. S. Roper, A. Park, editors. *The Living Forest: Non-Market Benefits of Forestry.* Proceedings of an International Symposium on Non-Market Benefits of Forestry. (United Kingdom Forestry Commission, 1999).

42. J. C. Jenkins, D. C. Chojnacky, L. S. Heath, R. A. Birdsey. National-Scale Biomass Estimators for United States Tree Species. *Forest Science.* 2003;49(1):12–35.

43. C. D. Oliver, B. C. Larson. *Forest Stand Dynamics*, updated edition. (John Wiley, 1996).

44. M. S. Ashton, M. L. Tyrrell, D. Spalding, B. Gentry. *Managing Forest Carbon in a Changing Climate.* (Springer, 2012).

45. R. L. Chazdon, E. N. Broadbent, D. M. Rozendaal, et al. Carbon Sequestration Potential of Second-Growth Forest Regeneration in the Latin American Tropics. *Science Advances.* 2016;2 (5):e1501639.

46. L. Poorter, F. Bongers, T. M. Aide, et al. Biomass Resilience of Neotropical Secondary Forests. *Nature.* 2016;530(7589):211–14.

47. N. L. Stephenson, A. Das, R. Condit, et al. Rate of Tree Carbon Accumulation Increases Continuously with Tree Size. *Nature.* 2014;507(7490):90–3.

48. R. E. McArdle, W. H. Meyer, D. Bruce. *The Yield of Douglas-Fir in the Pacific Northwest.* (USDA Forest Service, U.S. Department of Agriculture, 1961).

49. L. Davis, K. Johnson, P. Bettinger, T. Howard. *Forest Management*, fourth edition. (McGraw-Hill, 2001).

50. B. Lippke, E. Oneil, R. Harrison. Life Cycle Impacts of Forest Management and Wood Utilization on Carbon Mitigation: Knowns and Unknowns. *Carbon Management.* 2011;2(3): 303–33.

51. B. Lippke, J. Wilson, L. Johnson, M. Puettmann. *Life Cycle Environmental Performance of Renewable Materials in the Context of Building Construction, Phase Ii Research Report.* (The Consortium for Research on Renewable Industrial Materials, 2010).

52. B. Lippke, J. Wilson, J. Perez-Garcia, J. Bowyer, J. Meil. Corrim: Life-Cycle Environmental Performance of Renewable Building Materials. *Forest Products Journal.* 2004;54(6):8–19.

53. T. Stocker, D. Qin, G. Plattner, et al. *Climate Change 2013: The Physical Science Basis.* Intergovernmental Panel on Climate Change. (Cambridge University Press, 2013).

54. K. Naudts, Y. Chen, M. J. McGrath, et al. Europe's Forest Management Did Not Mitigate Climate Warming. *Science.* 2016;351(6273):597–600.

55. S. Bentsen, T. Nord-Larsen, S. Larsen, et al. *Letter Response To: Naudts et al. 2016. Europe's Forest Management Did Not Mitigate Climate Warming.* (Science eLetter: Science, American Association for the Advancement of Science, 2016 [Accessed July 28, 2016]). Available from: http://science.sciencemag.org/content/351/6273/597.e-letters.
56. IPCC. *Climate Change 2014: Synthesis Report.* (Intergovernmental Panel on Climate Change, 2014).
57. P. Smith, M. Bustamante, H. Ahammad, et al. Agriculture, Forestry and Other Land Use (AFOLU). *Climate Change 2014: Mitigation of Climate Change.* (Cambridge University Press, 2014).
58. R. A. Houghton, J. House, J. Pongratz, et al. Carbon Emissions from Land Use and Land-Cover Change. *Biogeosciences.* 2012;9(12):5125–42.
59. IPCC. *Terrestrial Carbon Processes: Background.* (Intergovernmental Panel on Climate Change, 2013). Available from: www.ipcc.ch/ipccreports/tar/wg1/099.htm.
60. N. Unger. Human Land-Use-Driven Reduction of Forest Volatiles Cools Global Climate. *Nature Climate Change.* 2014;4(10):907–10.
61. K. Covey, C. B. de Mesquita, B. Oberle, et al. Greenhouse Trace Gases in Deadwood. *Biogeochemistry.* 2016;130(3):215–26.
62. K. R. Covey, S. A. Wood, R. J. Warren, X. Lee, M. A. Bradford. Elevated Methane Concentrations in Trees of an Upland Forest. *Geophysical Research Letters.* 2012;39(15).
63. X. Lee, M. L. Goulden, D. Y. Hollinger, et al. Observed Increase in Local Cooling Effect of Deforestation at Higher Latitudes. *Nature.* 2011;479(7373):384–7.
64. O. Ndoye, M. R. Pérez, A. Eyebe. *The Markets of Non-Timber Forest Products in the Humid Forest Zone of Cameroon.* (Overseas Development Institute, 2016), 0968–2627.
65. C. Shackleton, S. Shackleton. The Importance of Non-Timber Forest Products in Rural Livelihood Security and as Safety Nets: A Review of Evidence from South Africa. *South African Journal of Science.* 2004;100(11–12):658–64.
66. J. H. De Beer, M. J. McDermott. *The Economic Value of Non-Timber Forest Products in Southeast Asia: With Emphasis on Indonesia, Malaysia and Thailand.* (Netherlands Committee for IUCN, 1989).
67. M. J. Balick, R. Mendelsohn. Assessing the Economic Value of Traditional Medicines from Tropical Rain Forests. *Conservation Biology.* 1992;6(1):128–30.
68. M. T. Rains, A. W. Rudie, T. H. Wegner. The Promise of Wood-Based Nanotechnology. *Consultant Magazine, Association of Consulting Foresters.* 2014.
69. UN FAO. *State of the World's Forests 2011.* (United Nations Food and Agriculture Organization, 2011).
70. A. Saxena, B. Guneralp, R. Bailis, G. Yohe, C. Oliver. Evaluating the Resilience of Forest Dependent Communities in Central India by Combining the Sustainable Livelihoods Framework and the Cross Scale Resilience Analysis. *Current Science.* 2016;110(7):1195.
71. T. Kozlowski. *Growth and Development of Trees: Vol. II: Cambial Growth, Root Growth, and Reproductive Growth.* (Academic Press, 1971).
72. P. Kramer, T. Kozlowski. *Physiology of Woody Plants.* (Academic Press, 1979).
73. A. L. Shigo. *Tree Anatomy.* (Shigo and Trees, Associates, 1994).
74. M. H. Zimmerman, C. L. Brown. *Trees: Structure and Function.* (Springer-Verlag, 1971).
75. N. Siddiqui, R. H. Mills, D. J. Gardner, D. Bousfield. Production and Characterization of Cellulose Nanofibers from Wood Pulp. *Journal of Adhesion Science and Technology.* 2011;25(6–7):709–21.
76. P. Fairley. Introduction: Next Generation Biofuels. *Nature.* 2011;474(7352):S2–S5.
77. J. R. Regalbuto. Cellulosic Biofuels – Got Gasoline? *Science.* 2009;325(5942):822–4.

29

Forest Distribution, Area, and Volume Changes

29.1 Forest Area and Timber Volume

Agriculture clearing has probably eliminated 11% of the forest land that would otherwise exist in the present climate (Table 15.2). Of the remaining forests (Figure 29.1), only about 40% of the timber was considered "accessible" in 1998 [1]. Accessible timber was defined as within 10 kilometers of a road or waterway for transport and not within a protected reserve, unlike "accessible land" in Figure 4.2 and Table 15.2.

Conifer forests are primarily found in the northern hemisphere (Figure 29.2) because most conifers are Neotropical species (Figure 15.1). Some southern hemisphere conifers (*Araucariaceae* family) are used for their wood. Northern hemisphere conifers have been established in some plantations in the southern hemisphere – especially slash (*Pinus elliottii*), loblolly (*P. taeda*), and Monterey (*P. radiata*) pines and Douglas fir (*Pseudotsuga menziesii*) (Figure 16.2b). Conifer forests are often found as pure (single-species) stands (forest communities, Chapter 14) or with only a few other species; consequently, many trees can be efficiently harvested and milled together.

Hardwood ("broadleaf" or "Angiosperm") forests are found throughout the world, with greatest concentrations in the African and South American tropics (Figure 29.2b). Hardwood stands often contain tens to hundreds of species together with only a few trees of each species on each hectare. Generally, only a few species have been promoted for commercial value; and processing machinery usually needs to be adjusted for each species. Consequently, hardwoods need to be harvested and processed with greater skill and care than conifers

A few hardwood species – poplars, aspens (*Populus*), and alders (*Alnus*) – commonly grow in pure species stands. *Dipterocarp* forests of southeastern Asia contain many species similar enough to be harvested and processed together.

The area of forest and standing volume of timber vary greatly among country groups (Figure 29.2), with six groups having most of the world's timber and only three having most of the conifers.

29.2 Forest Ownership

Most forests in the world are publicly owned and administered (Figure 29.3) [3]. This ownership strongly affects management.

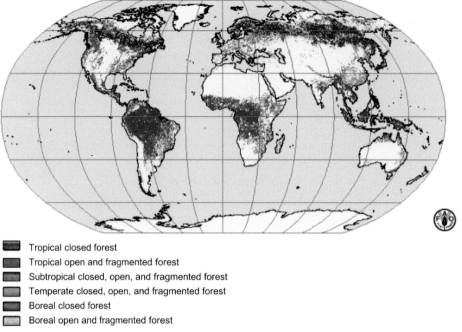

Tropical closed forest
Tropical open and fragmented forest
Subtropical closed, open, and fragmented forest
Temperate closed, open, and fragmented forest
Boreal closed forest
Boreal open and fragmented forest

Figure 29.1 World forests by major ecological domains (Figure 7.1). [1, 2]. [Also color plate 12, with legend.] Reproduced with permission, United Nations, Food and Agriculture Organization. (A black and white version of this figure will appear in some formats. For the color version, please refer to the plate section. Color plate 12.)

29.3 Area and Volume Changes

On average, very little of the world's wood growth is being harvested. The 385 billion cubic meters (m^3) of wood in the world's forests are potentially growing another 4.4% more wood each year, while the world is harvesting less than 1% of the standing volume and about 20% of the growth [3, 4]. The remaining 80% growth either burns, rots, or adds to the volume of wood in the forests.

Excess forest growth and increases in forest area are largely confined to developed regions (Figure 29.4). The small amount harvested is reflected in the crowded nature of many forests (Chapter 28).

Elsewhere, many forests are being overly harvested and the forest area is declining, albeit slowly. Of the 20% harvested of the world's wood growth, 55 is used for fuelwood – primarily inefficiently in open pits and crude stoves in developing countries in Africa, Central and South America, and Asia (Figure 29.5a). This subsistence fuelwood use contributes little to carbon dioxide sequestration (Figures 9.8, 28.5) and does little to avoid fossil fuels (Figures 24.1, 24.8). Another 16% is used for pulp to make paper. Only 29% is used to make longer-lasting materials such as buildings and furniture. Much more wood could be used in a variety of creative ways – such as building products – that could help the environment.

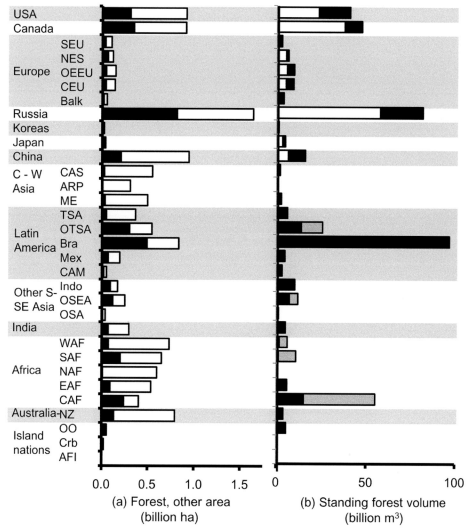

Figure 29.2 (a) Area and (b) volume of forests by country groups. (a) Black = forest; white = other land. (b) White = conifer; black = hardwoods; gray = unknown (probably hardwoods). [3]

Figure 29.5b shows the approximate average age of forests when harvested. For example, if 2% of the volume is harvested yearly, the average harvest age is roughly 50 years. (Accounting for growth would make it slightly older.) Where 4% is harvested, the age is roughly 25 years. India, North Africa, and East Africa probably harvest trees when young because they need the wood (Figure 29.5b), and they grow intensive plantations that are harvested when young.

The trends in forest area and volume change are for a variety of reasons (Figure 29.6). China has implemented an afforestation program. The United States

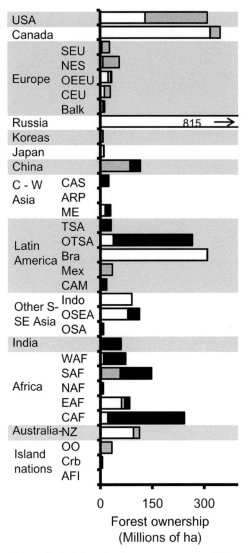

Figure 29.3 Forest ownership and administration by country groups. White = public; gray = private; black = unknown [3].

is increasing its wood volume much faster than its forest area, resulting in over-crowding that is reflected in catastrophic forest fires [6–8]. Brazil is decreasing its forest area and volume, a threat to biodiversity; on the other hand, Brazil has more area in forest reserves than any other global region (Figure 16.2). East Africa is rapidly reducing its forest area; and Indonesia and Central Africa are rapidly reducing their forest volume.

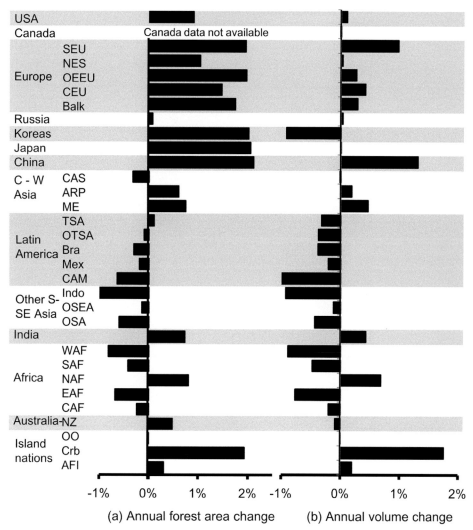

Figure 29.4 Annual percentage change in (a) forest area and (b) forest volume by country groups, 2010–15 [3, 5].

29.4 Forest Employment

Over 90% of the world's forest workers are employed to produce forest products [9]. Wood products contribute 65% of the global revenue obtained from forests, including solid wood products and wood for papermaking [9]. Timber production has also provided most of the machinery to shape the forests. Consequently, the present timber orientation has largely determined the forest stand structures and values.

Logging with large equipment employs few people in the woods and brings little value added (Figure 29.7). Secondary and tertiary processing of wood employs many more

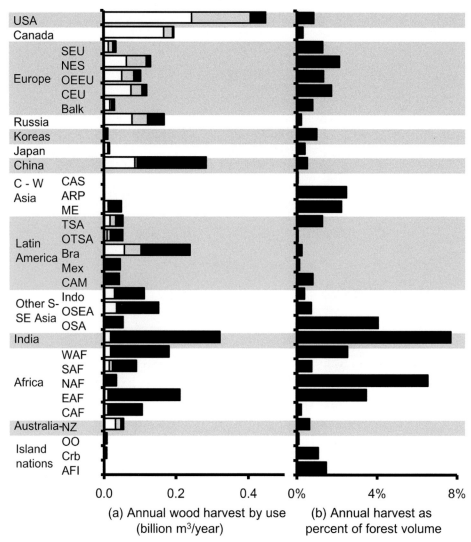

Figure 29.5 (a) Annual wood harvest by use by country groups. White = construction; gray = pulpwood; black = fuelwood. (b) Annual harvest as a percentage of standing volume [1].

people (Figure 29.7) [10] but is often done far from forests. Without secondary manufacturing in forest communities, there are few people and little turnover of money [11]. The few people mean little support for schools and retail establishments, so people become uninterested in living there (Figure 4.6).

    Large logging machinery is expensive, so loggers cannot use it seasonally; they need to operate year-round and rapidly repay the cost. Alternatively, logging and sawmilling could be done seasonally with small machinery and can complement farm, recreation, and other

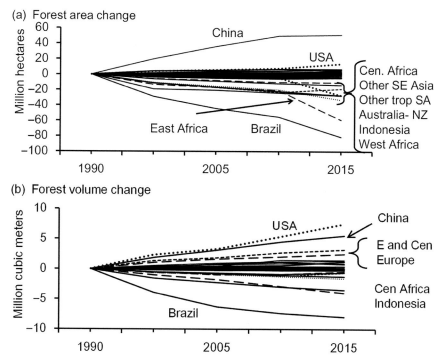

Figure 29.6 Change in forest area and standing volume (growing stock) for major countries and regions [3, 9].

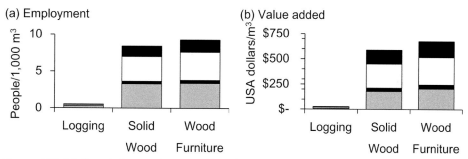

Figure 29.7 (a) Cumulative employment (b) and cumulative value added in logging, secondary processing of logs to lumber, and tertiary processing to furniture per forest volume sustainably harvested in Illinois state, USA. Bottom gray = direct employment; lower black = secondary production; white = secondary services; top black = financial services [10].

seasonal work. Saxena [12, 13] has found that, without the seasonal access to forests, people cannot maintain their livelihoods in some rural communities and need to migrate to cities.

New technologies are making it feasible to substitute smaller, less expensive, highly technical equipment in forestry for the large machines of the past. Innovations in precision, portable sawmills [14, 15] and roadless hauling techniques [16, 17] mean boards could be moved from the woods instead of heavy logs. Such smaller machinery costs less, but employs more skilled people in the forests. Costs of building roads are also saved. The savings by using small machinery are becoming greater than the costs of employing more people.

The low overhead of small equipment means the workers could work more carefully – wasting fewer trees and restoring the needed structures, similar to landcare workers [18–24]. The low overhead also means some work could be seasonal, since large monthly payments needed for large equipment are not necessary.

The global forestry sector employs about 13.7 million workers, 0.4% of the global labor force. [25, 26] for the 29% of the world's land in forests. Approximately 1.4% would be directly employed if conventional, large equipment were used and all of the world's wood growth were fully harvested and processed each year (based on Figure 29.7). A rough calculation suggests that four times as many people could be employed to harvest the same timber volume using small, high-technology equipment [14–17]. The rough calculation also shows that the small equipment would be more profitable.

References

1. UN FAO. *Global Forest Resource Assessment*. (United Nations, Food and Agriculture Organization, 2000).
2. UN FAO. *Forests 2000 by Major Ecological Domains*. (United Nations, Food and Agriculture Organization, 2017 [Accessed September 11, 2017]). Available from: www.fao.org/geonetwork/srv/en/metadata.show?id=1254&currTab=simple.
3. UN FAO. *FLUDE: The Forest Land Use Data Explorer*. (United Nations Food and Agriculture Organization, 2015 [Accessed May 25, 2016]). Available from: www.fao.org/forest-resources-assessment/explore-data/en.
4. C. D. Oliver, N. T. Nassar, B. R. Lippke, J. B. McCarter. Carbon, Fossil Fuel, and Biodiversity Mitigation with Wood and Forests. *Journal of Sustainable Forestry*. 2014;33 (3):248–75.
5. UN FAO. *Global Forest Resource Assessment*. (United Nations Food and Agriculture Organization, 2010).
6. D. C. Morton, M. E. Roessing, A. E. Camp. *Assessing the Environmental, Social, and Economic Impacts of Wildfire*. (Global Institute of Sustainable Forestry, School of Forestry and Environmental Studies, Yale University, 2003).
7. C. Oliver, D. Adams, T. Bonnicksen, et al. *Report on Forest Health of the United States by the Forest Health Science Panel*. (Center for International Trade in Forest Products, University of Washington, 1997).
8. R. N. Sampson, D. L. Adams, S. S. Hamilton, et al. Assessing Forest Ecosystem Health in the Inland West. *Journal of Sustainable Forestry*. 1994;2(1–2):3–10.
9. UN FAO. *State of the World's Forests 2011*. (United Nations Food and Agriculture Organization, 2011).
10. J. E. Henderson, I. A. Munn. *Forestry in Illinois: The Impact of the Forest Products Industry on the Illinois Economy: An Input-Output Analysis*. (Illinois Forestry Development Council, 2012).
11. D. Acemoglu, J. Robinson. *Why Nations Fail: The Origins of Power, Prosperity, and Poverty*. (Crown Business, 2012).
12. A. Saxena. Climate Change Resilience: Can Joint Forest Management Help Indian Forest and Communities? *Journal of Resources, Energy and Development*. 2011;8(2):75–88.

13. A. Saxena, B. Guneralp, R. Bailis, G. Yohe, C. Oliver. Evaluating the Resilience of Forest Dependent Communities in Central India by Combining the Sustainable Livelihoods Framework and the Cross Scale Resilience Analysis. *Current Science*. 2016;110(7):1195.

14. M. Matrix. *Lucas Mill*. (Lucas Mill, 2017 [Accessed June 8, 2017]). Available from: www .lucasmill.com.

15. *Wood-Mizer*. (Wood-Mizer LLC, 2017 [Accessed June 6, 2017]). Available from: www .woodmizer.com/us.

16. H. H. Acar, editor. *Timber Extraction by Cable Cranes, Monorail and Chute Systems in Turkish Forestry*. Working globally-sharing forest engineering challenges and technologies around the world; Coeur d'Alene. (Council on Forest Engineering, 2006).

17. W. Watts, T. Ward. *Zig-Zag Monocable Yarder: A Concept for Yarding Small Logs and Firewood*. (United States Forest Service, State and Private Forestry, 1989).

18. A. Campbell, G. Siepen. *Landcare: Communities Shaping the Land and the Future*. (ICON Group International, 1994).

19. J. Cary, T. Webb. Landcare in Australia: Community Participation and Land Management. *Journal of Soil and Water Conservation*. 2001;56(4):274–8.

20. J. Ellis-Jones. Poverty, Land Care, and Sustainable Livelihoods in Hillside and Mountain Regions. *Mountain Research and Development*. 1999:179–90.

21. NRM Group. *An Evaluation of Investments in Landcare Support Projects*. (URS Australia and Griffin, 2001).

22. E. Compton, R. B. Beeton. An Accidental Outcome: Social Capital and Its Implications for Landcare and the "Status Quo." *Journal of Rural Studies*. 2012;28(2):149–60.

23. A. Curtis, M. Lockwood. Landcare and Catchment Management in Australia: Lessons for State-Sponsored Community Participation. *Society and Natural Resources*. 2000;13(1):61–73.

24. R. Youl, S. Marriott, T. Nabben. *Landcare in Australia*. (SILC and Rob Youl Consulting Pty Ltd. 2006).

25. International Labour Organization. *International Year of Forests 2011: What About Labour Aspects of Forestry?* (International Labour Organization, 2011).

26. International Labour Organization. *Forestry, Wood, Pulp and Paper Sector*. (International Labour Organization, 2016 [Accessed October 21, 2016]). Available from: www.ilo.org/glo bal/industries-and-sectors/forestry-wood-pulp-and-paper/lang–en/index.htm.

30

Silviculture, Forest Degradation, Landscape Management

30.1 Background and Sustained Yield

During the relatively short 70 thousand years of modern human existence, reductions and expansions of global forest areas, species migrations, species extinctions, changes in dominant stand structures, and changes in landscape patterns have occurred with climate changes – and recently with human lifestyle changes. Windstorms, fires, floods, volcanic eruptions, and other disturbances have also altered stand structures and trajectories of change in all forests [1–5].

The approaches used to manage other resources have also been tried in forests: exploitation, infrastructure changes, preservation, sustained production ("sustained yield"), and free market supply-and-demand. Given the disproportionately large number of endangered forest species, little timber harvested, imbalance of stand structures, and little employment provided by forests (Chapter 29), none of these approaches is meeting current needs.

Sustained production provides the great promise for maintaining ecosystem services if the sustainability target is changed from timber to broader goals. A more general target could be to sustain forests of all native species in a diversity of structures across each forested landscape within each global ecological zone [6, 7]. The needed structures could be defined in scientifically agreed-upon ways for different forest types, such as Figure 14.5b or other classifications. Maintaining these structures would be a coarse filter approach (Chapters 14, 16). It would also allow all criteria for sustainable forestry to be maintained; and each region would contribute to both inter- and intra-generational equity emphasized in the Brundtland Report and elsewhere [8, 9].

Trees usually grow between 10 cm and 2 m per year in height when young. Trees can grow from 2 mm to 4 cm per year in diameter, slowing down as they age. They can grow in stem wood volume from 0.5 to over 8 m^3 per ha-year, with some intensive plantations growing over 20 m^3 per ha-year. Tree growth often appears slow and unnoticeable; however, each year's growth is added to previous years' and to every tree, so trees and forests become surprisingly larger and more crowded within a short time. Growth rates depend on soils; in some soils they can be nearly 50 meters tall by 50 years of age and in others can be less than 12 meters.

A stand's productivity is commonly measured by its "site index" – the height of the tallest trees in a stand at a given "index age" – usually 30, 50, or 100 years. Different species have different site indexes for the same area. Crowding of trees affects both individual tree growth and total volume growth of the area. In the same soil/climate conditions, some species can grow twice as much volume as others.

30.2 Silviculture Systems, Pathways, and Operations

Many centuries of experience – trial and error and research – throughout the world have led to an extensive body of knowledge about forestry. Silviculture is the management within a forest stand [10]. Specifically, it is "The art and science of controlling the establishment, growth, composition, health, and quality of forests and woodlands to meet the diverse needs and values of landowners and society on a sustainable basis" [11]. No single "correct" way exists to manage a stand, nor are sensitive operations such as "multi-aged" [12] "selection," and "clearcutting" [10] *good* or *bad* except in the context of their effect on the landscape pattern over time.

Silviculture had been expressed in terms of "silviculture systems" whose names are based on either the number of age classes within a stand or on the regeneration method used [11]. More recently silviculture is also being regarded in terms of silviculture "pathways" and "operations" [13, 14].

A stand development pathway is created as each stand's structure changes over time through growth and disturbances (Figure 14.5b) [13, 14]. A stand can have many potential pathways depending on tree growth rates, types and timings of disturbances, and regenerating species. Silviculturists design and direct each stand's pathway as an appropriate "silvicultural pathway" by thinning, clearcutting, or selection harvesting as well as weeding, planting, and other activities discussed below.

Silviculture involves relatively infrequent activities in an area compared to most resource management. A given stand is entered to do most operations for a few weeks only about once every decade except for annual or bi-annual underburning; and it is heavily harvested even less frequently. The occasional, short-duration activities that alter stand pathways are "silvicultural operations" [13]. They are generally grouped into types of operations and have several methods for doing each.

Operations often combat, mimic, protect, or recover from non-anthropogenic disturbances; or they copy agriculture techniques to ensure the desired species and structures are achieved. Each operation entails very intricate, scientific techniques, specialized tools and equipment, and knowledge. Peripheral subjects such as genetic improvement of trees, maintaining an inventory, growing tree seedlings in a nursery, maintaining an access system, and monitoring and extinguishing wildfires are also involved.

30.2.1 Stand Protection and Restoration

Forests are susceptible to disturbances with varying degrees of human influence – from windstorms or volcanoes that are almost entirely non-human to erosion, slope failure, fires that may have a human component in soil exposure, fuel buildup, or ignition source. Damage from disturbances can be minimized by putting stands in appropriate structures, combatted by direct actions against insects or wildfires, and mimicked by using forest harvesting or controlled burning as surrogates for windstorms and wildfires.

Not all disturbances can be minimized, and restoration efforts are needed to mitigate their impacts. Restoration at the stand level involves judicious use of the silvicultural operations described below to enable a stand to develop through an appropriate pathway [15, 16]. The appropriate pathway is best designed at the landscape level, discussed below.

30.2.2 Silviculture Systems

Silviculture systems have strong influences on the regeneration and future stand's pathway [10–12]. Systems include "clearcutting" all trees, leaving a few trees either temporarily or permanently ("two-aged methods" – "clearcutting with reserves," "seed tree," or "shelter-wood"), and "Uneven-aged (Selection)" – removing a limited proportion of trees in all sizes through "group selection," "single tree selection," and others [12]. Removing some trees to salvage crowded trees before they die while promoting growth of the remaining trees is referred to as "thinning." Thinning is not intended to promote regeneration.

Clearcutting removes all trees from a stand (Figure 14.14a) and is favored by loggers if all species can be profitably removed or if steep terrain mandates logging with cables. Concerned about a global wood shortage, silviculturists began favoring clearcutting in the 1950s when selection harvesting proved inadequate in sustaining timber growth [17]. Were carbon sequestration the only objective of forestry, clearcutting would probably be most appropriate in most cases (Figure 28.5b).

Other multiage silviculture systems are used to recruit a new generation of trees without clearcutting (Figure 30.1) [12, 18]. These strategies require expertise and may exclude

Figure 30.1 Partial cutting that successfully allows young, vigorous trees of many species to grow (Washington State, USA) [13, 14]. Reproduced with permission of Taylor and Francis Group.

Figure 30.2 (a) Previously clear-cut stand with valuable cherrybark oak trees growing straight. (b) Nearby stand on same soil but resulting from high grade harvesting, leading to crookedly growing sugarberry trees and no regenerating cherrybark oaks (Mississippi, USA) [13, 14]. Reproduced with permission of Taylor and Francis Group.

some tree species that need unshaded openings to survive (Figure 30.2) [19–21]. If applied to some stands across a landscape while clearcutting, thinning, and other systems are applied elsewhere on the landscape, they can promote a diversity of structures and greater biodiversity.

30.2.3 Silviculture Operations

Silviculture operations are generally divided into pre-regeneration "site preparation," "regeneration," post-regeneration "timber stand improvement (TSI)," and thinning activities. Overuse of a single operation in the past has led to large areas of little habitat diversity, even though the operation would be acceptable if not used excessively (Chapter 14).

Site preparation operations prepare a stand for regeneration. They can use different methods to mimic natural disturbances or facilitate planting of tree seedlings (young trees) depending on the regeneration mechanism of the desired species. Methods include "broadcast burning" the entire stand to eliminate excess biomass ("slash") from logging;

piling the slash (excess biomass such as tree limbs and leaves); using chemicals to kill excess vegetation; and sometimes creating ditches or plowing to drain or improve the soil.

Most forests regenerate without people planting seedlings or sowing seeds (Figure 16.2). Consequently, regeneration operations commonly can include protecting existing "advance regeneration" that was present before the harvest or preparing seedbeds for species that enter afterwards.

Seed sowing commonly requires collection of many seeds. Planting is costly and is used where wood is in high demand, where a desired species may not regenerate otherwise, or where genetically altered trees are desired. When planting seedlings instead of seeds is intended, tree seedlings are commonly grown in outdoor or greenhouse nurseries of various sophistication. They can be grown in individual plastic containers or directly in the soil.

Replanting genetically straight trees in place of crooked ones (Figure 16.1) can greatly increase the value of stands for wood products. Genetically modified trees "GMO" ("genetically modified organisms") have the same concerns as in agriculture (Chapter 21) [22].

"Timber Stand Improvement" (TSI) operations include weeding the stand by hand or with appropriate chemicals; cutting the excess trees that have no value ("thinning to waste" or "precommercial thinning"); applying fertilizers; and pruning lower tree limbs to obtain knot-free wood. Controlled burning is usually done by inducing a mild fire close to the ground to burn excess fuel and unwanted regenerating plants when desirable trees are large enough to withstand the fire. Inducing tree scars for animal nests and creating snags and down logs for other habitats are also TSI operations.

Thinning can keep trees from becoming physiologically weakened, spindly, and susceptible to insects and diseases or to falling over [23]. Usually the smaller, weaker trees are removed – "low thinning." Some dominant, vigorous trees can also be removed if they are malformed or in imminent danger of crowding other vigorous dominants. Overly crowded stands are subject to wildfires whether begun after partial cutting or clearcutting, with those beginning after selective harvesting creating "ladder fuels," where fires first ignite short trees in the understory and then preheat and ignite progressively taller trees until they reach the overstory.

30.3 Forest Degradation

Degradation is considered to be a diminished capacity to provide values. Each individual stand generally has diminished capacities to provide some values but provides others very well (Figure 28.2). Clearcutting is an acceptable practice that can cause degradation if used in the wrong situation or excessively (Figure 14.14a). A recent clearcut does not provide complex habitat; but it does provide habitats for butterflies, deer, and plants needing full sunlight very well (Figure 14.9c). However, the forest's capacity is not diminished if clearcut stands are complemented by stands of other structure providing other values across a landscape.

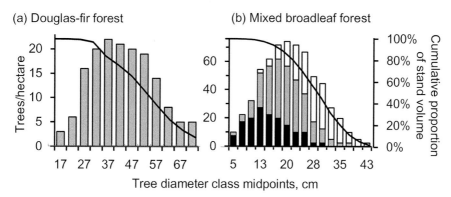

Figure 30.3 Diameter distributions of (a) single species stand and (b) mixed species stand showing that a large proportion of a stand's volume (line) are concentrated in the few, largest trees. Shades (right) represent different species [26, 27].

Even a few stands with overly crowded trees that pose insect or fire hazards would not degrade a large landscape because these stands would not imbalance the forest – place it "beyond the range of historical variability" [24] of stand structures across a landscape.

30.3.1 High-Grade Harvesting

Pretenses of selection thinning or uneven-aged regeneration methods [11, 12] have degraded forests by uniformly harvesting only the most healthy and valuable individuals – "high-grading" or "high-grade harvesting" (Figures 14.1c–d, 30.2b, 30.5) [25]. Removing the largest, most vigorous trees began with good intentions but incorrect science when the relay floristics (Figure 14.1a) paradigm was accepted. By removing the largest trees, it was assumed that the smaller, "younger" trees would replace them, growing straight and vigorously. Without silviculturists' supervision, loggers often still practice high-grading because they can harvest the large, valuable trees and not bother cutting the small, less valuable ones (Figure 30.3).

For the first few times that the most vigorous trees are removed – usually over several decades – the stand effectively maintains both closed forest habitats and timber production. However, removing these trees can decimate populations of the valuable tree species found only in the upper canopy in mixed species stands and lead them to extinction (Figure 30.2b). These upper canopy trees often need large disturbances and full sunlight to regenerate [19–21, 28, 29]. These stands can also mimic the rare, "pioneer loss" substructure of the complex structure [30, 31]. The pioneer loss structure would occur after all dominant (white) trees in the complex structure died ("white" trees of Figure 14.5b).

Most of a stand's volume and value are in a few large trees of a dominant species in both single- and mixed-species stands (Figure 30.3). Removing these trees can be effective at first because harvesting the large, valuable trees provides money to develop an

Figure 30.4 Stands whose most vigorous trees have been removed once, but still contain large trees that can be harvested, although crooked and of slightly inferior species. (a) Haryana, India. (b) Liberia.

infrastructure and then address the remaining forest. Usually enough initially unmerchantable species and crooked trees remain that can be profitably harvested once the infrastructure is in place (Figure 30.4). After several cutting entries, however, most remaining trees are too degraded – crooked, forked, and hollow – to have much use (Figure 30.5). The forest of few remaining species grows slowly; sometimes a few trees are harvested periodically just as they reach minimal value. These forests provide fewer wood products than if they contained more, vigorous trees growing larger before harvest. The slow growth and low tree volume also mean they are sequestering much less atmospheric carbon dioxide than would be possible in a different structure[32].

Mixed species stands can pose problems when harvesting is highly mechanized and timber prices are so low that the forest cannot be tended carefully, However, mixed species stands can actually provide more growth, biomass, timber quality, and biodiversity when management is skilled and profitable enough to take advantage of their benefits [33–37].

30.3.2 Degraded Forest Landscapes

As mentioned earlier, forests are considered degraded if they have reached a diminished capacity to provide an array of products and services such as the provision of timber and non-timber products, support to biodiversity, carbon storage, and others [38]. A large landscape without a variety of structures could be referred to as degraded. Degradation can result from environmental or anthropogenic influences. Many past well-intended ways

Figure 30.5 Stands whose most vigorous trees have been removed multiple times, and the remaining few species and weak trees grow slowly and provide little wood or carbon dioxide sequestration. (a) Idaho, USA. (b) Madhya Pradesh, India.

to manage forests – and some not so well-intended – have degraded forests in different patterns.

Degraded landscapes can contain many stand structures that had been rare across the landscape in the past, such as the high graded stands described above, overly crowded stands, stands with diminished use because of genetic de-selection (Figure 16.1a), stands of exotic species of little value, and stands where only the large trees and most valuable species have been removed (Figures 14.14d, 30.2b, 30.5) [30, 31]. Landscape degradation can also occur if there is a shortage of some structures across the landscape. Because of fire prevention, there are probably less areas of the previously common open and savanna structures in many forests in the world (Figure 14.11).

30.4 Landscape Management and Sustainability

Sustainability of forests has been a concern for over 500 years. In 1346 in France, the Waters and Forests Administration had the Brunoy executive order for foresters "to inspect all woods and ensure that they can perpetually sustain themselves in good condition." In 1713, H. C. von Carlowitz wrote, "Sylvicultura oeconomica, oder haußwirthliche Nachricht und Naturgemäße Anweisung zur Wilden Baum-Zucht. (Remagen-Oberwinter, Germany, Verlag Kessel.Dupuy" [39].

Providing a sustainable flow of timber from each stand by selective harvesting proved unsuccessful, but organizing many stands across a landscape to provide timber sustainably is feasible and done in many places [40]. Techniques have been adopted for providing a sustainable flow of other values from a forest landscape [41–43]. The key is to ensure some stands within the landscape are in each structure and species composition as needed. Thus, some stands will provide part of the non-commodity and non-timber forest product commodity values and complement stands in other structures that provide other values (Figure 28.2).

There is no single template for an appropriate "balance" of structures, but many forest landscapes are completely lacking some structures or have extremely little area of other structures. Many restoration efforts will be needed to move landscapes away from extreme imbalances of structures.

Forest stands do not remain static, however. As they grow to a different structure, they will lose their capacity to provide some values but will provide others [44]. Another management challenge to sustainability is to ensure that some stands are always changing into each structure and species composition as other stands change out of them.

Computer and satellite technologies are developing that will enable forests to be planned and monitored to ensure it is being managed sustainably [41, 42, 45, 46].

30.5 Broad Temporal and Spatial Scales and Land Use

Many forests have been interspersed with grasslands, lakes, deserts, sand dunes, rock outcrops, and other non-forest areas, further complicating the distributions of structures. Now, the forests are also interspersed with human-built landscape features – cities, farms, suburbs, highways, levees, and others. These human-built structures often have greater impact than is suggested by the area of forest they displace if they make movement among different areas and structures more difficult for plants and animals (Chapter 14).

Land use planning/zoning and landscape management may reverse these losses. Land use zoning is done in many countries and is increasingly including wildlife corridors and habitats [47–49]. Forest reserve areas are increasing worldwide and currently comprise over 15% of the forests (Figure 16.2). The distribution of reserves may not be the most appropriate, with areas such as Idaho, Montana, and Alaska, USA, having many reserves but relatively few species, and areas such as Georgia, USA, having many species but few reserves. For biodiversity, reserves are probably most justifiable to maintain the complex and old savanna structures. Clearcutting and multi-age silviculture can create many features in relatively young stands; and multiage silviculture is helpful if complex and old savanna forests are lacking. However, structural features of old trees such as large upper limbs and trunk cavities come primarily with age. The issue of how to address a complex forest reserve if a windstorm or fire destroys it needs to be prepared for.

30.6 Agro-Forestry and Intensive Plantations

"Agro-forestry" systems mix growing trees for timber and agriculture crops on the same area. The systems enable landowners to diversify the products grown on their land. They are especially effective where the trees and other crops occupy such different niches that they do not strongly inhibit each other's productivity; such as when winter wheat is grown beneath deciduous poplar trees before the poplar leaves emerge and cast shade in the spring and before the trees are large (Figure 30.6). Agroforestry systems are also useful in developing countries where land, wood, and money to import wood or its substitutes are limited because of a high PPPA (Chapter 6). In such cases, they have a similar role to intensive plantations.

"Intensive forest plantations" are commonly composed of a few species such as Monterey (*Pinus radiata*) or slash pines (*Pinus elliottii*), *Eucalyptus*, or poplars (*Populus*) that are

Figure 30.6 Agroforestry, with winter wheat growing beneath poplar (*Populus* sp.) trees (Haryana, India).

planted and cultured with almost farm-like intensity. The result is rapid growth of uniform trees which are scheduled for harvest at a young age [50].

Intensive forest plantations are often promoted to relieve the "pressure" to harvest other forests, to meet some anticipated wood shortage, to provide efficiencies of harvesting uniform wood with machinery, or to ensure a cheap raw product for mills. They are also promoted for fuel wood in countries with insufficient natural forests and an insufficiently strong economy to import fuelwood or alternative energy. Some intensive plantations are economically successful, such as Eucalyptus market pulp in South America or poplar plantations in India

The young age of harvesting intensive plantations means trees mostly contain juvenile wood that is unsuitable for some purposes. The rapid growth, high investment, and uniform growth create a financial and biological "window" for profitable harvest of very few years. Little or no profit is made if the world's abundant wood happens to be at a low price at that time [51].

30.7 Enhancing Biodiversity and Rural Employment

A sophisticated approach that maintains all structures (Figure 14.5b) across the landscape will probably be most successful in protecting forest biodiversity (Chapter 16). It could be done through a combination of some preserved areas and other areas actively managed for a variety of structures [52] as well as corridors across highways, farmlands, and similar obstacles. Local people could do appropriate silvicultural operations in the various stands to maintain a dynamic diversity of structures, removing wood products and non-timber forest products and providing other ecosystem services (Chapter 28).

The world needs social drivers, especially to employ more people in rural areas (Chapter 6). Forestry could become a social driver to provide new foci for employment, more biodiversity, reduction of fossil fuel use, and to avoid carbon dioxide emissions. Processing wood within the forest using more labor and smaller, highly technical tools could provide more employment and stimulate indirect employment – schoolteachers, shopkeepers, bus drivers, repair people, public administrators – so that small cities could established in currently rural areas. Such cities could have a critical mass of people to support retail shops with a variety of amenities (Figure 4.6). Such rural prosperity could relieve overly crowded cities and reduce the pressure to convert forests to other uses.

References

1. C. D. Canham, P. Marks. The Response of Woody Plants to Disturbance Patterns of Establishment and Growth, in Pickett S., White P., editors, *The Ecology of Natural Disturbance and Patch Dynamics*. (Academic Press, Inc., 1985): pp. 197–216.
2. V. H. Dale, A. E. Lugo, J. A. MacMahon, S. T. Pickett. Ecosystem Management in the Context of Large, Infrequent Disturbances. *Ecosystems*. 1998;1(6):546–57.

3. J. R. Runkle. Disturbance Regimes in Temperate Forests, in Pickett S., White P., editors, *The Ecology of Natural Disturbance and Patch Dynamics*. (Academic Press, Inc., 1985): pp. 17–33.

4. P. S. White. Pattern, Process, and Natural Disturbance in Vegetation. *The Botanical Review*. 1979;45(3):229–99.

5. P. S. White, A. Jentsch. The Search for Generality in Studies of Disturbance and Ecosystem Dynamics. In *Progress in Botany*. (Springer, 2001): pp. 399–450.

6. S. G. Boyce. *Landscape Forestry*. (John Wiley and Sons, 1995).

7. C. D. Oliver. A Landscape Approach: Achieving and Maintaining Biodiversity and Economic Productivity. *Journal of Forestry*. 1992;90:20–5.

8. G. Brundtland, M. Khalid, S. Agnelli, et al. *Our Common Future*. (Oxford University Press, 1987).

9. C. D. Oliver. Sustainable Forestry: What Is It? How Do We Achieve It? *Journal of Forestry*. 2003;101(5):8–14.

10. M. S. Ashton, M. J. Kelty. *The Practice of Silviculture: Applied Forest Ecology*, tenth edition. (John Wiley and Sons, 2018).

11. D. Adams, J. Hodges, D. Loftis, et al. *Silviculture Terminology with Appendix of Draft Ecosystem Management Terms*. (Society of American Foresters, 1994).

12. K. L. O'Hara. *Multiaged Silviculture: Managing for Complex Forest Stand Structures*. (Oxford University Press, 2014).

13. C. D. Oliver, K. L. O'Hara. Effects of Restoration at the Stand Level, in Stanturf J. A., Madsen P., editors, *Restoration of Boreal and Temperate Forests*. (CRC Press, 2004): pp. 31–59.

14. C. D. Oliver, K. L. O'Hara, P. J. Baker. Effects of Restoration at the Stand Level, in Stanturf J. A., Madsen P., editors, *Restoration of Boreal and Temperate Forests*. (CRC Press, 2015).

15. J. A. Stanturf. *Restoration of Boreal and Temperate Forests*. (CRC Press, 2015).

16. J. A. Stanturf, B. J. Palik, R. K. Dumroese. Contemporary Forest Restoration: A Review Emphasizing Function. *Forest Ecology and Management*. 2014;331:292–323.

17. S. G. Boyce, C. D. Oliver. The History of Research in Forest Ecology and Silviculture. In *Forest and Wildlife Science in America: A History*. (Forest History Society, 1999). pp. 414–53.

18. K. L. O'Hara. Silviculture for Structural Diversity: A New Look at Multiaged Systems. *Journal of Forestry*. 1998;96(7):4–10.

19. C. D. Oliver, E. C. Burkhardt, D. A. Skojac. The Increasing Scarcity of Red Oaks in Mississippi River Floodplain Forests: Influence of the Residual Overstory. *Forest Ecology and Management*. 2005;210(1):393–414.

20. L. K. Snook. Catastrophic Disturbance, Logging and the Ecology of Mahogany (*Swietenia Macrophylla* King): Grounds for Listing a Major Tropical Timber Species in Cites. *Botanical Journal of the Linnean Society*. 1996;122(1):35–46.

21. L. K. Snook. Regeneration, Growth, and Sustainability of Mahogany in Mexico's Yucatan Forests, in Lugo A., Figueroa-Colon J., Alayon M., editors, *Bigleaf Mahogany: Genetics, Ecology, and Management*. (Springer-Verlag, 2003): pp. 169–92.

22. K. M. Gartland, R. M. Crow, T. M. Fenning, J. S. Gartland. Genetically Modified Trees: Production, Properties, and Potential. *Journal of Arboriculture*. 2003;29(5):259–66.

23. C. D. Oliver, B. C. Larson. *Forest Stand Dynamics*, update edition. (John Wiley, 1996).

24. P. Morgan, G. H. Aplet, J. B. Haufler, et al. Historical Range of Variability: A Useful Tool for Evaluating Ecosystem Change. *Journal of Sustainable Forestry*. 1994;2(1–2):87–111.

25. P. Catanzaro, A. D'Amato. *High Grade Harvesting*. (University of Massachusetts Extension, 1998).

26. R. E. McArdle, W. H. Meyer, D. Bruce. *The Yield of Douglas-Fir in the Pacific Northwest*. (USDA Forest Service, U.S. Department of Agriculture, 1961).

27. G. L. Schnur. *Yield, Stand, and Volume Tables for Even-Aged Upland Oak Forests*. (US Department of Agriculture, 1937).

28. P. J. Baker, S. Bunyavejchewin, C. D. Oliver, P. S. Ashton. Disturbance History and Historical Stand Dynamics of a Seasonal Tropical Forest in Western Thailand. *Ecological Monographs*. 2005;75(3):317–43.

29. J. Fernandez-Vega, K. R. Covey, M. S. Ashton. Tamm Review: Large-Scale Infrequent Disturbances and Their Role in Regenerating Shade-Intolerant Tree Species in Mesoamerican Rainforests: Implications for Sustainable Forest Management. *Forest Ecology and Management.* 2017;395:48–68.
30. J. F. Franklin, T. A. Spies, R. Van Pelt, et al. Disturbances and Structural Development of Natural Forest Ecosystems with Silvicultural Implications, Using Douglas-Fir Forests as an Example. *Forest Ecology and Management.* 2002;155(1):399–423.
31. P. S. Park, C. D. Oliver. Variability of Stand Structures and Development in Old-Growth Forests in the Pacific Northwest, USA. *Forests.* 2015;6(9):3177–96.
32. N. Sasaki, F. E. Putz. Critical Need for New Definitions of "Forest" and "Forest Degradation" in Global Climate Change Agreements. *Conservation Letters.* 2009;2(5):226–32.
33. C. D. Oliver. Even-Aged Development of Mixed-Species Stands. *Journal of Forestry.* 1980;78 (4):201–3.
34. D. Piotto. A Meta-Analysis Comparing Tree Growth in Monocultures and Mixed Plantations. *Forest Ecology and Management.* 2008;255(3):781–6.
35. D. Piotto, D. Craven, F. Montagnini, F. Alice. Silvicultural and Economic Aspects of Pure and Mixed Native Tree Species Plantations on Degraded Pasturelands in Humid Costa Rica. *New Forests.* 2010;39(3):369–85.
36. F. Montagnini, D. Piotto. Mixed Plantations of Native Trees on Abandoned Pastures: Restoring Productivity, Ecosystem Properties, and Services on a Humid Tropical Site. *Silviculture in the Tropics.* (Springer, 2011). p. 501–11.
37. F. Montagnini, E. Gonzalez, C. Porras, R. Rheingans. Mixed and Pure Forest Plantations in the Humid Neotropics: A Comparison of Early Growth, Pest Damage and Establishment Costs. *The Commonwealth Forestry Review.* 1995:306–14.
38. D. Schoene, W. Killmann, H. von Lupke, M. LoycheWilkie. *Definitional Issues Related to Reducing Emissions from Deforestation in Developing Countries.* (Food and Agriculture Organization of the United Nations, 2007).
39. F. Schmithüsen. Three Hundred Years of Applied Sustainability in Forestry. *Unasylva.* 2013;64 (1):3–11.
40. L. Davis, K. Johnson, P. Bettinger, T. Howard. *Forest Management*, fourth edition. (McGraw-Hill, 2001).
41. J. B. McCarter, J. S. Wilson, P. J. Baker, J. L. Moffett, C. D. Oliver. Landscape Management through Integration of Existing Tools and Emerging Technologies. *Journal of Forestry.* 1998;96 (6):17–23.
42. C. D. Oliver, J. B. McCarter, K. Ceder, C. S. Nelson, J. M. Comnick. Simulating Landscape Change Using the Landscape Management System, in Millspaugh J. J., Thompson F. R., editors, *Models for Planning Wildlife Conservation in Large Landscapes.* (Elsevier and Academic Press, 2009): pp. 339–66.
43. N. L. Crookston. Suppose: An Interface to the Forest Vegetation Simulator. In Teck, R.; Moeur, M.; Adams, J. *Proceeding Forest Vegetation Simulator Conference.* (US Forest Service General Technical Report INT-GTR-373, 1997).
44. M. C. Duguid, E. H. Morrell, E. Goodale, M. S. Ashton. Changes in Breeding Bird Abundance and Species Composition over a 20 year Chronosequence Following Shelterwood Harvests in Oak-Hardwood Forests. *Forest Ecology and Management.* 2016;376:221–30.
45. J. M. Marzluff, J. J. Millspaugh, K. R. Ceder, et al. Modeling Changes in Wildlife Habitat and Timber Revenues in Response to Forest Management. *Forest Science.* 2002;48(2):191–202.
46. J. Millspaugh, F. R. Thompson. *Models for Planning Wildlife Conservation in Large Landscapes.* (Academic Press, 2011).
47. T. Nakamura, K. Short. Land-Use Planning and Distribution of Threatened Wildlife in a City of Japan. *Landscape and Urban Planning.* 2001;53(1):1–15.
48. L. C. Weaver, P. Skyer. Conservancies: Integrating Wildlife Land-Use Options into the Livelihood, Development and Conservation Strategies of Namibian Communities. in *Conservation and Development Interventions at the Wildlife/Livestock Interface: Implications*

for wildlife, livestock and human health. (Animal Health and Development (AHEAD) Forum, 2003);30:89–104.

49. J. Randolph. *Environmental Land Use Planning and Management.* (Island Press, 2004).
50. W. C. Price, N. Rana, A. Sample. *Plantations and Protected Areas in Sustainable Forestry.* (CRC Press, 2006).
51. C. D. Oliver, R. Mesznik. Investing in Forestry: Opportunities and Pitfalls of Intensive Plantations and Other Alternatives. *Journal of Sustainable Forestry.* 2006;21(4):97–111.
52. R. S. Seymour, M. L. Hunter, Jr. Principles of Ecological Forestry, in Hunter M. L., Jr., editor. *Maintaining Biodiversity in Forest Ecosystems.* (Cambridge University Press, 1999): pp. 22–61.

Part XI

Perspective

31

Integrating the Environment, Resources, and People

31.1 Managing Dynamic Socioenvironmental Systems

The environment, resources, and human societies are constantly changing but at different rates. Some changes are gradual but continuous and others are abrupt between periods of stability. Most are irreversible. The world's climates have been unusually regular for the past 10 thousand years (Chapter 9), but the human population, areas inhabited, and lifestyles have altered dramatically (Chapter 5). Changes often created hardships, especially if people were not socially or technically prepared. To endure and even thrive with changes, people need to be informed, cooperative (Chapter 6), and equally or more flexible than the environmental and resource systems they are confronted with ("law of requisite variety," Chapter 3 [1]).

The recent millennia of stable climates were a "Golden Age" where societies became more prosperous than during all previous 60 thousand-plus years of modern human existence. Besides a population and knowledge expansion, there were a "Green Revolution" and science, energy, technology, communication, transportation, and health revolutions. Climate change is currently most talked about, but other aspects of life will also be different. People's "Golden Age" could continue or could decline, depending on the choices they make.

Resources, the environment, and human societies are best understood if viewed as complex, dynamic systems with emergent properties (Chapter 2). They are based on simple "agents" and building blocks that create systems not completely explainable from their components – water from hydrogen and oxygen, living creatures from a mixture of elements and energy, various human societies from a mixture of people's behaviors. Components and the systems as a whole are constantly changing, reorganizing, often reforming into dramatically different behaviors. Subcomponents also reorganize at different rates, maintaining a vibrancy that gives these systems their longevities.

People are now appreciating the dynamic nature of the Earth. They realize that to sustain human life, the various changes need to be managed. Consequently, "sustainability" does not mean a "static" world, but one in which changes are recognized and proactively addressed (Chapter 3). This book has examined nine systems – three environmental, five resource, and human societies. Simplistically, the interfaces among these create $9 \times 9 = 81$ interactions. Each of the systems has many internal variations that generate more interactions, so that a very large number of interactions occur – with as many possible futures for people and the world. All interactions cannot be controlled. Managing just one factor such as climate or biodiversity will not create an ideal world. And there is neither perfect knowledge nor perfect ignorance about the interactions.

People have learned much about these systems and how to address them. Their outcomes are best considered probabilistic, not deterministic (Figure 2.1). The many likely and impactful events are generally addressed, with contingencies for slightly less likely ones. People can learn to identify, monitor, understand, anticipate, and manage potential changes (Chapter 3). The systems cannot be addressed by a few highly educated people. Instead, it needs most of society to understand and willingly participate in appropriate lifestyles – lifestyles that can become quite fulfilling.

Important are to look for opportunities in anticipated changes; work cooperatively, promote knowledge and knowledge, but keep a diversity of ideas; maintain a communication infrastructure; ensure decision-makers are concerned about the people's welfare; and share resources instead of hoarding them (Chapter 6). Knowledge and awareness can be improved with science and education.

Management is situational: a good decision at one time/place may be bad at another. New decisions may be needed to rectify past actions that are no longer effective. Mental and computer models and case studies can "game" possible future scenarios.

People have learned that some environmental and resource systems are best managed by preserving them intact, others by modifying them, and others by replacing the system with a new one. In three resources systems – energy, water, and minerals – the total amounts do not change, but their uses can be expanded by circulating them – the "multiplier" effect. Biological systems – biodiversity, agriculture, and forestry – can be increased or reduced; and people have learned quite a bit about how to do it.

Models help identify leverage points when/where/how efficient actions will be effective. Leverage points can be used to remain in or transform to a desirable condition and anticipate and preadjust to avoid or mitigate an undesirable one.

The understanding of environmental and resource system behaviors is a never-ending task: the more that is understood, the more people can benefit. Presently, the world needs more diffuse, leapfrog technologies instead of current centralized, infrastructure-intense ones; more social drivers and "inclusive economies" to increase and diversify employment (Chapter 6); and "critical population small cities" to repopulate rural areas and reduce crowded cities (Chapter 4).

People can share resources without war, as evidenced by the many cross-boundary water treaties (Chapter 18). People can transform societal behaviors with taxes, regulations, and other means. Finally, there are enough examples of stupid human decisions not to be complacent: the depletion of forests on Easter Island [2], selling phosphate reserves without investing sustainably in Nauru [3], wars, and others [4, 5].

31.2 Future Climate Scenarios

The short-term future climate is relatively predictable, but the longer-term may change in either of two ways, discussed below. Both scientists and the public accept weather predictions as probabilistic, rather than demanding certainty before taking action. For example, they do not wait for certainty before preparing for a blizzard, drought, hurricane, or tornado. It is now highly probable both that changing climates and a net global warming are occurring and that carbon dioxide emissions, largely from fossil fuels, contribute. Given this high probability coupled with the global wars and strife accompanying fossil fuel use and the ease of transitioning away from fossil fuels (Chapter 25), it is prudent to work globally and collectively toward a fossil fuel-free world.

31.2.1 Short-Term Future Climate

The recent climate trajectory and global temperature increase will probably continue in the short term – the next few decades and possibly centuries (Chapters 8 and 9). Weather patterns will remain or become more erratic; equatorial regions will become hotter and drier, and upper latitudes may heat less or even become cooler. These are already transforming some grasslands to deserts and may transform some forests to grasslands. Sea levels will probably rise a few centimeters or meters.

The above short-term changes are common to both of two highly possible climate and geography scenarios that will begin anytime between now and a few thousand years hence. This is not a long time considering that most large modern cities in Europe, Asia, and Africa are much older than 1 thousand years.

31.2.2 Scenario 1: Long-Term Climate Changes Without Glaciers Advancing

A strong probability is that heat from the high atmospheric carbon dioxide levels will prevent glacial advances, as predicted as "virtually certain" for the next 1 thousand years by the Intergovernmental Panel on Climate Change (IPCC) [6]. If the present warming trajectory continues, the heat would become difficult for people, crops, and many other biota to live near the tropics. As a worst-case scenario, many plant and animal species could go extinct before they could migrate poleward if the warming rate does not slow. Climates at upper latitudes would probably remain as cool as now, and possibly even cooler because they will receive less sunlight as the Milankovitch Cycle progresses (Chapter 9). Also as a worst-case scenario, people would migrate – peacefully or otherwise – and agriculture could be confined to a narrow region at mid-latitudes between cool upper latitudes and hot lower latitudes. Unlike times of glacial advances, the lower latitudes would not be cooled by the ice and dry atmosphere. The sea level would also not lower, so productive coastal plain landforms would not be exposed as in glacial periods (Figure 9.5).

31.2.3 Scenario 2: Long-Term Climate Changes with Glaciers Advancing

If global warming from greenhouse gases were not to prevail now or in a thousand years, glaciers would return over several thousand years to similar positions but at first not covering as much area toward the equator than they did 18 thousand years ago (Figures 9.2, 9.5a). The polar regions would be cold and some would be covered with glaciers or permafrost (Figure 10.8); equatorial regions would be dry but possibly cool; and the sea level would drop by about 60 meters – about half of its drop of 18 thousand years ago (Chapter 9). Expanses of land would emerge where current coastal plains are, but not as much as before (Figure 9.5a).

People would probably migrate away from the poles and inhabit suitable areas in temperate and tropical zones and onto the newly exposed land. Species would adjust by migrating, as they always have. These migrations would be slower than if Scenario 1 prevailed. People could fight each other over migrations and land and water rights; or, they could work cooperatively and tap glaciers for water and hydroelectric power, shipping the water to dry regions through pipes and/or canals (Chapter 19). They could make use of the dry low latitudes for Concentrated Solar Power energy generation (Chapter 25). If the world's people were scientifically and technically informed and worked cooperatively, the changes could be satisfactory.

31.3 Population Changes

If it continues, the recent trend toward a replacement fertility rate could stabilize the world's population in about 70 years (Figure 5.13). A fertility rate slightly below replacement could reduce the world population to a level less susceptible to problems associated with congestion (Chapter 5). During the next few decades, a surge of young people currently under 35 years old will be in their prime working years. They will be more numerous than their predecessors and will strain the shelter/ transportation infrastructures and the job market – and probably many will be left out and bitter.

In some developed countries, recent low fertility rates are leaving few young people as the population ages. Other countries have high fertility rates that may lead to extreme surges in regional populations, with attendant poverty. Each of these situations can be treated with understanding and careful action; however, each is highly susceptible to panic and violence by xenophobes who insist their "race" outnumber others.

The world's population is increasingly being concentrated in large cities (Chapter 5). If climate changes bring food, water, and job shortages, the cities may be stressed even more. Social drivers that entice people to jobs elsewhere may relieve the pressures. Migrations could occur to milder temperate zones for technical farm or forest work (discussed below) and/or to deserts for technical industrial work (also discussed below). Education, training, and appropriate accommodations would help this transformation.

31.4 Constructing Infrastructures

Infrastructures of bridge, highways, dams, and mid-rise buildings in much of the developed world are declining (Chapter 10). They may be replaced with modern infrastructures that accommodate new technologies. Some people in crowded cities may be dispersed elsewhere while outdated buildings are converted to open spaces that make cities more livable. Technical agriculture (Chapter 22) and forestry (Chapter 30) social drivers that help reverse depopulation of rural areas could utilize infrastructures of "critical population small cities" (Chapter 4) with enough people to provide retail, entertainment, and service amenities such as book stores, restaurants, sports teams, theaters, schools, and hospitals. These could be made energy efficient – mass timber high-rises, renewable energy and water, and good insulation. A shift to industrial production in deserts could also utilize these "critical population small cities" (Chapter 25).

31.5 Biodiversity

Enhancing biodiversity – number and viability of species and their habitats – will be critical for people's well-being (Chapters 13 and 16). All causes for native species extinctions are not known. Known causes – chemicals, loss and fragmentation of habitats, and exotic pests – will need to continue being addressed; as will newly discovered causes. Recent establishment of reserves as well as monitoring species and avoiding trade of endangered species parts (for example, furs and ivory) would need to continue. Prohibiting the shipping of raw organics overseas would help avoid invasive exotic pests that eliminate species (Chapter 16); concomitantly, it would force secondary manufacture in the country growing the crop and so should promote more inclusive economies (Chapter 6) [7]. In any case, the world will increasingly and inevitably have "novel communities" of species from different realms that have not lived together until recently. It may be prudent to continue "rewilding" – introducing analogous species or genes from kindred species where the native ones are lost. Increasingly, animals and plants are adapting to modern human environments, although people can aid by promoting corridors and habitats for desirable species (Chapter 16).

31.6 The Resources

Water may become scarcer in many parts of lower latitudes in the next few decades and centuries as atmospheric circulation patterns change. Different crops may be grown or croplands abandoned. Crops raised for biofuels will become less efficient because of their water use (Chapter 22). Aquifers currently being overdrawn will need to be tended to – with reductions in irrigation, other water uses, and impermeable surfaces. Storm drains, ditches, and other infrastructures that rapidly send water to rivers and oceans instead of to groundwater may need to be altered to conserve useable water in aquifers (Chapter 19). Recycling, desalinization, and rooftop rainwater harvest could be widely expanded. Some existing and newly built dams, levees, and other water infrastructures will continue to be needed and used for flood control, hydroelectricity, and irrigation; however, they could be designed or retrofitted with environmentally friendly features for silt and fish passage (Chapter 25).

Food is now grown so efficiently and sufficiently that the agriculture area occupies only 11% of the terrestrial Earth but can feed all people, save for distributional problems (Chapters 4 and 22). With increasingly erratic weather, more detailed attention will be needed to avoid pest, frost, flood, drought, and other problems. Intensive, broadcast applications of pesticides, fertilizers, and excess water to flush them are increasingly being replaced with precise, "spot" applications and organic-style intercropping techniques using Geographic Information Systems (GIS), Global Positioning Systems (GPS), drones, and other technologies (Chapter 21). Changes in energy and agricultural chemical application policies could accelerate this transformation. Knowledgeable agriculture labor to work with this technology could thus become a social driver for rural communities, saving water, chemicals, fertilizer and energy, and money (Chapter 22).

Agricultural land is displacing forests and their biodiversity as cropland biofuels are used (Chapter 24), meat is eaten in more countries and is raised in feedlots (Chapter 22). Alternatively, raising range-fed animals could save forests – as could changed diets. Biofuels could be made from forest and agriculture/food waste rather than from human-edible food.

Renewable energy sources could continue to be adopted rapidly – probably as rapidly as fossil fuels were adopted in the last century (Chapter 25). The result will be less global strife, a significant reduction in atmospheric carbon dioxide. Renewable energy technologies are less centralized and so will avoid clumsy energy infrastructures where not already present. The rapid introduction of renewable energy sources will enhance lifestyles in developing countries and need not detract from lifestyles in developed countries – especially if combined with energy savings by avoiding generating electricity from fossil fuels, improving building insulation, and using mass timber (wood) construction to replace fossil fuel-consuming concrete, brick, and steel (Chapter 25). Deserts near oceans may emerge as new industrial centers where Concentrated Solar Power (CSP) can be used for industrial energy and to desalinate water for industrial and other uses. With air and rail passenger and freight movement needing less energy and fewer highways, the efficacy of superhighways may decline, although roads for local traffic would not. Current superhighways may be retrofit for rail traffic or other innovative transportation mechanisms.

Minerals, and especially elements, will probably continue to be in demand in new uses, types, and volumes in high technologies such as computers, photovoltaic solar panels, and other systems where special properties of specific elements are needed (Chapter 27). The scarcity of some critical elements and their known presence in only a few countries could mean conflicts and wars over critical elements could occur, similar to those over fossil fuels. Hopefully, international protocols for resolving issues amicably will develop; and cool heads will prevail. Meanwhile, systems for recycling elements from discarded material or tailings could be implemented much more aggressively. Scientists are also constantly seeking technologies that do not need extremely scarce elements.

Forests occupy about 30% of the terrestrial land but are greatly underutilized and neglected – with very little wood harvested and few people employed (Chapters 28 and 30). Fewer forests would burn up if silviculturally tended in ways that keep fuel loads smaller and fires less intense. The currently erratic climates will continue to contribute to forest fires. Forest reserves now exist to protect ecosystem services somewhat, but far more proactive management is needed both to restore ecosystem services and to expand wood product availability (Chapter 29). A skilled forest workforce could become a social driver for rural areas and further populate the "critical population small cities" described above. This workforce could combine landcare and logging activities, using small, highly technical small equipment that employs more people than conventional logging (Chapter 30).

Most of the environmental movement since the 1960s has been filled with alarmist rhetoric, anecdotes, and single-issue problems and solutions. These activities may have been necessary, although frustrating; as a result, people are now much more sensitive to the environment and resource limits than before. This book takes a different, necessary direction. It looks at the socioenvironmental system holistically and calmly, and describes the current condition. It suggests likely and unlikely future events. Faced with a probabilistic and uncertain future, it is important to understand what *not* to do. It is important not to remain static because the environment, resources, and societies will not. There is no single correct course of action, but some will be more likely to give better outcomes than others. Rather than an alarming, gloomy, fearful outlook that causes people to resist change, an optimistic, adventurous, cooperative attitude is necessary, but is not sufficient. Also necessary will be many people, at various technical and scientific levels, to manage the resources described in Chapters 17 through 30. The public in all walks of life will need to appreciate and be engaged with the analytical intricacies and tradeoffs involved.

References

1. W. R. Ashby. Requisite Variety and Its Implications for the Control of Complex Systems. *Cybernetics*. 1958;1(2):83–99.
2. J. Diamond. *Collapse: How Societies Choose to Fail or Succeed*. (Penguin, 2005).
3. J. M. Gowdy, C. N. McDaniel. The Physical Destruction of Nauru: An Example of Weak Sustainability. *Land Economics*. 1999;75(2):333–8.
4. B. W. Tuchman. *The March of Folly: From Troy to Vietnam*. (Random House, 2011).
5. J. H. Bernstein. The Data-Information-Knowledge-Wisdom Hierarchy and Its Antithesis. *NASKO: North American Symposium on Knowledge Organization*. 2011;2(1):68–75.
6. T. Stocker, D. Qin, G. Plattner, et al. Climate Change 2013: The Physical Science Basis. in *Intergovernmental Panel on Climate Change*. (Cambridge University Press, 2013).
7. D. Acemoglu, J. Robinson. *Why Nations Fail: The Origins of Power, Prosperity, and Poverty*. (Crown Business, 2012).

Appendix I

Country Groups Used in Analyses

Key to Figure 1.4: Country group numbers (first column), group names (second column), and included countries (third column)

Column numbers			
1	2	1	2
1	Africa – small islands	4	North Africa (continued)
	1. Cape Verde		3. Libyan Arab Jamahiriya
	2. Comoros		4. Morocco
	3. Mauritius		5. Tunisia
	4. Reunion		6. Western Sahara
	5. Sao Tome and Principe	5	Southern Africa
	6. Seychelles		1. Angola
2	Central Africa		2. Botswana
	1. Burundi		3. Lesotho
	2. Cameroon		4. Madagascar
	3. Central African Republic of the Congo		5. Malawi
	4. Democratic Republic of the Congo		6. Mozambique
	5. Gabon		7. Namibia
	6. Equatorial Guinea		8. South Africa
	7. Rwanda		9. Swaziland
3	East Africa		10. Zambia
	1. Djibouti		11. Zimbabwe
	2. Eritrea		12. Saint Helena (not shown)
	3. Ethiopia	6	West Africa
	4. Kenya		1. Benin
	5. Somalia		2. Burkina Faso
	6. Sudan		3. Chad
	7. Uganda		4. Côte d'Ivoire
	8. United Republic of Tanzania		5. Gambia
4	North Africa		6. Ghana
	1. Algeria		7. Guinea-Bissau
	2. Egypt		8. Guinea

492

(*cont.*)

Column numbers			
1	2	1	2
6	West Africa (continued)	12	Middle East (continued)
	9. Liberia		8. Iraq
	10.Mali		9. Israel
	11.Mauritania		10. Jordan
	12.Niger		11. Lebanon
	13.Nigeria		12. Pakistan
	14.Senegal		13. Syrian Arab Republic
	15.Sierra Leone		14. Turkey
	16.Togo		15. West Bank
7	Arabian Peninsula	13	Other South Asia
	1. Bahrain		1. Bangladesh
	2. Kuwait		2. Bhutan
	3. Oman		3. Maldives
	4. Qatar		4. Nepal
	5. Saudi Arabia		5. Sri Lanka
	6. United Arab Emirates	14	India
	7.Yemen	15	Other Southeast Asia
8	Central Asia		1. Brunei Darussalam
	1. Kazakhstan		2. Cambodia
	2. Kyrgyzstan		3. Timor-Leste
	3. Mongolia		4. Lao People's Democratic Rep.
	4. Tajikistan		5. Malaysia
	5. Turkmenistan		6. Myanmar
	6. Uzbekistan		7. Philippines
9	China		8. Singapore
10	Japan		9. Thailand
11	Korean Peninsula		10. Viet Nam
	1. Dem. People's Rep. of Korea	16	Indonesia
	2. Republic of Korea	17	Balkans
12	Middle East		1. Albania
	1. Afghanistan		2. Bosnia and Herzegovina
	2. Armenia		3. Bulgaria
	3. Azerbaijan		4. Croatia
	4. Cyprus		5. Romania
	5. Gaza Strip		6. Slovenia
	6. Georgia		7. The Former Yugoslav Rep. of
	7. Iran (Islamic Republic of)		Macedonia

(*cont.*)

Column numbers			
1	2	1	2
18	Central Europe	24	Caribbean
	1. Austria		1. Antigua and Barbuda
	2. Belgium and Luxembourg		2. Bahamas
	3. France		3. Barbados
	4. Germany		4. Bermuda
	5. Ireland		5. British Virgin Islands
	6. Liechtenstein		6. Cayman Islands
	7. Netherlands		7. Cuba
	8. Switzerland		8. Dominica
	9. United Kingdom		9. Dominican Republic
19	Other Eastern Europe		10. Grenada
	1. Belarus		11. Guadeloupe
	2. Czech Republic		12. Haiti
	3. Estonia		13. Jamaica
	4. Hungary		14. Martinique
	5. Latvia		15. Montserrat
	6. Lithuania		16. Netherland Antilles
	7. Poland		17. Puerto Rico
	8. Republic of Moldova		18. Saint Kitts and Nevis
	9. Slovakia		19. Saint Lucia
	10. Ukraine		20. Saint Vincent / the Grenadines
20	Russian Federation		21. Trinidad and Tobago
21	Northern Europe		22. United States Virgin Islands
	1. Denmark	25	Central America
	2. Finland		1. Belize
	3. Iceland		2. Costa Rica
	4. Norway		3. El Salvador
	5. Sweden		4. Guatemala
22	Southern Europe		5. Honduras
	1. Andorra		6. Nicaragua
	2. Greece		7. Panama
	3. Italy	26	Mexico
	4. Malta	27	United States
	5. Portugal	28	Australia and New Zealand
	6. San Marino		1. Australia
	7. Spain		2. New Zealand
23	Canada		

(*cont.*)

Column numbers			
1	2	1	2
29	Other Oceania	30	Temperate South America
	1. American Samoa		1. Argentina
	2. Cook Islands		2. Chile
	3. Fiji		3. Uruguay
	4. French Polynesia		4. Falkland Islands (Malvinas)
	5. Guam	31	Brazil
	6. Kiribati	32	Other Tropical South America
	7. Marshall Islands		1. Bolivia
	8. Micronesia		2. Colombia
	9. Nauru		3. Ecuador
	10. New Caledonia		4. French Guiana
	11. Niue		5. Guyana
	12. Northern Mariana Islands		6. Paraguay
	13. Palau		7. Peru
	14. Papua New Guinea		8. Suriname
	15. Samoa		9. Venezuela
	16. Solomon Islands		
	17. Tonga		
	18. Vanuatu		

Appendix II

Elements, Chemical Symbols, Atomic Numbers, and Masses

Element symbols, names, atomic numbers (number of protons), and atomic masses (average protons and neutrons) [1]. Figure 26.1 shows the Periodic Table.

Symbol	Name	Atomic number	Atomic mass	Symbol	Name	Atomic number	Atomic mass
Ac	Actinium	89	227	Na	Sodium	11	23
Ag	Silver	47	108	Nb	Niobium	41	93
Al	Aluminum	13	27	Nd	Neodymium	60	144
Am	Americium	95	243	Ne	Neon	10	20
Ar	Argon	18	40	Ni	Nickel	28	59
As	Arsenic	33	75	No	Nobelium	102	259
At	Astatine	85	210	Np	Neptunium	93	237
Au	Gold	79	197	O	Oxygen	8	16
B	Boron	5	11	Os	Osmium	76	190
Ba	Barium	56	137	P	Phosphorus	15	31
Be	Beryllium	4	9	Pa	Protactinium	91	231
Bh	Bohrium	107	264	Pb	Lead	82	207
Bi	Bismuth	83	209	Pd	Palladium	46	106
Bk	Berkelium	97	247	Pm	Promethium	61	145
Br	Bromine	35	80	Po	Polonium	84	209
C	Carbon	6	12	Pr	Praseodymium	59	141
Ca	Calcium	20	40	Pt	Platinum	78	195
Cd	Cadmium	48	112	Pu	Plutonium	94	244
Ce	Cerium	58	140	Ra	Radium	88	226
Cf	Californium	98	251	Rb	Rubidium	37	85
Cl	Chlorine	17	35	Re	Rhenium	75	186
Cm	Curium	96	247	Rf	Rutherfordium	104	261
Co	Cobalt	27	59	Rg	Roentgenium	111	272
Cr	Chromium	24	52	Rh	Rhodium	45	103
Cs	Cesium	55	133	Rn	Radon	86	222
Cu	Copper	29	64	Ru	Ruthenium	44	101
Db	Dubnium	105	262	S	Sulfur	16	32

(*cont.*)

Symbol	Name	Atomic number	Atomic mass	Symbol	Name	Atomic number	Atomic mass
Ds	Darmstadtium	110		Sb	Antimony	51	122
Dy	Dysprosium	66	163	Sc	Scandium	21	45
Er	Erbium	68	167	Se	Selenium	34	79
Es	Einsteinium	99	252	Sg	Seaborgium	106	266
Eu	Europium	63	152	Si	Silicon	14	28
F	Fluorine	9	19	Sm	Samarium	62	150
Fe	Iron	26	56	Sn	Tin	50	119
Fm	Fermium	100	257	Sr	Strontium	38	88
Fr	Francium	87	223	Ta	Tantalum	73	181
Ga	Gallium	31	70	Tb	Terbium	65	159
Gd	Gadolinium	64	157	Tc	Technetium	43	98
Ge	Germanium	32	73	Te	Tellurium	52	128
He	Helium	2	4	Th	Thorium	90	232
Hf	Hafnium	72	178	Ti	Titanium	22	48
Hg	Mercury	80	201	Tl	Thallium	81	204
Ho	Holmium	67	165	Tm	Thulium	69	169
Hs	Hassium	108	277	U	Uranium	92	238
I	Iodine	53	127	Uub	Ununbium	112	
In	Indium	49	115	Uuh	Ununhexium	116	
Ir	Iridium	77	192	Uuo	Ununoctium	118	
K	Potassium	19	39	Uup	Ununpentium	115	
Kr	Krypton	36	84	Uuq	Ununquadium	114	
La	Lanthanum	57	139	Uus	Ununseptium	117	
Li	Lithium	3	7	Uut	Ununtrium	113	
Lr	Lawrencium	103	262	V	Vanadium	23	51
Lu	Lutetium	71	175	W	Tungsten	74	184
Md	Mendelevium	101	258	Xe	Xenon	54	131
Mg	Magnesium	12	24	Y	Yttrium	39	89
Mn	Manganese	25	55	Yb	Ytterbium	70	173
Mo	Molybdenum	42	96	Zn	Zinc	30	65
Mt	Meitnerium	109	268	Zr	Zirconium	40	91
N	Nitrogen	7	14				

Reference

1. Los Alamos National Laboratory. *Periodic Table of the Elements*. (US Los Alamos National Laboratory Chemistry Division, 2016 [Accessed December 10, 2016].) Available from: http://periodic.lanl.gov/images/periodictable.pdf.

Index

logging, secondary, and tertiary processing, 463–6
middlemen, 453, *See also* forest characteristics, work force
seasonal labor, 168, 453, 465
forest land uses, 254
forest loss prevention. *See* carbon dioxide atmospheric reduction
forest management
degradation, 472, *See also* high grade harvesting, degraded forest stands, degraded forest landscapes
free market supply and demand, 468
landscape management. *See* landscape management
silviculture. *See* Silviculture
sustained yield, 468, *See also* sustainability, Criteria for Sustainable Forestry
forest stand protection, 470
forest stand structures, 216, 222, 446, 447, 463
forest values, 445–6, *See also* Criteria for Sustainable Forestry, forest stand structures
non-commodity forest values, 453, *See also* ecosystem services
non-timber forest products, 453
timber products, 453–5
Forrester, J.W., 18, 21, 25, 82
fossil aquifer, 300
fossil fuel, 108, 110, 377, 378, 379, 388, 400, 439, *See also* country troups:energy; produced, imported, exported
"true" cost of fossil fuels, 395
clean fuel VS dirty fuel, 379
global trade, 395
reduction scenarios, 128–31, 397–402, 403
world reserves, 378, 398
fossil water stocks, 268, *See also* fossil aquifer
founder crops, 327
founder effect, 195, 201
fracking, 288, 420, 422, *See* mining
fractionate, 410, 417
framework, xxiii, 12, 23, 26, 27
free range, 343
frost avoidance, 131
frost heaving, 185, 268
frost pockets and frost drains, 90, 342

gamma-ray wavelengths. *See* energy, radiation wavelengths
gangs of disks (plowing), 336
Gaud, W.S., 329
General System Theory, 12
genetic differentiation, 195, 199
genetic drift, 167, 195, 198, 201
genetic reproduction, 194
genetic selection, 195, 198, 201
genetic variations, 194, 320
genetically modified organisms, 329, 472, *See also* GMOs

genotypes, 202, 204, 231, 251, 329, 343
geoengineering, 131
giardia, 267
glacial periods, 112, 118, 182, 187
glacier accumulation zone, 118, 123
glacier equilibrium line, 118, 123
glacier extent, 120, *See also* Maps, global land cover changes and landform, recent glaciation
global dimming, 111
global ecological zones, 88
global vegetation cover types, 87
global warming, 4, 106, 110, 111, 112, 127, 128, 187, 232, 379, 402, 452, 488
temperature graph, 113
global warming potential (GWP), 105
GLOFs, 168, 184, 185, 295, 305, *See also* jokuloups
GMOs. *See also* genetically modified organisms
Gondwanaland, 116
Goulburn-Broken Catchment, 25, 31, 143, 274, 337
granite, 116, 157–9, 163, 171, 418, 420, 424, 432
grasslands, 35, 55, 121, 163, 173, 174, 214–18, 225–32, 238–42, 270, 276, 277, 322, 356–8, 476, 487
gravitation water, 277
grazing, 26, 140, 144, 152, 157, 159, 163, 165, 167, 168, 174, 206, 214, 216, 220, 223, 241, 242, 251, 280, 293, 295, 299, 302, 334, 342, 343, 358
Greek fire, 375
Green Revolution, 58, 63, 326, 327, 329, 344, 362, 485
greenhouse gases, 99, 103, 105–12, 118–27, 128, 265, 369, 488, *See also* carbon dioxide, methane, nitrogen oxides, CFC, HFC, HCFC
carbon dioxide equivalents, 106
global warming potential, 105
radiative forcing, 105
water, 265
greenhouses, 313, 334, 342
Gross Domestic Product per capita (GDPpc). *See* Indexes of country well-being
Gross Domestic Product per capita (Purchasing Power Parity) (GDPpcPPP). *See* Indexes of country well-being
Gross National Income per capita (GNIpc). *See* Indexes of country well-being
ground fires, 225
groundwater, 141, 143, 144, 163, 265, 268, 272, 274, 275, 276, 281, 288, 293, 337, 447, 489
groupthink, 76–7
growing space, 203, 212–15, 219, 221, 230
guano, 339
Gulf Stream, 95

Haber–Bosch method, 339
Hadley cells, 92, 121
halo of fresh water, 273, 300

Printed in the United States
by Baker & Taylor Publisher Services